BRITISH GEOLOGICAL SURVEY

J G REES and
A A WILSON

Geology of the country around Stoke-on-Trent

Memoir for 1:50 000 Geological Sheet 123
(England and Wales)

CONTRIBUTORS

Stratigraphy
B M Besly
R G Crofts
A S Howard

Geophysics
J D Cornwell
Z K Dabek
D J Evans

Palaeontology
B Owens
W H C Ramsbottom
N J Riley
G Warrington

Petrography
R J Merriman

Geochemistry
H W Haslam

Structure
S M Corfield

Hydrogeology
R A Monkhouse

Engineering geology
T P Gostelow

London: The Stationery Office   1998

ISBN 0 11 884537 3

The grid used on the figures is the National Grid taken from the Ordnance Survey map © Crown copyright reserved Ordnance Survey licence no. GD 272191/1998.

Bibliographical reference

REES, J G, AND WILSON, A A. 1998. Geology of the country around Stoke-on-Trent. *Memoir of the Geological Survey, Sheet 123 (England and Wales).*

Authors

J G Rees, BSc, PhD
British Geological Survey, Keyworth

A A Wilson, BSc, PhD
formerly British Geological Survey

Contributors

J D Cornwell, MSc, PhD
R G Crofts, BSc
Z K Dabek, BSc
D J Evans, BSc, PhD
T P Gostelow, MSc, PhD
H W Haslam, MA, PhD
A S Howard, BSc, PhD
R J Merriman, BSc
R A Monkhouse, MA
B Owens, BSc, PhD
N J Riley, BSc, PhD
G Warrington, DSc
British Geological Survey

B M Besly, BSc, PhD
Keele University

S M Corfield, BSc, PhD
Manchester University

W H C Ramsbottom MA, PhD
formerly British Geological Survey

Printed in the UK for The Stationery Office
J 39951 C6 3/98

Other publications of the Survey dealing with this district and adjoining districts

BOOKS

British regional geology
Central England (3rd edition), 1969
The Pennines and adjacent areas (3rd edition) 1954

Memoirs
The geology of the area around Chester and Winsford, Sheet 109, 1986
Geology of the country around Macclesfield, Congleton, Crewe and Middlewich, Sheet 110, 1968
Geology of the country around Buxton, Leek and Bakewell, Sheet 111, 1985
Geology of the country around Nantwich and Whitchurch, sheet 122, 1966
Geology of the country around Ashbourne and Cheadle, Sheet 124, 1988
Geology of the country around Burton upon Trent, sheet 140, 1982

Reports
A standard nomenclature for the Triassic formations of the Ashbourne district, 81/14, 1982

The conglomerate resources of the Sherwood Sandstone Group of the country east of Stoke-on-Trent, Staffordshire, Mineral Assessment Report No. 91, 1982

Stoke-on-Trent: a geological background for planning and development, WA/91/01, 1992

MAPS

1:625 000
Solid geology (UK south sheet)
Quaternary geology (UK south sheet)
Aeromagnetic anomaly (UK south sheet)
Bouguer gravity anomaly (UK south sheet)

1:250 000 (Solid)
Sheet 52N 04W (Mid-Wales and Marches) 1990
Sheet 52N 02W (East Midlands) 1983
Sheet 53N 04W (Liverpool Bay) 1978
Sheet 53N 02W (Humber-Trent) 1983

1:250 000 (Solid and Drift)
Sheet 53N 04W (Liverpool Bay) 1978

1:100 000
Hydrogeology of Clwyd and the Cheshire Basin, 1989

1:50 000 (Solid)
Sheet 109 (Chester) 1986
Sheet 123 (Stoke-on-Trent) 1995

1:50 000 or 1:63 360 (Solid and Drift)
Sheet 109 (Chester) 1965
Sheet 110 (Macclesfield) 1968
Sheet 111 (Buxton) 1978
Sheet 122 (Nantwich) 1967
Sheet 123 (Stoke-on-Trent) 1995
Sheet 124 (Ashbourne) 1983
Sheet 138 (Wem) 1967
Sheet 139 (Stafford) 1974
Sheet 140 (Burton upon Trent) 1982

Geology of the country around Stoke-on-Trent

In a region exceptionally rich in natural resources, and with a long history of mineral extraction, up-to-date geological information is essential as a foundation for the planning of future land use and development. This memoir is intended to satisfy that basic need, and at the same time to provide an overview of the geology for amateur and professional geologists alike.

The district described in this memoir lies at the south-western end of the Pennines, where Carboniferous strata are flanked by Permo-Triassic cover rocks to the south and west. The Potteries conurbation, which includes Stoke-on-Trent and Newcastle-under-Lyme, forms a built-up tract on the Carboniferous outcrop. This contrasts sharply with surrounding rural areas, which include parts of Cheshire and Shropshire as well as Staffordshire.

The Carboniferous rocks include the Coal Measures of the Potteries and Shaffalong coalfields. These strata contain probably the thickest sequence of workable coals in Britain and have been of major economic importance to the district, as have some of the clays of overlying Carboniferous red-bed formations.

The Permian and early Triassic rocks are mainly sandstones, which provide important sand and gravel resources as well as acting as major aquifers. The younger Triassic rocks consist mostly of mudstones, but also contain major deposits of rock salt.

The southern part of the district is traversed by a suite of dykes of Palaeogene age, the most south-easterly members of the Hebridean Tertiary Igneous Province.

Most of the district is overlain by extensive spreads of Pleistocene glacial and postglacial deposits.

Cover photograph
Pottery kilns, Gladstone Pottery Museum, Longton. (MN 27961)

Upper Coal Measures, High Lane Opencast site. The strata, dipping to the east, are dislocated by a zone of westerly dipping faults. The slope at the base of the photograph is the mudstone floor of the Great Row Coal, the seam behind the rubble. The prominent thick seam is the Spencroft Coal which is split into three leaves and is overlain by a sandstone-rich sequence. This is in turn overlain by the Peacock Coal, and the coals and ironstones of the 'Blackband Group'. The thickness of the sequence between the Great Row and Spencroft coals is about 17 m.

CONTENTS

PLATES

Cover photograph Pottery kilns, Gladstone Pottery Museum, Longton
Frontispiece Upper Coal Measures, High Lane Opencast site

FIGURES

TABLES

ACKNOWLEDGEMENTS

In this memoir the following chapters were written by the authors and contributors listed on the title page. **Pre-Carboniferous**: JGR, JDC, ZKD. **Carboniferous**: Dinantian JGR, NJR, RJM; Namurian AAW, JGR; Westphalian (Coal Measures) JGR, NJR; Westphalian (Barren Measures) BMB, AAW; ?Stephanian BMB. **Permian**: JGR, JDC, ZKD. **Triassic**: Sherwood Sandstone Group JGR, AAW; Mercia Mudstone and Penarth groups AAW, JGR, GW. **Palaeogene**: JGR. **Quaternary**: AAW, JGR, JDC. **Structure**: JGR, SMC, JDC, ZKD, DJE. **Applied Geology**: Economic Geology JGR; Hydrogeology RAM; Geological hazards TPG, JGR. The memoir was edited by T J Charsley and J I Chisholm.

We gratefully acknowledge information and assistance generously provided by numerous organisations during the course of the survey. These include: the North Staffordshire group of the Geologists' Association for their unlimited support, British Coal and Redland Brick, particularly for access to quarries, Staffordshire County Council, Stoke-on-Trent City Council, Newcastle Borough Council, Stoke-on-Trent City Museum, the National Federation of Clay Industries and many geotechnical companies (especially Wardell Armstrong). We also acknowledge the access and help given by numerous landowners and farmers throughout the district during the course of the survey. The survey was supported in part by the Department of the Environment. Thanks are also due to the many individuals who have given us comments, constructive criticism and support during the survey and drafting of this memoir, especially D B Thompson (Keele University), J O'Dell (British Coal) and E L Parry and A Peacock (National Rivers Authority) and W T C Sowerbutts (Manchester University). British Coal, Shell UK, and Hamilton Brothers plc are all thanked for permission to reproduce geophysical data. Research by B M Besly was supported by Natural Environment Research Council grants GT4/76/G5/48 and GR3/6961. Research by S M Corfield was supported by a grant from Conoco UK Ltd.

NOTES

Throughout the memoir the word 'district' refers to the area covered by the 1:50 000 geological sheet 123 (Stoke-on-Trent).

National Grid references are given in square brackets; all lie within 100 km square SJ, unless otherwise stated.

The location and registration number of all boreholes and shafts referred to are given in Appendix 1.

Numbers preceded by the letter E refer to the BGS sliced rock collection (see Appendix 2).

Enquiries concerning geological data for the district should be addressed to the Manager of the National Geological Records Centre at BGS, Keyworth.

PREFACE

An understanding of geology is essential for the exploration of earth-based resources and the avoidance of ground hazards, and is therefore of basic importance in land-use planning. In recognition of this the British Geological Survey is funded by central Government to improve understanding of the geology of the UK through a programme of data collection, geological interpretation, publication and archiving. One aim of this programme is to ensure coverage of the UK land area by modern 1:50 000 geological maps, mostly with explanatory memoirs. The memoir for Stoke-on-Trent is part of the output from that programme.

The economic development of few districts of Britain can have been as greatly influenced by geology as that described here. The very name Stoke-on-Trent is synonymous with the pottery industry, which became established in this area because of the presence of suitable geological resources. Other industries, notably coal mining and iron and steel manufacture, also had their roots in the local geology. Many of these industries are now in decline, and the redevelopment of former colliery sites, claypits and quarries in and around the urbanised areas has created its own problems, which can be assessed only by reference to modern geological maps.

The information presented in this memoir incorporates the results of detailed mapping by the British Geological Survey, as well as a digest of subsurface information from the coal, hydrocarbon, water and site investigation industries. It is intended to be of practical value to a wide range of users including specialists in the earth sciences or related disciplines, planning authorities concerned with land-use planning and development, and those who require geological information as a basis for mineral exploration.

The memoir contains much hitherto unpublished material, and the many specialist contributions highlight the increasing emphasis placed by the Survey on a multidisciplinary approach to geological mapping. The memoir represents a major advance in the understanding of the geological history of the district, and I am confident it will play its part for many years to come in serving the needs of the scientific, planning and commercial communities.

David A Falvey, PhD
Director

British Geological Survey
Keyworth
Nottingham
NG12 5GG

HISTORY OF SURVEY

The district covered by the Stoke-on-Trent (123) sheet of the 1:50 000 geological map of England and Wales was originally surveyed by E Hull and W W Smyth, and the results published on one-inch Old Series sheets 72 and 73 in 1857.

The primary six-inch survey of the district was carried out between 1892 and 1902 by G Barrow, C E de Rance, W Gibson, T I Pocock and C B Wedd, and between 1920 and 1926 by T Robertson and T H Whitehead.

Part of the district was resurveyed at the six-inch scale by F W Cope and J R Earp between 1941 and 1946.

The present resurvey was carried out between 1954 and 1992. It was mapped at the six-inch scale between 1954 and 1966 by R A Allender, G S Boulton, W Clyne, W B Evans, D Price, B J Taylor, A J Wadge and A A Wilson; between 1976 and 1980 by T J Charsley, J I Chisholm, D P Piper, I P Stevenson and R J Tappin; and at the 1:10 000 scale between 1988 and 1992 by J G Buchanan, R G Crofts, A S Howard, J G Rees and A A Wilson. This final phase of work was funded partly by the Department of the Environment.

Geological six-inch or 1:10 000 scale National Grid maps included wholly or in part in 1:50 000 sheet 123 (Stoke-on-Trent) are listed below, together with the initials of the geological surveyors and last date of survey. In the case of marginal sheets, only those surveyors who mapped within the area covered by the 1:50 000 sheet are listed.

Copies of the maps have been deposited in the BGS libraries at Keyworth and Edinburgh for public reference and may also be inspected in the BGS London Information Office in the Natural History Museum Earth Galleries, South Kensington, London. Copies may be purchased directly from BGS as black and white dyeline sheets.

SJ 63 NE	AAW, JGR	1991
SJ 64 NE	RAA, AAW	1992
SJ 64 SE	AAW	1992
SJ 65 NE	BJT	1956
SJ 65 SE	RAA, JGB	1991
SJ 73 NW	GSB, AAW, JGR	1991
SJ 73 NE	GSB, JGR	1991
SJ 74 NW	RAA, AAW	1991
SJ 74 NE	AAW	1991
SJ 74 SW	GSB, AAW	1991
SJ 74 SE	AAW, JGR	1991
SJ 75 NW	WBE	1957
SJ 75 NE	WBE	1957
SJ 75 SW	RAA, WBE, AAW, JGB	1991
SJ 75 SE	RAA, WBE, AAW	1990
SJ 83 NW	AAW, JGR	1991
SJ 83 NE	WBE, JGR	1991
SJ 84 NW	AAW	1989
SJ 84 NE	JGR	1988
SJ 84 SW	AAW, JGR	1991
SJ 84 SE	RGC	1988
SJ 85 NW	BE	1960
SJ 85 NE	RAA, WBE	1960
SJ 85 SW	AAW, RGC	1989
SJ 85 SE	AAW, JGB	1988
SJ 93 NW	JGR	1990
SJ 93 NE	TJC, AJW	1977
SJ 94 NW	ASH	1988
SJ 94 NE	JIC, DPP, AAW	1977
SJ 94 SW	JGR, ASH	1989
SJ 94 SE	WBE	1979
SJ 95 NW	AAW	1959
SJ 95 NE	DP	1958
SJ 95 SW	AAW	1960
SJ 95 SE	JIC, AAW	1978

In the course of the last resurvey the following open-file reports were produced. They contain information additional to that presented on the 1:10 000 maps, and details not included in this memoir. They can be consulted at BGS libraries or purchased from the same outlets as the dyeline maps.

CORNWELL, J D, and DABEK, Z K. 1993. Geophysical investigations in the Stoke-on-Trent District. *British Geological Survey Technical Report* WK/94/04.

CROFTS, R G. 1990. Geology of the Trentham district: 1:10 000 Sheet SJ84SE. *British Geological Survey Technical Report* WA/90/06.

CROFTS, R G. 1990. Geology of the Kidsgrove district: 1:10 000 Sheet SJ85SW. *British Geological Survey Technical Report* WA/90/07.

HASLAM, H W. 1993. Geochemistry of Carboniferous sediments from the Sidway Mill Borehole, Staffordshire. *British Geological Survey Technical Report* WP/93/04.

HOWARD, A S. 1990. Geology of the Werrington district: 1:10 000 Sheet SJ94NE. *British Geological Survey Technical Report* WA/90/09.

REES, J G. 1990. Geology of the Hanley district: 1:10 000 Sheet SJ84NE. *British Geological Survey Technical Report* WA/90/05.

REES, J G. 1990. Geology of the Longton district: 1:10 000 Sheet SJ94SW. *British Geological Survey Technical Report* WA/90/10.

REES, J G, and CLARK, M C. 1992. Geology of the Tunstall district: 1:10 000 Sheet SJ85SE. *British Geological Survey Technical Report* WA/90/08.

WAINE, P J, HALLAM, J R, and CULSHAW, M E. 1990. Engineering Geology of the Stoke-on-Trent area. *British Geological Survey Technical Report* WN/90/11.

WILSON, A A. 1990. Geology of the Silverdale district: 1:10 000 Sheet SJ84NW. *British Geological Survey Technical Report* WA/90/04.

WILSON, A A, REES, J G, CROFTS, R G, HOWARD, A S, BUCHANAN, J G, and WAINE, P. 1992. Stoke-on-Trent: A geological background for planning and development. *British Geological Survey Technical Report* WA/90/01.

ONE

Introduction

About three quarters of the Stoke-on-Trent district lies in Staffordshire, the remainder being in Cheshire and a small part of Shropshire. Topographically, geologically and economically the district falls into two broad areas. In the west is the Cheshire Plain, a region of subdued topography formed by drift-covered mudstones of Triassic age, the Mercia Mudstone Group. This fertile area is largely given over to agriculture and settlements are sparse and small. The remainder of the district forms higher ground, rising to over 280 m in the north-east. This area represents the south-western extremity of the Pennines, and is underlain by Carboniferous rocks with a local Triassic cover, mainly sandstones of the Sherwood Sandstone Group, and a patchy veneer of drift. The Carboniferous rocks have provided an abundance of raw materials for human use, of which coal, ironstone and clay have been the most extensively worked. The availability of these, particularly of pottery clays, was responsible for the growth of several towns, including Longton, Fenton, Hanley, Burslem and Tunstall. They all form part of the city of Stoke-on-Trent, named after the ecclesiastical and administrative centre of the area. The conurbation of these towns, along with the older town of Newcastle-under-Lyme, has become informally known as 'the Potteries' and can rightly claim to be the centre of the British ceramics industry.

HISTORY OF RESEARCH

The district has attracted the attention of geologists for over two centuries. By the middle of the 19th century a commendable map existed of the economically important Carboniferous rocks (Cope, 1852), and by the 1860s papers relating to the geology of the district were published in most years. The latter part of the 19th century and early part of the 20th century saw much detailed work by local geologists, notably W Hind, J T Stobbs and J Ward. It was at this time, when the geological resources were of greatest value to the local economy, that the Geological Survey was most active in the district, producing detailed maps and a succession of memoirs (Gibson and Wedd, 1902, 1905; Gibson, 1905). The main influence of the First World War was to stimulate the search for oil in the district, at the suggestion of the local oil pioneer, Lord Cadman (Rowland and Cadman, 1960; Torrens, 1994). Less geological work was done in the late 1920s and during the 1930s, but the last edition of the Stoke-on-Trent memoir was published (Gibson, 1925). Renewed interest in local coal resources during the Second World War brought about a considerable amount of new geological work (and subsequent publication), notably by R V Melville, J O' N Millot and F W Cope. The

last 30 years have seen fluctuating interest in geological research, largely mirroring the health of local extractive industries. This memoir comes at a time when large-scale mining in the district is all but over, and a new geological focus has emerged, directed towards an understanding of the environmental impact of human activities, and to the recognition and protection of resources for the future.

Most of the major geological works relating to the Stoke-on-Trent district are included in the reference list in this memoir. For fuller lists of published local work the reader is referred to Gibson (1905), Challinor (1978a) and Bentley (1983).

GEOLOGY

A summary of the near-surface geological sequence in the district is given on the inside of the front cover, and the outcrop of the main stratigraphical divisions is shown in Figure 1.

The pre-Carboniferous basement is not exposed, nor, apparently, has it been penetrated by boreholes, but its gross structure has been elucidated by geophysical work (Chapter 2). The district is located close to the apex of a triangular-shaped area of Proterozoic (Late Precambrian) continental crust, the Midlands Microcraton, formed of two microplates. The boundaries of the microplates are marked by faults that have had long and complex histories. During several periods, from the late Proterozoic onwards, these faults have separated areas of different depositional history, or of different response to earth movements.

The main part of the memoir (Chapters 3, 4 and 5) describes the sedimentary rocks preserved at or near the surface in the district. These were deposited between the beginning of the Carboniferous period, about 360 million years ago (Leeder, 1988), and the end of the Triassic period, about 205 million years ago (Forster and Warrington, 1985). During this time the landmass of which Britain formed a part drifted from just south of the equator northwards across the tropics. The depositional environments that prevailed during this migration reflect their low-latitude position. Inland and upland areas were commonly arid or semi-arid, with seasonal fluvial systems that were prone to desiccation; where high water tables existed, as in river floodplains and coastal swamps, lush equatorial rain forests were established; and where marine waters submerged the land, limestone seas developed. These environments, particularly during the Carboniferous, were also affected by sea-level changes, caused by expansion and contraction of ice sheets over the south pole. The other main influence on depositional environments was tectonic, and associated with the mechanisms that controlled Britain's

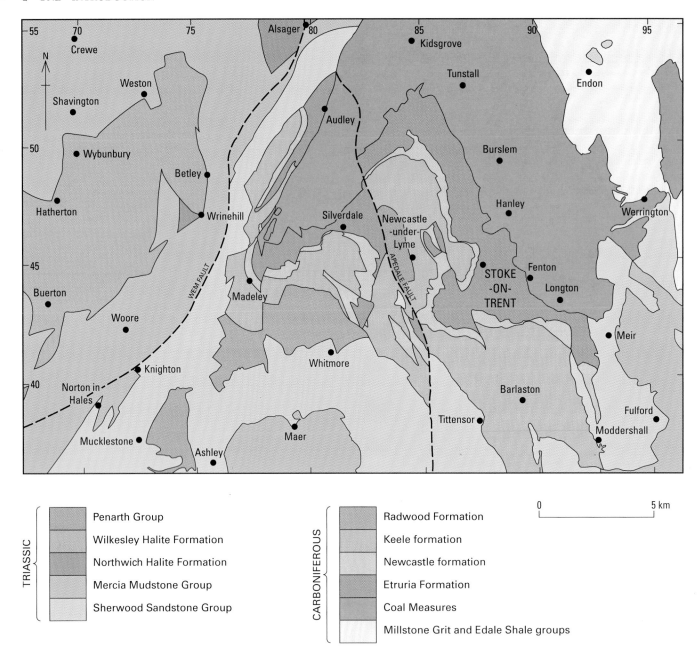

Figure 1 Distribution of the major stratigraphical divisions at outcrop in the district.

northward migration. During the Carboniferous, Permian and Triassic the tectonic setting of the district changed several times, with profound effects on depositional environment. The only major break in sedimentation during this period was brought about by compressional regimes associated with the Variscan Orogeny, which caused a regional uplift.

The only rocks of post-Triassic age are doleritic dykes of Palaeogene age (described in Chapter 6). These are notable as the most south-easterly intrusions of the Hebridean Tertiary Igneous Province.

In contrast to the environments in which the bedrock (solid) formations were deposited, the overlying superficial (drift) sediments, which are less than 2 million years

old, were deposited during or after the Pleistocene glaciations. Most of the drift deposits (described in Chapter 7) were thus laid down under, or on the margin of, thick ice sheets.

As already noted, structures such as faults have had a notable effect on the evolution of the district. Chapter 8 describes the major structures and summarises their influence on the deposition, burial and deformation of strata.

Many parts of the geological sequence provide valuable resources in themselves or act as reservoirs for hydrocarbons or water; others are notable for the geological hazards they may pose. The applied geology of the district is described in Chapter 9.

TWO

Pre-Carboniferous

In the Stoke-on-Trent district no rocks of pre-Carboniferous age crop out, and it appears unlikely that they occur in any of the boreholes reported to date (see below). Consequently the stratigraphy and structure of the basement can only be determined by geophysical methods (detailed by Cornwell and Dabek, 1994) within the known regional context of pre-Carboniferous geology.

GEOPHYSICS

Geophysical data

The Bouguer gravity anomaly data for the Stoke district were collected by the Anglo-American Oil Company and the Geological Survey, and form part of the BGS National Gravity Databank. The regional coverage of gravity stations is about 1 per 1.1 square kilometres. This was increased locally by detailed gravity traverses, including those of Brooks (1961) in the north-western part of the district. Data from the latter were put into digital form (Cornwell and Dabek, 1992) and the results incorporated into the dataset for processing. The observed Bouguer gravity map for the Stoke district (Figure 2a) is dominated by a gravity low over the Cheshire Basin and by flanking gradients associated with the Red Rock and Hodnet faults and, to a lesser extent, with the Wem Fault. The significance of the pronounced Cheshire Basin gravity anomaly has been discussed by several authors since the original work of White (1948), most recently by Abdoh et al. (1990). In the south of the map smaller-amplitude gravity anomalies are associated with other low density Permo-Triassic rocks in the Stafford and Needwood basins. Within the Stoke-on-Trent district, a local gravity high occurs over the Western Anticline and, to the east, the values again rise towards a gravity high over the Dinantian rocks of the Derbyshire Dome.

The aeromagnetic data for the district also form part of the BGS geophysical databank. The results were first obtained in analogue form by surveys flown in the district in 1955 (east of Grid line 70E) and 1960 (west of this line). The 1955 survey was flown at 305 m mean terrain clearance with east–west flight lines at a spacing of 1.61 km and north–south tie lines with a spacing of 9.66 km. The 1960 survey was flown with the same ground clearance along north–south flight lines 2 km apart, and east–west tie lines 10 km apart. The analogue data were subsequently converted to a digital form (Smith and Royles, 1989) and used to provide the contour map in Figure 2b in which the aeromagnetic data have been adjusted to a vertical magnetisation (reduction to the pole). The most pronounced anomaly is the north-north-east-trending high over the Western Anticline, which is believed to be due to the volcanic

rocks proved in the Apedale Borehole. Most of the other, longer wavelength anomalies are probably associated with more deep-seated magnetic basement rocks. Apart from the Palaeogene Butterton–Swynnerton dykes, none of the surface rocks would be expected to have significant magnetisation.

The data on physical properties were derived largely from borehole logs (Smith et al., 1980). The densities of the main rock types used for interpretations are listed in the caption to Figure 3b. Little information is available on the magnetic properties of the rocks in the district but it is expected that all the sedimentary rocks have negligible magnetisation. Some magnetic susceptibility measurements were made on a small amount of material available from volcanic rocks in Apedale No. 2 Borehole (Chapter 3). Accurate measurements were not possible because of the small size of the fragments but the results indicate significant values (more than 0.01 SI units), particularly for samples from a depth of about 451 m. Measurements of the remanent magnetisation (information from P B Sadler) indicate that the intensities are too low to influence the total magnetisation (Q-values of about 0.05).

The Triassic rocks in the west of the district have been investigated intensively using seismic reflection surveys because of their hydrocarbon potential, and a few such surveys have been carried out over the Carboniferous rocks. Most of the results remain confidential, although some are illustrated by Corfield (1991) and Evans et al. (1993). Seismic surveys have added little to our knowledge of pre-Carboniferous rocks in the district.

The general significance of the regional geophysical data in districts to the east and north-east of the Stoke-on-Trent district are discussed by Cornwell (in Chisholm et al., 1988, and Aitkenhead et al., 1985).

Geophysical interpretation methods

Qualitative interpretation of the geophysical data was carried out using contour maps (Figure 2; see also small-scale coloured diagrams on the published 1:50 000 geological map) and various colour and shaded relief plots of the observed data and its processed forms. Quantitative interpretation of the geophysical data was based largely on the modelling of profiles using the GRAVMAG program (Busby, 1987). Three-dimensional modelling of the gravity data was undertaken for the part of the district west of the main north-easterly gravity lineament (the Pontesford Lineament on Figure 3a).

Some of the main results of the geophysical interpretations have been compiled in Figure 3a, including the locations of geophysical lineaments. In some cases these are clearly related to known faults, but the full significance of others is not clear.

Figure 2
Bouguer gravity
anomaly and
aeromagnetic
maps of the
district.

a

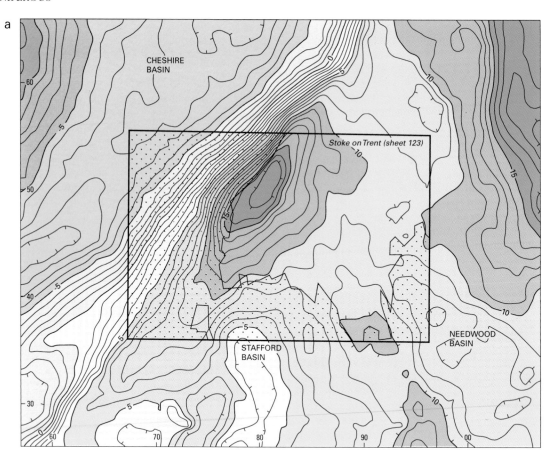

b

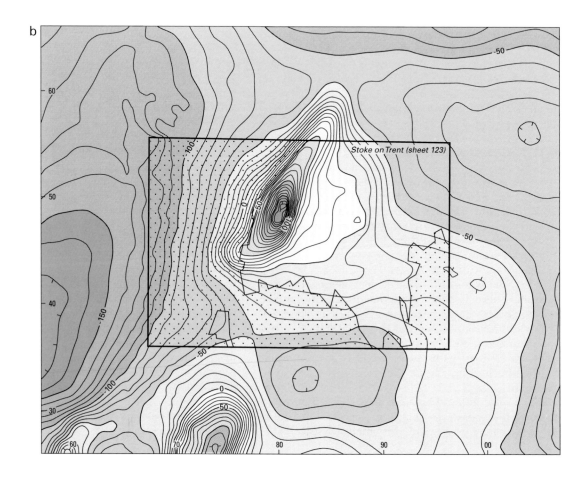

INTERPRETATION OF PRE-CARBONIFEROUS STRUCTURE

Magnetic anomalies in the region (Figure 3) generally indicate the existence of magnetic basement rocks buried under thick developments of non-magnetic sedimentary rocks. However, the main anomaly in the Stoke district, centred on Apedale, is exceptional not only for its well-defined character but also because its source apparently includes volcanic rocks in the Carboniferous sequence as well as in the basement. Similar anomalies to the south-west of the district (on the southern margin of Figure 2b) are probably due solely to Proterozoic (Uriconian) rocks in the basement. Apedale No. 2 Borehole, drilled almost on the peak of the Apedale anomaly, proved over 840 m of volcanic rocks buried by less than 450 m of sedimentary rocks. The location and magnetisation of these are consistent with their being the main source of the observed magnetic anomaly, and a Dinantian age has been assigned to them on the basis of petrological characteristics (Chapter 3). The form and subsurface extent of these Carboniferous volcanic rocks is discussed here as their magnetic effects cannot be easily distinguished from those of the basement.

The Apedale magnetic anomaly is elongated in a north-north-easterly direction. The central part of the anomaly is almost certainly all associated with the proven volcanic rocks, indicating that these are closest to surface over a distance of 2 km about the site of the borehole. To the north-north-east the anomaly diminishes in amplitude, suggesting either an increase in depth or a decrease in magnetisation or thickness (these effects cannot readily be distinguished). There is, however, evidence for a local culmination just north of the district near Mow Cop. The magnetic anomaly decreases abruptly to the north-north-east of this, near Astbury, where outcropping Dinantian rocks contain only minor volcanic horizons (Chapter 3).

The anomaly, which closely follows the line of the Red Rock Fault (Figure 43) for about 18 km, can be divided into two main sections approximately south and north of Apedale (Figures 2b and 3a). To the south the anomaly is broad and extends with decreasing amplitude to the west-south-west beneath the Permo-Triassic rocks west of the Red Rock, Wem and Hodnet faults (Figure 43). North of Apedale the anomaly is characterised by a narrower form and by an abrupt change in gradient along its west-north-west flank. The latter feature is interpreted as indicating magnetic source rocks at two different depths, with the Apedale volcanic rocks being truncated to the west-north-west at, or close to, the Red Rock Fault. There is some evidence that the Apedale volcanic rocks continue to the east of the main anomaly, though with decreased thickness (Figure 3b). Their eastward margin could be indicated by an increase in magnetic gradient on Figure 2b [at 390 350], a feature shown as a lineament in Figure 3a.

The deeper magnetic body appears to lie, probably at a depth of a few kilometres, in the basement. The depth is uncertain because of overlapping effects of the near-surface magnetic body. To the east (Figure 3, profile iii)

the magnetic basement is deeper, though still at less than 8 km. The western limit of shallow magnetic basement is defined by the Pontesford Lineament (Lee et al., 1990). This is an extensive zone of steep gravity gradients which probably represents a deep-seated fault, or faults. Near the surface the lineament is represented by faults of similar trend, including the Hodnet and Red Rock faults (Figure 43). West of the Pontesford Lineament the magnetic basement descends to considerable depths or could be absent altogether.

From regional evidence (Pharaoh et al., 1987) it can be seen that the Stoke-on-Trent district occurs approximately at the northern apex of a triangular, structurally stable area of late Proterozoic rocks, the Midlands Microcraton. This is formed of western and eastern microplates which have had different stratigraphical, magmatic and structural histories, and are separated by a suture zone of Avalonian age. The western microplate contains a basic volcanic suite, generally referred to as the Uriconian group, which crops out in Shropshire and forms The Wrekin. Several magnetic anomalies in Shropshire, including the two anomalies at the southern margin of Figure 2b, are interpreted as fault-controlled blocks of concealed Uriconian rocks. The Uriconian rocks in Shropshire are interpreted to occur at the north-western margin of the Midlands Microcraton. Their elevated position with respect to Uriconian rocks towards the centre of the microcraton is suggested by Lee et al. (1990) to have been caused by their relative uplift in flower structures generated along the margin of the microcraton.

Similarities with aeromagnetic anomalies over Uriconian rocks in Shropshire suggest that the shallow magnetic basement east of the Pontesford Lineament (MB in Figure 3b) is a northward extension of the same basement type. It is also probable that the magnetic basement immediately east of the Pontesford Lineament has been elevated relative to the deeper magnetic basement farther east by uplift in flower structures similar to those proposed by Lee et al. (1990) for the Uriconian rocks of Shropshire.

The considerable depth of the magnetic basement north-west of the Pontesford Lineament (Figure 3) suggests that the western boundary of the Midlands Microcraton in the Stoke-on-Trent district broadly coincides with the lineament. This suggests that the area to the north-west consists of the same basement as that in Wales. Such a model is supported by several geophysical features, including the major gradient zones in the area (Figure 3a) which show a north-easterly trend characteristic of Lower Palaeozoic rocks in Wales. By analogy with other districts to the south, the area west of the Pontesford Lineament saw sedimentation of a considerably thicker Lower Palaeozoic sequence than that on the microcraton to the east, which acted as a stable shelf area (Lee et al., 1990).

The eastern part of the Stoke-on-Trent district probably lies on the eastern microplate of the Midlands Microcraton. This is suggested by several geophysical lineaments and faults with the approximate north-west trend typical of the eastern microplate (Figures 3a and 43). Where the suture zone between the two microplates occurs in the district is unclear; it may correspond with a system of

3a

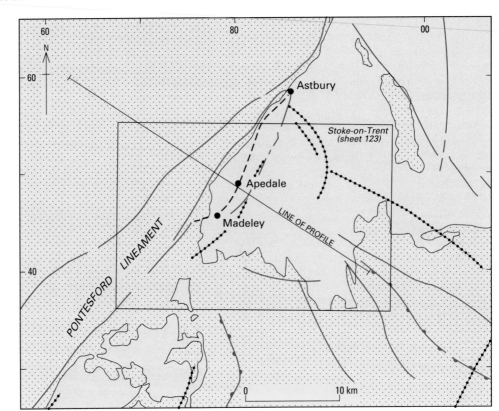

Gravity
lineament

Axis of
gravity low

Magnetic
lineament

Axis of Apedale
magnetic anomaly

Triassic outcrop

Carboniferous
outcrop

3b i Magnetic Profile

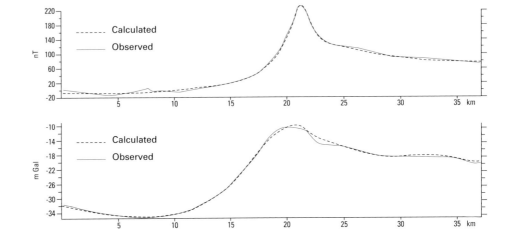

ii Gravity Profile

iii Profile Model

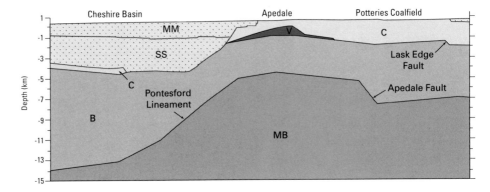

Figure 3a *(opposite)* Compilation map with main geophysical and geological features and location of aeromagnetic and gravity profile. The map shows the main gravity and magnetic lineaments referred to in the text, and margin of the Permo-Triassic outcrop (stippled).

Figure 3b *(opposite)* Observed aeromagnetic and Bouguer gravity profiles and model producing calculated profiles. Location of profile shown in Figure 3a. Background fields: aeromagnetic 145nT; gravity 25 mGal.

Key to the abbreviations on model profile (iii): (with density in Mg/m^3 and magnetic susceptibility in SI units)

MM	Mercia Mudstone:	(2.35,0)
SS	Sherwood Sandstone	
	(and Permian sandstones):	(2.48,0)
C	Carboniferous sedimentary rocks	(2.50,0)
V	Volcanic rocks (Apedale):	(2.80,0.02)
B	Basement (undifferentiated):	(2.70,0)
MB	Magnetic basement:	(2.70,0.02)

near-northerly trending faults, including the Lask Edge, Sandon and Hopton faults (Chapter 8) which represent the northern extent of the Malvern Lineament of the south Midlands (Lee et al., 1990).

During Caledonian deformation, in the mid-Devonian Acadian Orogeny, the microcraton probably acted as an indenter during the closure of the Iapetus and Tornquist oceans (Soper et al., 1987). This had a profound effect upon the deformation of the thick Lower Palaeozoic successions flanking the microcraton, which gained an arcuate fabric around its apex. The metamorphic grade of these flanking successions is commonly much higher than that of the Lower Palaeozoic rocks preserved on the microcraton (Lee et al., 1990). However, the rim of the microcraton, represented by areas underlain by shallow basement, as in Shropshire and at Apedale, is also likely to have a substantially higher metamorphic grade than rocks near its centre. The faults of the Pontesford and Malvern lineaments had reverse displacements during the Caledonian movements.

THREE

Carboniferous

Following the Acadian Orogeny (Chapter 8) sedimentation resumed over much of northern Britain in the late Devonian or early Carboniferous. Structures that had formed during Caledonian compression were extensionally reactivated at this time by crustal back-arc stretching, associated with closure of the Rheno-Hercynian basin far to the south (Leeder, 1988). Sedimentary basins developed during the Dinantian, with trends controlled largely by the reactivated Caledonian faults. Thicker sequences were deposited in the basins than on the intervening 'block' areas. One of these basins, the North Staffordshire Basin (Trewin and Holdsworth, 1973), existed in the north-east of the present district, bounded on the west by the Lask Edge Fault system (Figure 4; Corfield, 1991) and on the north-east by the Derbyshire Block. The area west of the basin, at least as far as the Wem–Red Rock Fault system, is known as the Market Drayton Horst and was part of a block which effectively formed a northern promontory of the Wales–Brabant High. This block is in its western part underlain at a depth of only a few kilometres by magnetic Proterozoic rocks (Chapter 2). Another basin may have lain to the west of the Wem–Red Rock fault system; the Apedale volcanism implies that the fault system was active in the Dinantian, and the component faults (part of the Pontesford Lineament, Figure 3a) dip to the north-west. In the early Carboniferous regime of crustal extension the downthrow would have been in the same direction, giving basinal conditions there. However, Trewin and Holdsworth (1973) inferred the presence of a block in this area, to provide the source for westerly derived clastics in the early Namurian.

The active differentiation of blocks and basins by crustal extension had largely ceased by Namurian times, and overall subsidence of northern Britain continued as a result of post-rift thermal sagging of the crust (Chapter 8). Subsidence gave rise to the Pennine Basin, a broad structure bounded to the north by the Southern Uplands block of southern Scotland, and to the south by the Wales–Brabant High; the present district was located near to its southern margin (Figure 4). In detail, Namurian and early Westphalian subsidence patterns within the Pennine Basin continued to be influenced by the same blocks and basins that had controlled Dinantian sedimentation. Later in the Westphalian, the subsidence patterns changed and new structures began to control sedimentation. These were actively developing folds, initiated by the earliest phases of Variscan compression (Chapter 8).

During the Carboniferous, between 360 and 300 million years ago (Leeder, 1988), Britain occupied a broadly equatorial position, drifting northwards across the equator in the late Carboniferous. In the course of Dinantian extension northern Britain was sharply differentiated into hot and arid inland areas, coastal areas which saw widespread production and deposition of carbonates, and submarine basins where much of the carbonate was redeposited. This differentiation changed as extension ceased and thermal subsidence began, forming the Pennine Basin. The latter was progressively filled, from the early Namurian onwards, by large deltaic systems on which extensive rain-forests developed. The Westphalian coals were formed in this environment. Uplift of parts of the Pennine Basin, from the mid Westphalian onwards, resulted in a rejuvenation of the drainage system and was accompanied by a return to increasingly arid conditions.

DINANTIAN ROCKS

The Dinantian rocks comprise limestones, mudstones and sandstones, interbedded with tuffs, and in this district have been proved only in boreholes on the western side of the Potteries Coalfield (Figure 4) at Bowsey Wood, Bittern's Wood, and probably Apedale (see below). The information from these boreholes is of limited value, and the account given here therefore includes data from adjacent districts.

The Dinantian sequences of the North Staffordshire Basin and the Market Drayton Horst differ in age, thickness and depositional facies; the relationship between the successions is shown in Figure 4.

North Staffordshire Basin

No boreholes have penetrated the basinal sequence in the Stoke-on-Trent district. Most information comes from the Nook's Farm and Gun Hill boreholes (Figure 4), in the adjoining Macclesfield and Buxton districts, approximately 10 km and 20 km respectively north-east of Stoke-on-Trent (Hudson and Cotton, 1945; Aitkenhead et al., 1985; Rees et al., 1996). The three formations recognised in the Gun Hill Borehole belong to the North Staffordshire Basin sequence and probably extend into in the Stoke-on-Trent district. They are only briefly outlined here; Aitkenhead et al. (1985) described them in greater detail. The Dinantian sequence that developed in the basin is thicker, and was deposited in deeper water, than the equivalent sequence on the Market Drayton Horst. Sedimentation is also likely to have commenced earlier (perhaps in the late Devonian) in the basin than on the horst.

MILLDALE LIMESTONES

This formation (redefined by Aitkenhead and Chisholm, 1982) is dominated by dark bioclastic calcisiltites and calcilutites, which were probably deposited in deep water.

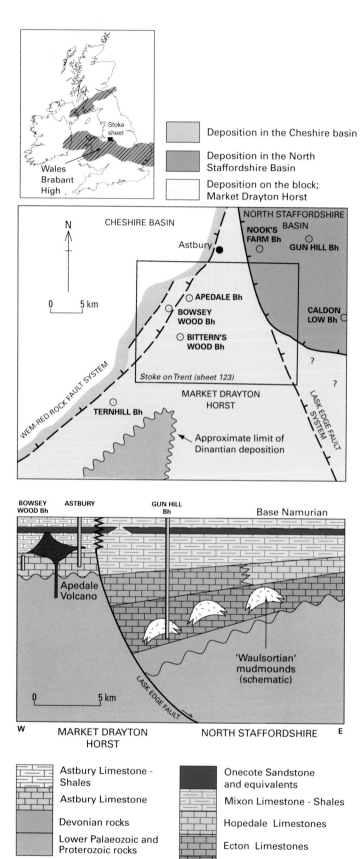

Figure 4 Dinantian setting of the district.

'Waulsortian' carbonate mudmounds ('knoll reefs') also occur. The formation includes 61.5 m of 'green tuff and ?lava' in the Gun Hill Borehole (Hudson and Cotton, 1945). Volcanic rocks were also proved at Nook's Farm Borehole. The limestones range between Courceyan and Holkerian in age (Aitkenhead et al., 1985), the diachronous top being drawn at the base of the Ecton Limestones. Other Dinantian (or late Devonian) formations proved below the Milldale Limestones in the Caldon Low Borehole (Chisholm et al., 1988) may also extend into the present district. The lowest demonstrably Dinantian strata at Caldon Low are the Rue Hill Dolomites, of peritidal and shallow marine origin; a similar depositional environment may be inferred for the sandy dolomitic limestones proved in the lowest 76 m of the Gun Hill Borehole. The Redhouse Sandstones, unfossiliferous pebbly deposits of 'Old Red Sandstone' facies, underlie the Rue Hill Dolomites at Caldon Low and may occur also in the present district.

ECTON LIMESTONES

The Ecton Limestones (Hudson and Cotton, 1945; Aitkenhead and Chisholm, 1982) consist of medium to dark grey thinly bedded and bioclastic limestones, interbedded with fissile mudstones. Most sharp-based limestones are interpreted as turbidites, while conglomeratic limestones are interpreted as debris flows. One such conglomerate marks the base of the formation in the Gun Hill Borehole; the top of the formation is defined by the base of the Mixon Limestone-Shales. The formation probably passes laterally into the Astbury Limestone on the Market Drayton Horst and into the Hopedale Limestones at the south-west margin of the Derbyshire Block. At Gun Hill the formation is 182.5 m thick and is of Holkerian to Asbian age.

MIXON LIMESTONE-SHALES

This formation, named by Hudson and Cotton (1945), was redefined by Aitkenhead and Chisholm (1982). It is dominated by mudstones, which are interbedded with graded limestones, sandstones and tuffs. The base of the formation is placed at the base of the mudstone-rich sequence overlying the Ecton Limestones, and the top at the base of the Lask Edge Shales. The formation passes laterally into the Astbury Limestone-Shales (see below). Limestones of the Mixon Limestone-Shales are mostly bioclastic, and are commonest in the lower part of the formation. Where they are interbedded with tuffs (discussed below) they too are tuffaceous. A laterally extensive sandstone member, the Onecote Sandstones, is recognised towards the top. The formation is interpreted to have been deposited in moderately deep water, probably below fairweather wave-base, during the latest Asbian and Brigantian. The formation is 209.7 m thick in the Gun Hill Borehole (Evans et al., 1968).

Market Drayton Horst

The thickness of the Dinantian succession on the Market Drayton Horst is difficult to estimate, as only the upper part of the succession has been proved. This is exposed at Astbury (north of the district) and was drilled in the Ternhill Borehole (south-west of the district). The succession probably thins southwards, towards the axis of the the Wales–Brabant High.

ASTBURY LIMESTONE

The lowest beds known at Astbury are pale, thickly bedded limestones of Asbian age, the Astbury Limestone (Evans et al., 1968). Coral and brachiopod faunas from the formation suggest association with the 'reef' (mudmound) facies of Derbyshire (Evans et al., 1968). Similar faunas, and lithologies typical of mudmound associations, have been found in the Bowsey Wood and Bittern's Wood boreholes, and it is likely that these provings, also of Asbian age in the case of Bowsey Wood Borehole, are of Astbury Limestone. The base of the formation has not been proved and it is not known whether the formation overlies older Dinantian formations or rests unconformably upon Lower Palaeozoic or Proterozoic rocks. In the northern part of the district the top of the formation is defined by the base of the Astbury Limestone-Shales, but in Bowsey Wood and Bittern's Wood boreholes the formation appears to be unconformably overlain by Namurian rocks. The formation probably passes laterally northwards and eastwards into the Ecton Limestones of the North Staffordshire Basin. The pale massive limestones present at Astbury (as described by Evans et al., 1968) resemble the contemporaneous Bee Low and Kevin limestones of the shelf areas to the east (Aitkenhead et al., 1985; Chisholm et al., 1988). The similarity suggests that the three formations had similar depositional environments although they developed on blocks that were separated by the North Staffordshire Basin.

Bowsey Wood Borehole (Earp and Calver, 1961) cored 13.5 m of very fine- to very coarse-grained, slightly dolomitised packstones containing shell fragments with micritic (cyanobacterial) coatings, and calcareous mudstones. The fauna is of late Asbian age and includes the foraminifera *Archaediscus* sp., cf. *Bradyina* sp., *Earlandia* sp., *Endothyra* sp., *Endothyranopsis* sp., *Eostaffella* sp., *Gigasbia* sp., cf. *Koskinobigenerina* sp., *Mediocris* sp., *Millerella* sp., cf. *Neoarchaediscus* sp., *Palaeonubecularia* sp., *Paraarchaediscus stilus*, P. sp., *Planoarchaediscus concinnus*, *Priscella* sp., *Pseudoammodiscus volgensis*, *Pseudoglomospira* sp., *Spinobrunsiina* sp. and *Valvulvinella* sp.; the bryozoan *Rhopalonaria* sp.; the corals *Rotiphyllum* sp. and *Siphonodendron martini*; the brachiopods *Dictyoclostus* sp., *Echinoconchus* sp., *Gigantoproductus* sp., *Krotovia?* sp., *Orbiculoidea nitida*, *Spirifer* sp. and *Tornquistia* sp.; the pectinoid bivalve *Pseudamussium* sp.; and the trilobite *Cummingella* sp. The calcareous microflora contained calcispheres, *Draffania biloba*, kamaeniids, *Koninckopora inflata*, *Scalebrina* sp. and stacheiinids.

The 1.5 m of limestones recovered in Bittern's Wood Borehole comprise white to buff limestone of partly oolitic and partly fine crystalline character. Recovery was poor and no fauna was recorded.

ASTBURY LIMESTONE-SHALES

This formation is known only in the Dinantian inlier at Astbury (Figure 4) in the adjoining Macclesfield district, where it is less than 100 m thick (Evans et al., 1968). It consists of mudstones interbedded with limestones, sandstones and tuffs. The base of the formation is drawn at the base of the mudstone-rich sequence overlying the Astbury Limestone and the top is defined by the base of the overlying Lask Edge Shales. The formation passes laterally to north and east into the Mixon Limestone-Shales of the North Staffordshire Basin, which were probably deposited in deeper water. The presence of coals in the Astbury Limestone-Shales suggests that the formation was deposited partly under emergent conditions (Evans et al., 1968).

A coarse-grained feldspathic sandstone member, the Astbury Sandstone (Evans et al., 1968), is probably of shallow-water origin. It is unlikely to be equivalent to the non-feldspathic Onecote Sandstones, present in the Mixon Limestone-Shales at about the same level (see above).

It is probable that the Astbury Limestone-Shales were deposited over most of the Market Drayton Horst. Their local absence in Bowsey Wood and Bittern's Wood boreholes results probably from late Dinantian erosion related to an episode of uplift (Gutteridge, 1989).

DINANTIAN VOLCANIC ROCKS

Apedale No.2 Borehole did not encounter Dinantian limestones or shales below the Namurian sequence, but proved 847 m of tuffs. These are of Carboniferous aspect (see below), and their occurrence here suggests Dinantian vulcanicity in the Audley–Apedale area, probably centred on the Wem–Red Rock Fault system.

The tuffs are largely basic lapilli tuffs, consisting of highly vesicular pumice, microcrystalline lava fragments and broken feldspar crystals. Olivine basalt is the most common lava type, but many fragments are too altered for definite identification. Epiclastic material, including quartz-chlorite-schist, occurs in the upper part of the sequence. All the tuffs are strongly altered to intergrowths of secondary chlorite, calcite, white mica and haematite (E35040–35048, Appendix 2).

Because of the alteration, major element geochemistry can give only limited information on the primary composition of the tuffs. Immobile trace elements are more useful for this purpose, and the ratio of Nb/Y: Zr/TiO$_2$ suggests that the parental lavas were highly alkaline basalts and basanites. This composition is also supported by the relative proportions of Nb, Zr and Y. Together these data indicate that the parental lavas were derived from fertile mantle of ocean island basalt type, and erupted in a within-plate (?rift) volcanic province.

These tuffs are compositionally unlike those of Uriconian (Proterozoic) type, but very like those of known Dinantian age within the Pennine Basin (Sutherland, 1982). Their character is very similar to those of late Asbian to early Brigantian age at Astbury, but unlike the older Dinantian tuffs in the North Staffordshire Basin (Rees et al., 1996), suggesting that they too are of later Dinantian age.

NAMURIAN ROCKS

Namurian rocks crop out in the north-eastern part of the district, where a sequence of mudstones, siltstones and sandstones almost 1 km thick was deposited in the North Staffordshire Basin. The sequence thins south-westwards under the Coal Measures towards the west side of Stoke-on-Trent, where it is known at depth in the Bowsey Wood, Bittern's Wood and Apedale boreholes. The lowest Namurian rocks are probably absent in the attenuated sequence in this western area, due to south-westward onlap on to the Market Drayton Horst below Stoke-on-Trent. Namurian thickness variations (Figures 5 and 6) largely reflect the infilling of the relict block and basin topography inherited from the Dinantian (Figure 4).

In 1905 Gibson, following Hind and Howe (1901), referred the Namurian rocks beneath the Chatsworth Grit (known at the time as the Third Grit) to the Pendle-side Series, and those up to the Rough Rock were referred to as the Millstone Grit. Subsequently, Gibson (1925) placed the entire Namurian sequence within the Millstone Grit Series, a scheme maintained by Hester (1932). However, Gibson made a fundamental distinc-tion between the sequence containing quartzitic sand-stones, below the base of the Chatsworth Grit, and that containing feldspathic sandstones, above it. This lithol-ical distinction is the basis for the classification used here (Table 1). The sequence from the base of the lowest feldspathic sandstone (usually the Chatsworth Grit, though locally the Roaches Grit) up to the Subcrenatum Marine Band, is referred to as the Millstone Grit Group. The underlying sequence, featuring protoquartzitic sand-stones, is referred to as the Edale Shale Group. The nomenclature of the protoquartzitic sandstones used here is that of Chisholm et al. (1988). In both groups, mudstones and other finer- grained lithologies occurring between named sandstone units have not normally been given formal names. The classification used by Evans et al. (1968) in the Macclesfield district, which includes the northern part of the North Staffordshire Basin, is also shown in Table 1.

The protoquartzites were derived from the weathered terrain of the Wales–Brabant High to the south (Trewin and Holdsworth, 1973; Chisholm et al., 1988). The felds-pathic sandstones were transported into the district from a larger, more distant source area situated to the north or north-east (Collinson, 1988; Leeder, 1988). In the North Staffordshire Basin, in Kinderscoutian (R_1) and Marsdenian (R_2) times, feldspathic sandstone deposition in the north was contemporaneous with protoquartzite deposition in the south (Evans et al., 1968, fig. 5).

The sedimentology of the Namurian Pennine Basin fill has been reviewed by Collinson (1988). The strongly cyclical nature of the sequence has long been recognised (Ramsbottom, 1977). At a simplistic level, mudstones containing marine faunas, known as marine bands, inter-digitate with lithologies devoid of marine faunas. Glacially induced fluctuations of sea level seem the most likely cause of the faunal and lithological variations (Leeder, 1988; Maynard and Leeder, 1992), and the effects of the fluctuations on sedimentation have been widely discussed (Bristow, 1988; Holdsworth and Collinson, 1988; Collinson et al., 1992; Maynard, 1992; Read, 1991). A general review of Namurian cyclicity, with particular reference to the North Staffordshire Basin, is given by Holdsworth and Collinson (1988).

NAMURIAN BIOSTRATIGRAPHY

The Namurian sequence is divided into zones and stages by means of faunas contained in discrete bands, which represent the marine phases of successive cycles. The marine bands are usually less than a few metres thick, yet some extend laterally for many thousands of kilometres

Table 1 Previous and current Namurian strati-graphical classifi-cation in this district.

GROUPS	NAMED FORMATIONS (mudstones between sandstone formations are not named)	STAGES	NAMES USED IN MACCLESFIELD DISTRICT (EVANS et al. 1968)
MILLSTONE GRIT GROUP	Rough Rock (G_1b)	YEADONIAN (G_1)	Rough Rock Group
	Chatsworth Grit (R_2c)	MARSDENIAN (R_2)	Middle Grit Group
	Roaches Grit (R_2b)		
EDALE SHALE GROUP	Brockholes Sandstones (R_2b)		
	Kniveden Sandstones (R_1c)	KINDERSCOUTIAN (R_1)	Upper Churnet Shales and Kinderscout Grit Group
	Cheddleton Sandstones (H_1-H_2)	ALPORTIAN (H_2)	Middle Churnet Shales
		CHOKIERIAN (H_1)	
	Hurdlow Sandstones (E_2c)	ARNSBERGIAN (E_2)	Lower Churnet Shales
	Minn Sandstones (E_1b-E_2a)		Minn Beds
	Lask Edge Shales (E_1a-b)	PENDLEIAN (E_1)	Lask Edge Shales

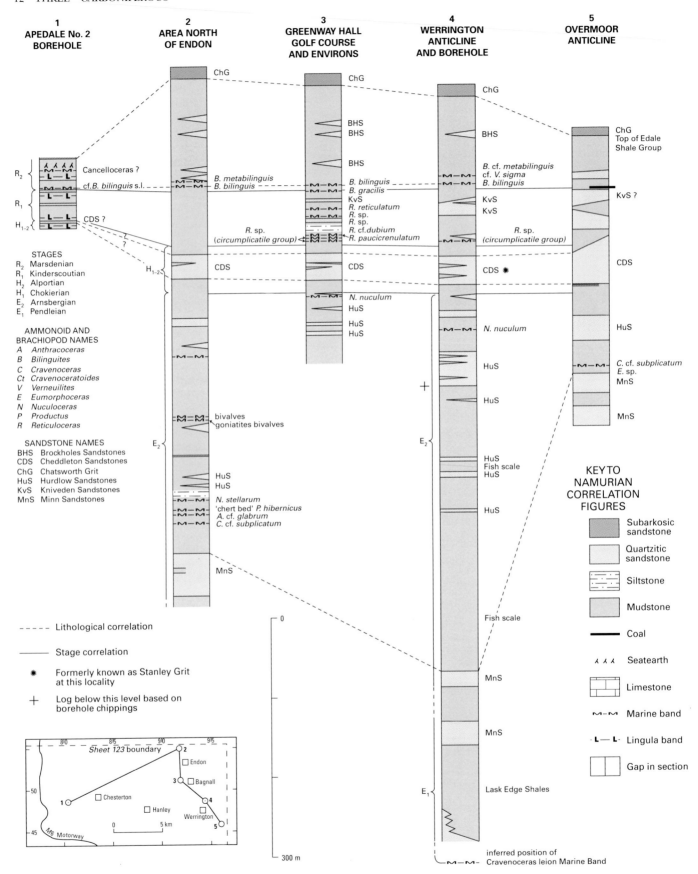

Figure 5   Correlation of the Edale Shale Group in the district.

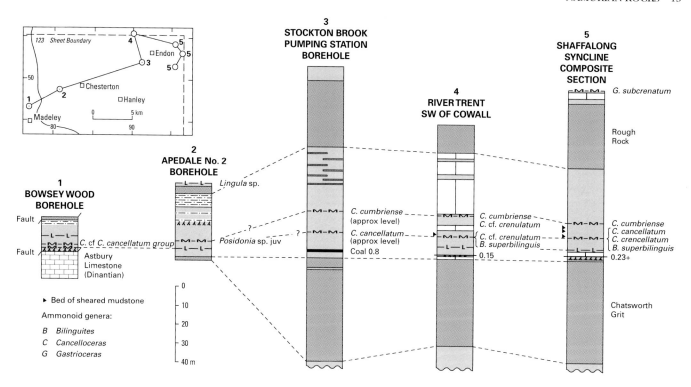

Figure 6 Correlation of the Millstone Grit Group in the district. For key see Figure 5.

(Ramsbottom, 1969). They consist of dark grey to black mudstones, commonly pyritic, with limestone or dolomite beds. Their faunas reflect water depth and substrate oxygenation, with an ammonoid (goniatite) facies representing the environment farthest offshore (Holdsworth and Collinson, 1988; Ramsbottom, 1962, 1969, 1980). Individual marine bands may contain a range of facies in vertical section, reflecting the initial transgression, highstand and then basinward regression of the shoreline. Such changes have been illustrated in detail by Wignall (1987) in respect of the Gastrioceras cumbriense Marine Band. The ammonoids provide a refined stratigraphical zonation; its application to the present district was pioneered by Hester (1932). A summary of Namurian marine band stratigraphy was given by Holdsworth and Collinson (1988). Faunas of Chokierian and Alportian age have rarely been seen, as mudstones of these stages are poorly exposed.

In boreholes for which only chipping samples are available, such as Apedale No.2 Borehole, biostratigraphical data have been enhanced by palynological information. Namurian palynology is based on Owens et al. (1977), though local work on Namurian coals (Millot, 1939) and marine bands (Turner et al., 1994) has been published.

Edale Shale Group

The Edale Shale Group comprises mudstones and siltstones, with subordinate protoquartzitic sandstones. The group is subdivided on the basis of sandstone formations within it. The intervening mudstones, which comprise the major part of the group, are unnamed, apart from those below the Minn Sandstones, which are called Lask Edge Shales. The sandstones include the Minn, Hurdlow, Cheddleton, Kniveden and Brockholes sandstones, which tend to form strong features, with intervening tracts of subdued terrain developed on the softer mudstones. The type area is in Edale, some 35 km to the north, where sandstones are rare (Stevenson and Gaunt, 1971).

In the eastern part of the district the group is probably conformable with the Dinantian sequence, the base being drawn at the base of the Cravenoceras leion Marine Band. In the western part of the district the group is unconformable on probable Dinantian rocks and the onset of sedimentation was significantly later than in the east; at Apedale it was probably delayed until Chokierian times. The top of the group is defined by the conformable, but diachronous, incoming of feldspathic sandstones at the base of the Millstone Grit Group.

The mudstones of the group are grey to dark grey and most contain nonmarine faunas such as fish scales and probable bivalve fragments, as found in chippings from the Werrington Borehole. In the less common marine mudstones most fossils are preserved as flattened impressions; undeformed specimens in calcareous nodules (bullions) are rare. Potassium bentonites, representing thin layers of volcanic ash, are present at several horizons (Trewin, 1968; Trewin and Holdsworth, 1972, 1973; Ashton, 1974; and Bolton, 1978). Carbonate rocks occur in the form of beds or nodules of sideritic ironstone, dolomite or limestone (Holdsworth, 1966). A distinctive calcareous siltstone lithology was described by Trewin and Holdsworth (1973) in the lowest part of the group.

Most of the sandstones in the group are hard, pale coloured and quartzitic, and were previously known informally as 'crowstones' (Holdsworth, 1964). Detailed petrographic descriptions of these have been provided by Evans et al. (1968), Trewin and Holdsworth (1973) and Chisholm et al. (1988). Some samples described by them originate from the Stoke-on-Trent district. Variations within and between sandstone formations are small. The following generalised description is based on many samples of the Minn, Hurdlow, Cheddleton, Kniveden and Brockholes sandstones (Appendix 2). The sandstones are quartz arenites (protoquartzites of Trewin and Holdsworth, 1973), composed mostly of quartz of igneous and metamorphic origin; lithic grains include acidic to intermediate volcanic rocks, chert, and quartzitic lithologies; feldspars are rare, but potassium feldspar and plagioclase both occur. Heavy minerals commonly include zircon, tourmaline, anatase and rutile, and are often concentrated on laminae or bedding planes. Grains usually have sutured contacts, and most pores have been largely filled by quartz cements.

The Minn and Hurdlow sandstones are mainly turbidites; individual beds tend to be graded, with sole marks, and are laterally persistent. The Cheddleton, Kniveden and Brockholes sandstones are more varied in origin; some may be turbidites, but the majority are shallow-water fluviodeltaic deposits (Chisholm et al., 1988).

The thickness of the group in the north-eastern part of the district appears to be at least 930 m, as suggested by evidence from the Werrington Borehole. From Werrington the group thins southwards, through the area of the Overmoor Anticline, towards the margin of the basin (Figure 5). The thinning is accompanied by lithological change, as exposures near the anticline are chiefly in sandstone. At the southern end of the anticline several sandstones, of likely Chokierian to Kinderscoutian age, appear to converge, cutting out intervening mudstone. The same southward transition was noted in the Ipstones Edge area of the Ashbourne district (Chisholm et al., 1988), where the amalgamated sandstones are called the Ipstones Edge Sandstones.

The greatest thinning of the Edale Shale Group is to the west of Stoke-on-Trent. In Apedale No.2 Borehole, equivalent mudstones with thin sandstones are only 88 m thick and rest unconformably on older strata. Palynology and limited ammonoid evidence from mudstone chippings suggest that most of this thickness is of Marsdenian and Kinderscoutian age (Figure 5). However, the palynomorphs Bellisporites nitidus, ?Cirratriradites saturni and Ibrahimspores brevispinosus suggest that the lowest 11 m may be early Chokierian in age. In Bittern's Wood Borehole (Figure 8) the estimated thickness of the Namurian (most of it likely to be Edale Shale Group) was 108 m. The thinning of the Edale Shale Group to the west of Stoke-on-Trent probably reflects the influence of the Market Drayton Horst.

LASK EDGE SHALES (E_{1a-b})

This formation, first named in the Macclesfield district (Evans et al., 1968), comprises mudstones and calcareous siltstones between the base of the Cravenoceras leion Marine Band and the base of the Minn Sandstones.

The Werrington Borehole penetrated 110 m of the formation (from 686 to 796 m depth). Chippings are dominantly of mudstones, pyritic in part, with fish scales, plant remains and some sideritic nodules. Thin sandstones are also present, especially towards the base. As in the equivalent sequence at Limekiln Brook, Astbury (Evans et al., 1968), the marine faunas are mainly in the upper half of the sequence.

MINN SANDSTONES

The term Minn Beds was introduced by Evans et al. (1968) for the protoquartzites and interbedded mudstones that form the prominent hills of Bosley Minn and Wincle Minn in the Macclesfield district. Aitkenhead et al. (1985) modified the name to Minn Sandstones and included all protoquartzitic sandstones, with their interbedded mudstones and calcareous siltstones, of E_{1b}, E_{1c} and E_{2a} age. The sedimentology and petrography of these were described by Trewin and Holdsworth (1973). The thickness of the the formation in the Werrington Borehole is 93 m.

Trewin and Holdsworth (1973) identified three sandstone units in the North Staffordshire Basin. The upper units occur near the southern end of the Blackwood Anticline, where blocks of fine-grained quartzite from the top unit occur at surface [9259 5482]. A borehole at Knowles Farm [9257 5489] appears to have penetrated mudstones between the upper units; to the north of the district these mudstones are of E_{2a} age, containing E. grassingtonense and E. erinense (Evans et al., 1968). Two fine-grained protoquartzitic sandstones, each 24 m thick, are exposed in the axial region of the Overmoor Anticline [961 461]. These sandstones, depicted as a single bed on the 1:50 000 map, occur below mudstones of low E_{2b} age (see below), and so probably lie near the top of the Minn Sandstones.

BEDS BETWEEN THE MINN AND HURDLOW SANDSTONES

The beds are exposed in stream sections north of Endon. A marine band containing Cravenoceras cf. subplicatum, Anthracoceras sp. and Posidonia corrugata identified in one stream [9241 5335] is probably the same as that found on the Overmoor Anticline [9612 4618], which yielded C. cf. subplicatum and Eumorphoceras sp. A marine band exposed in another stream [9219 5414] is likely to be at a slightly higher horizon. It contains Cravenoceratoides sp., possibly Ct. edalensis, and is overlain by 19 m of mudstones with Posidonia corrugata. Above are 0.2 m of cherty ferruginous siltstone with crinoid debris, Productus hibernicus, Rugosochonetes and Cravenoceras sp. This lithology and faunal assemblage are associated with the Cravenoceratoides nititoides Marine Band across much of north-western Europe. In a notable section at Holehouse Lane [9210 5480 to 9181 5487] the probable nititoides band is overlain by some 12 m of mudstones with scattered P. corrugata and Anthracoceras sp., succeeded in turn by 0.22 m of soft grey mudstone with abundant Nuculoceras stellarum, Anthracoceras sp., Posidonia corrugata and Posidoniella aff. vetusta [9202 5483]. Between beds with N. stellarum and attenuated representatives of the Hurdlow Sandstones are mudstones with beds of sideritic siltstone.

HURDLOW SANDSTONES

The bulk of the sequence from the top of the Minn Sandstones up to the base of the Cheddleton Sandstones consists of mudstones, but there are also some proto-quartzitic sandstones, collectively called the Hurdlow Sandstones. The thicker beds form ridges at outcrop.

The most persistent of the Hurdlow Sandstones forms extensive features east and west of Endon. It lies at a horizon probably below the lowest of the Nuculoceras nuculum bands which occur in the late Arnsbergian sequence. In quarries in Henridding Wood [9143 5425] 10 m of thickly bedded fine-grained protoquartzite are overlain by 0.9 m of fine-grained platy sandstone with plants, annelid trails and U-shaped burrows. Probably the same sandstone forms the ridge between Manor Farm, Endon [935 536] and Knowsley Common [944 513]. North of Endon, three thin sandstones are recognised within mudstones above the Nuculoceras stellarum Band. It is likely that at least one of the fine-grained quartzitic sandstones on the Overmoor Anticline [9621 4628] belongs to the Hurdlow Sandstones, since it lies close to the Minn Sandstones, yet overlies the band containing *C. subplicatum*. Mudstones between this sandstone and some higher leaves of Hurdlow Sandstones expose some of the *N. nuculum* bands near Moor Hall, Bagnall. At one locality [9491 5118], 0.3 m of calcareous mudstone yielded *N. nuculum*, *Cravenoceras?* and *Posidonia corrugata*, whilst an infilled pylon pit [9451 5095] yielded debris of *N.* cf. *nuculum*.

BEDS BETWEEN THE HURDLOW AND CHEDDLETON SANDSTONES

In the adjacent Ashbourne district three marine bands with *N. nuculum*, two with *Isohomoceras subglobosum*, and a *Lingula* band, were recognised in these beds (Chisholm et al., 1988). In the present district some Hurdlow Sandstones are locally present near Bagnall among beds containing *N. nuculum* (see above), but elsewhere no sandstones have been mapped. On Stanley Moor, an exposure [9210 5148] about 30 m below the Cheddleton Sandstones yielded *?N. nuculum*, *Aviculopecten* sp., *Posidonia corrugata* and *Posidoniella* sp. No mudstones containing *Isohomoceras subglobosum* have been found in the present district.

CHEDDLETON SANDSTONES

The Cheddleton Sandstones are laterally persistent proto-quartzitic sandstones of Chokierian and Alportian age, which were identified and defined in the Ashbourne district where the associated marine bands are well known (Chisholm et al., 1988). In the Stoke-on-Trent district the Cheddleton Sandstones are between 25 and 70 m thick. The sandstones form prominent features west of Endon, around the box-fold at the southern end of the Lask Edge Anticline at Stanley Moor [922 514], and around the Werrington Anticline. The name Stanley Grit was formerly used by the Geological Survey for the sandstone in this area (Gibson and Wedd, 1902; 1905; Gibson, 1905; 1925).

The best exposures of the Cheddleton Sandstones are in two quarries near Endon. In the more northerly of these [9197 5280], 18 m of thickly bedded fine- and medium-grained sandstone contain three siltstones between 0.6 and 1.35 m thick. A bedding plane rich in large plant stems is present

2.7 m below the top. In the quarry 200 m to the south, mudstones and siltstones are more numerous. Two leaves of sandstone occur south of the valley at Endon. The upper leaf is about 21 m thick and is exposed at Broughton Wood [9245 5125] where a plant-rich bedding plane occurs near the top of a 10.7 m section. The lower leaf is about 12 m thick, and the two leaves are separated by siltstones and mudstones, formerly exposed in a trial pit [9200 5141]. The village of Stanley is built on Cheddleton Sandstones, containing minor folds and bounded by faults. Two leaves of sandstone are recognised near Stanley Pool, where they are locally rich in plant remains. In a quarry below the pool [9290 5203] the lower leaf comprises 9 m of thickly bedded fine-grained sandstone overlain by 2.45 m of fine- to medium-grained sandstone with an undulating channelled base, 1.2 m in amplitude. The mudstone overlying the lower leaf was penetrated to 4.75 m in a borehole on the dam [9301 5197]. The sandstones can be traced southwards to Bagnall Grange, where they are estimated to be 28 m thick. The lower sandstones [9382 5230] contain quartz pebbles, as are found in equivalent sandstones at Sharpcliffe Rocks in the Ashbourne district (Chisholm et al., 1988). Sporadically exposed sandstones with up to two feature-forming mudstone or siltstone partings, about 25 m thick, can be traced round the Werrington Anticline. The outcrop [945 489] is repeated by the Armshead Fault.

On the eastern limb of the Overmoor Anticline four feature-forming sandstones, converging southwards, probably represent the Ipstones Edge Sandstones (H_1–R_1 inclusive) of the Ashbourne district (Chisholm et al., 1988). The lowest of these, exposed in the stream south of Heywood Grange [9625 4554] is a probable equivalent of the Cheddleton Sandstones. Exposures include fine-grained ripple-laminated sandstones with partings of purple and green siltstone and mudstone. The status of the higher sandstones is less clear, though they are possibly of Kinderscoutian age, and thus may be equivalents of the Kniveden Sandstones (see below). One is incompletely seen in the stream near Heywood Grange [9622 4559] in a 2 m exposure of thickly bedded very coarse-grained sandstone with small quartz pebbles; another is overlain by traces of coal [9636 4488].

BEDS BETWEEN THE CHEDDLETON AND KNIVEDEN SANDSTONES

Sections near Cheddleton on the western edge of the Ashbourne district (Chisholm et al., 1988) and east of Biddulph in the Macclesfield district (Evans et al., 1968) show that the Cheddleton Sandstones are overlain by the Homoceratoides prereticulatus Marine Band (H_{2c}). In this district mudstones of Alportian age overlying the Cheddleton Sandstones are seldom exposed. Younger mudstones, of Kinderscoutian age, are better exposed, and reach some 70 m in thickness.

The Kinderscoutian mudstones are exposed in a stream flowing into Stanley Pool [9360 5151], and contain *Reticuloceras* sp. (*circumplicatile* group), as does an exposure farther south [9397 5095]. Another section through the Kinderscoutian mudstones between the Cheddleton and Kniveden sandstones is exposed at Greenway Hall Golf Course, Baddeley Green [9186 5127 to 9189 5106]. It contains several marine bands, with faunas indicating the presence of the R_{1a} and R_{1c} zones. Thin quartzitic sandstones are interbedded with intervening mudstones of presumed R_{1b} age. The sandstones are unnamed in the present district but between Leek and Buxton they are called the Blackstone Edge Sandstones (Aitkenhead et al., 1985).

Stream section at Greenway Hall Golf Course

		Thickness m
Beds between the Kniveden and Brockholes sandstones		
Mudstone with *Bilinguites bilinguis* (R$_{2b}$)		1.60
Mudstone, grey		1.20
Gap	about	3.10
Mudstone with *Bilinguites gracilis* (R$_{2a}$)		0.90
Gap	about	8.00
Mudstone, fissile, grey		2.10
Kniveden Sandstones		
Sandstone, fine-grained, quartzitic	about	2.70
Beds between the Cheddleton and Kniveden sandstones		
Mudstone, grey		0.30
Gap	about	4.00
Mudstone, fissile, dark grey with *Anthracoceras* sp. or *Dimorphoceras* sp., *Homoceras* sp., *Reticuloceras reticulatum* and bivalves (R$_{1c}$)		1.80
Gap	about	1.20
Mudstone, grey		5.50
Mudstone, fissile, grey, sparse bivalves		0.23
Sandstone, medium-grained, with trace fossils		0.15
Mudstone, fissile, grey, with *Reticuloceras* sp. and bivalves		0.45
Mudstone, fissile, grey		2.40
Sandstone, very fine-grained, thinly bedded, with siltstone laminae in lowest 1.20 m		4.00
Mudstone, grey, with two 0.10 sandstone beds near base		1.50
Mudstone, fissile, grey, with thin beds of ferruginous siltstone		10.70
Mudstone, grey, fissile, earthy in top 0.10 m; *Reticuloceras* cf. *dubium*, *R.* sp. (nodosum group) (R$_{1a}$)		1.45
Mudstone; grey; *R. paucicrenulatum* (R$_{1a}$)		0.40
Mudstone; grey; *R.* sp. (*circumplicatile* group) (R$_{1a}$); *R.* sp. in basal 0.60 m		3.20
Mudstone, fissile, grey; pale grey in basal 1.00 m		2.90
Gap	about	3.00
Mudstone, grey; *R.* sp. (*circumplicatile* group) (R$_{1a}$)		0.30
Gap	about	14.00
Cheddleton Sandstones		

KNIVEDEN SANDSTONES

Fine-grained protoquartzites of late Kinderscoutian (R$_{1c}$) age are termed Kniveden Sandstones, following usage in the Buxton and Ashbourne districts (Aitkenhead et al., 1985; Chisholm et al., 1988). The sandstones thin northwards from the Werrington Anticline, where they are about 23 m thick and form persistent features; they are absent in the district to the north (Evans et al., 1968). The stratigraphy of the Kinderscoutian sequence in the North Staffordshire Basin, including the Kniveden Sandstones, was described by Ashton (1974).

Exposures on the Werrington Anticline [9392 4987] reveal fine- to coarse-grained, thin- to medium-bedded sandstone with linguoid current ripples. Several beds have strongly bioturbated tops, *Isopodichnus*-like arthropod tracks, 2–10 mm wide, being the most common trace. One of the sandstones is exposed in the Greenway Hall Golf Course stream section [9186 5127] (see above).

BEDS BETWEEN THE KNIVEDEN AND BROCKHOLES SANDSTONES

These strata, mainly mudstones, are well exposed in the stream section on Greenway Hall Golf Course (see above). The occurrence of *Bilinguites bilinguis* in the section probably marks the level of the B. bilinguis (early form) Band. *B. bilinguis* has also been recorded in a stream draining to the River Trent [9066 5519] and east of Cowall [9069 5516] where the ammonoid occurs abundantly, accompanied by *Homoceratoides* aff. *divaricatus* and *Caneyella*.

The B. bilinguis Band (*sensu stricto*) was also recorded near Bagnall [9374 5023] where it contained abundant *B. bilinguis*, *Anthracoceras* or *Dimorphoceras* sp., *Caneyella rugata* and *Dunbarella speciosus*. Some 10 m higher in the sequence [9364 5053] medium dark grey mudstones containing *B.* cf. *metabilinguis*, cf. *Verneuilites sigma*, *Caneyella rugata* and an orthocone nautiloid, may represent the B. metabilinguis Band.

BROCKHOLES SANDSTONES

The Brockholes Sandstones were defined in the adjoining Ashbourne district as protoquartzites of R$_{2b}$ age (Chisholm et al., 1988). Impersistent ridge features, which probably represent sandstones at this level in the sequence, occur on Brown Edge and on Greenway Hall Golf Course. Sandstone fragments associated with the ridges are quartzitic and very fine grained. In the north-eastern corner of the district the Brockholes Sandstones interdigitate with easterly derived feldspathic sandstones of the Roaches Grit. The relationship is most clearly demonstrated in a ravine [959 559] near Great Longsdon in the adjoining Macclesfield district (Chisholm et al., 1988).

Millstone Grit Group

The group is characterised by immature feldspathic sandstones, interbedded with grey mudstones and siltstones. It crops out around the margins of the Potteries and Shaffalong coalfields, in the north-east of the district. The sandstones form well-marked laterally persistent ridges, with intervening areas of lower ground underlain by mudstones or siltstones. Exposures are mostly in old quarries or natural scarps.

The base of the group is placed at the base of the lowest feldspathic sandstone in the Namurian sequence, and coincides generally with the base of the Chatsworth Grit, or with the base of the Roaches Grit where this is present. The base is thus diachronous, and in the north-eastern part of the district protoquartzites characteristic of the Edale Shale Group interdigitate with feldspathic sandstones of the Roaches Grit (see above). The top of the Millstone Grit Group is drawn at the base of the Sub-crenatum Marine Band, which marks the base of the Coal Measures.

In the Stoke-on-Trent district the group comprises mainly subarkosic sandstones, the Roaches Grit, Chatsworth Grit and Rough Rock. These consist of quartz, dominantly of igneous origin, and up to 25 per cent of feldspar. Orthoclase and microcline predominate greatly over plagioclase. The sandstones mostly form laterally extensive sheets, with cross-stratified, locally pebbly sandstone overlying interbedded siltstones and sandstones. The sandstones are commonly overlain by seatearths, with or without thin coals. The mudstones between the sandstones contain marine faunas in their lower parts.

The sandstones of the group represent the repeated progradation of deltaic systems across the Pennine Basin from a distant northern source. In the North Staffordshire Basin these feldspathic sandstones were being deposited simultaneously with protoquartzites entering from the south (Evans et al., 1968). The Roaches Grit represents the last deep-water infill of the North Staffordshire Basin (Jones, 1980) and the succeeding sandstones represent shallow-water, fluviodeltaic systems. The thickness of the group is between 100 and 170 m in the area of the North Staffordshire Basin but decreases to less than 50 m in the area west of Stoke-on-Trent (Figure 6).

ROACHES GRIT

The formation was referred to as the Third Grit or Roaches Grit by Hind (1910), and as the Roches Grit by Challinor (1921). The Stoke-on-Trent district is on the south-western limit of its depositional area. Compared with its development in the type area, to the north-east, it is thin and mostly fine grained. The sandstones form ill-exposed ridges at the north-eastern corner of the district [968 550], where they interdigitate with the Brockholes Sandstones (see above). The petrography of the Roaches Grit was described by Evans et al. (1968), Jones (1980), and Chisholm et al. (1988). Jones showed that the Roaches Grit was deposited by a west-north-west-migrating delta that was supplied from the same northern source as the other Millstone Grits.

BEDS BETWEEN THE ROACHES GRIT AND CHATSWORTH GRIT

These beds are very poorly exposed in the Stoke-on-Trent district. In the northern part of the Shaffalong Syncline they appear to be about 45 m thick and to consist of mudstone (Evans et al., 1968). In the adjoining Ashbourne district the lowest mudstones, containing marine bands with *Bilinguites superbilinguis* and *Verneuilites sigma*, are overlain by barren mudstones which pass upwards into siltstones and the Chatsworth Grit (Chisholm et al., 1988).

CHATSWORTH GRIT

Chatsworth Grit is the name given to the major feldspathic sandstones in R_{2c} (Smith et al., 1987) in the south Pennine area. The Chatsworth Grit commonly forms a prominent ridge, as at Brown Edge [908 539], Baddeley Edge [916 516] and around the Werrington Anticline. The sandstones are mostly medium to very coarse grained. They are rich in angular and fresh orthoclase and microcline, and well-rounded quartz, of mainly igneous origin. Heavy minerals include zircon, anatase and rutile. Pore spaces are filled with secondary quartz, illite, sericite and kaolinite (the latter often replacing orthoclase) (Appendix 2). Rounded, gravel-sized clasts of vein quartz are common, with more angular pebbles con-

centrated on erosional set boundaries. The sandstones are commonly reddish brown due to reddening associated with the Permo-Triassic unconformity. The sandstones are interpreted as the deposits of rivers which flowed from the east or north-east (Mayhew, 1966; Kerey, 1978; Chisholm et al., 1988).

The greatest thickness of the formation, 60 m, was penetrated in a borehole at Stockton Brook Pumping Station. It appears to thin southwards to about 30 m on the eastern fringe of the district near Caverswall [962 440], and to the west it virtually dies out, with indications only of a sandstone about 1 m thick in chippings in Apedale No. 2 Borehole. The most complete surface section, in a cutting on the A53 road near Longsdon [955 543], exposes about 34 m of cross-bedded and densely jointed coarse-grained sandstone. The best quarry sections are at Moss Hill [917 522] near Endon, and in Bagnall [9300 5086] where some 27 m of medium-grained cross-bedded sandstone may be seen.

BEDS BETWEEN THE CHATSWORTH GRIT AND ROUGH ROCK

Beds at this level are mainly mudstones and silty mudstones with a thickness of 33.5 to 57 m. Generally they form ill-exposed depressions between bounding sandstone ridges. The most complete section is in a disused brick pit [9644 5322] near Wall Grange, where a coal 0.2 m thick is overlain by mudstones with three marine bands, successively containing *Lingula*, *Cancelloceras cancellatum* and *C. cumbriense* (Figure 6, column 5). Between the ammonoid bands are three horizons of highly sheared mudstone resembling the 'crozzle' beds described by Cope (1946) at a similar stratigraphical level west of Buxton, and known also in Lancashire. The mudstones above the *C. cumbriense* Band become increasingly silty upwards, to the base of the Rough Rock. The borehole at Stockton Brook proved a coal 0.8 m thick, with overlying mudstones in which both ammonoid bands were present (Figure 6). Chippings from Apedale No. 2 Borehole yielded marine fossils at the inferred levels of the two ammonoid bands. In Bowsey Wood Borehole (Figure 6) a band with *Cancelloceras* cf. *C. cancellatum* group is overlain by one with *Lingula mytilloides*, which may be equivalent to the *C. cumbriense* Band.

ROUGH ROCK

The Rough Rock is the name applied to major, laterally persistent subarkosic sandstones of Yeadonian age in the Pennine Basin. In the present district it forms a single bold feature around the margins of the Potteries and Shaffalong coalfields. It is typically medium to coarse grained and moderately to poorly sorted. Most coarse grains and resistates are rounded. The quartz is of both igneous and metamorphic origin, and the feldspars are dominated by orthoclase and microcline. Rock particles include chert. Zircon is a common heavy mineral. Quartz, kaolinite, illite and sericite often fill pores (Appendix 2). Bristow (1988) showed that the quartz/feldspar ratio of the Rough Rock increases towards the south of the Pennine Basin (the highest ratio he recorded was in the North Staffordshire Basin). The sandstones are often red stained. Cross-bedding is common; Bristow (1988) noted planar cross-stratification at Knypersley Reservoir [900 553] showing a consistent westward palaeocurrent flow there. The Rough Rock of the Pennine area has been interpreted as the deposits of a braided river system flowing across an alluvial plain from the same northern source as the other Millstone Grits (Bristow, 1988). The thickness of the formation in the Stockton Brook Borehole was 34.1 m (corrected for dip), though the full thickness of Rough Rock in the district may reach 40 m. The minimum thickness of the formation at outcrop appears to be about 25 m.

The best quarry exposures are on Baddeley Edge in some 30 m of thickly bedded medium and coarse-grained sandstone [9138 5096, 9143 5125] and at Wall Grange Brick Pit [9644 5322] where the basal 18 m of medium-grained sandstone are exposed. The best natural exposures are at High Tor, Brown Edge [9059 5392] in 12 m of medium-grained cross-bedded sandstone, and at Wetley Rocks [966 492], 250 m into the adjoining Ashbourne district (Chisholm et al., 1988, plate 7).

BEDS BETWEEN THE ROUGH ROCK AND THE SUBCRENATUM MARINE BAND

These beds were cored in the Ridgeway Borehole. They consist of sandy seatearth with sideritic nodules, immediately overlain by the Subcrenatum Marine Band. The Six Inch Mine Coal of Lancashire is not present here (Magraw, 1957).

WESTPHALIAN ROCKS

Westphalian rocks crop out over much of the central and eastern parts of the district. They fall into two groups, the Coal Measures, which contain numerous coals, and the overlying Barren Measures, which include few coals, and comprise the Etruria, Newcastle, Keele and Radwood formations. The uppermost part of the Barren Measures, the Radwood Formation, is of probable Stephanian age.

The Westphalian sequence in the Stoke-on-Trent district provides some of the best evidence in Britain for syndepositional crustal stress changes during the Upper Carboniferous. The evidence comes principally from the thickness trends of sequential parts of the Westphalian sequence, as shown in isopachyte maps (Figure 7). The maps show that the influence of the Market Drayton Horst and North Staffordshire Basin, which had been important in the Dinantian and Namurian, declined in the early part of the Westphalian, and that, from the mid-Westphalian onwards, Variscan compression played an increasingly important role in controlling deposition. Details of the basin evolution are given in Chapter 8.

Coal Measures

The Coal Measures of North Staffordshire consist of a grey, mudstone-dominated sequence with numerous coal seams. They form three separate coalfields, each named after the syncline in which they are preserved: the Potteries Coalfield, the Shaffalong Coalfield and the Cheadle Coalfield. The Potteries Coalfield extends northwards into the Macclesfield district, and that part has been described by Evans et al. (1968). The Cheadle Coalfield, to the east, lies in the Ashbourne district and was described by Chisholm et al. (1988). The present account covers the whole of the Potteries and Shaffalong coalfields (Figure 7). All the North Staffordshire coalfields were described by Gibson (1905), who gave details of many sections which have not been reproduced here. The thickness of the Coal Measures in the Potteries Coalfield ranges between about 900 and 1600 m; only the lowest 200 m or so are preserved in the Shaffalong Coalfield.

The Coal Measures are of Langsettian, Duckmantian and Bolsovian age (Westphalian A to C). The base of the Langsettian Stage is marked by the Subcrenatum Marine Band, that of the Duckmantian Stage by the Vanderbeckei (Seven-Feet Banbury) Marine Band, and that of the Bolsovian Stage by the Aegiranum (Gin Mine) Marine Band.

At outcrop the sandstones in the Coal Measures tend to make ridges, or form higher ground than less resistant intervening mudstones. The Coal Measures and Etruria Formation generally form lower ground than the Newcastle and Keele formations, which contain several major sandstones. The outcrops of the latter formations also tend to be less scarred by features relating to former mining activities.

The generalised gamma and sonic logs of Bittern's Wood Borehole, showing the geophysical response of the Coal Measures and other Carboniferous formations, are shown in Figure 8.

DEPOSITIONAL ENVIRONMENTS OF THE COAL MEASURES

The Coal Measures comprise claystones, mudstones, siltstones, sandstones, ironstones, coals and rare limestones, which were deposited in delta or alluvial plain settings in the Pennine Basin. The sedimentology of the Coal Measures has been described by Fielding (1984a; 1984b; 1986) and Guion and Fielding (1988). A broad division of depositional environments into 'upper delta plain' and 'lower delta plain' was proposed for the Pennine

Figure 7 *(opposite)* Map showing principal collieries and opencast coal pits in the Potteries Coalfield and isopachyte maps for intervals in the Westphalian sequence.

a) locations of collieries and pits; b–l) isopachytes (in metres) between specified markers in the Westphalian sequence; the outcrop of each interval is shaded: Map i) approximates the thickness of the Etruria Formation after the effects of the diachronous base have been removed; map j) approximates the thickness of the Newcastle formation after the effects of the diachronous top have been removed; map k) approximates the thickness of the Keele formation after the effects of the diachronous base have been removed; map l) demonstrates the diachronous base of the Etruria Formation. In maps i) and j) stratigraphical thicknesses in the south and west of the district are derived from borehole geophysical logs; in k) data are wholly derived from borehole geophysical logs.

a.

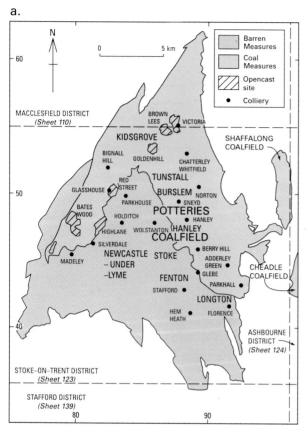

N

0   5 km

Barren
Measures

Coal
Measures

Opencast
site

• Colliery

MACCLESFIELD DISTRICT
(Sheet 110)

SHAFFALONG
COALFIELD

BROWN
LEES   VICTORIA

KIDSGROVE

BIGNALL
HILL   GOLDENHILL

CHATTERLEY
WHITFIELD

TUNSTALL

RED
STREET

GLASSHOUSE

BURSLEM

NORTON

PARKHOUSE   SNEYD

POTTERIES

BATES
WOOD   HOLDITCH   HANLEY

HIGHLANE   WOLSTANTON   HANLEY

SILVERDALE   COALFIELD

NEWCASTLE   STOKE   BERRY HILL

MADELEY   –UNDER   ADDERLEY
            –LYME   GREEN
                        GLEBE

FENTON

STAFFORD   PARKHALL

LONGTON

HEM   FLORENCE
HEATH

CHEADLE
COALFIELD

ASHBOURNE
DISTRICT
(Sheet 124)

STOKE–ON–TRENT DISTRICT
(Sheet 123)

STAFFORD DISTRICT
(Sheet 139)

b.

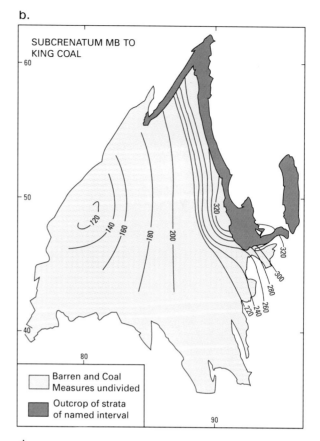

SUBCRENATUM MB TO
KING COAL

Barren and Coal
Measures undivided

Outcrop of strata
of named interval

c.

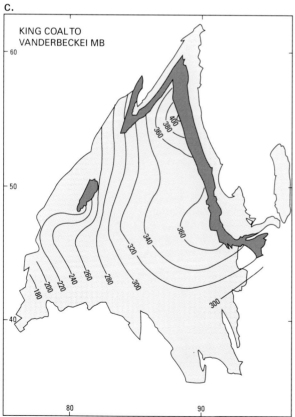

KING COAL TO
VANDERBECKEI MB

d.

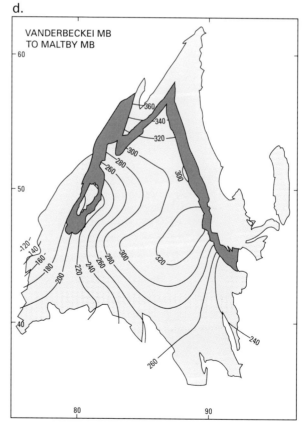

VANDERBECKEI MB
TO MALTBY MB

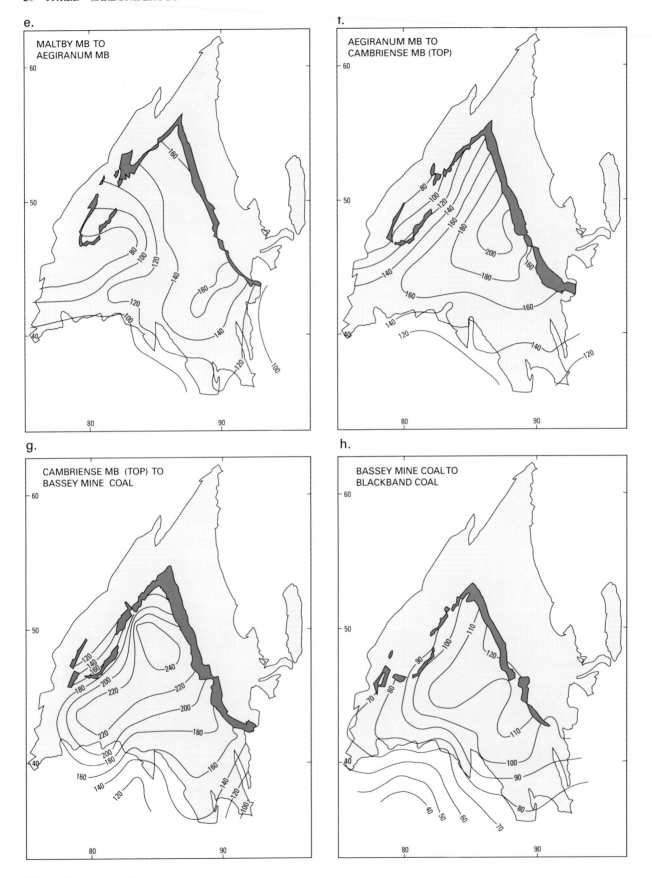

Figure 7 *(continued).*

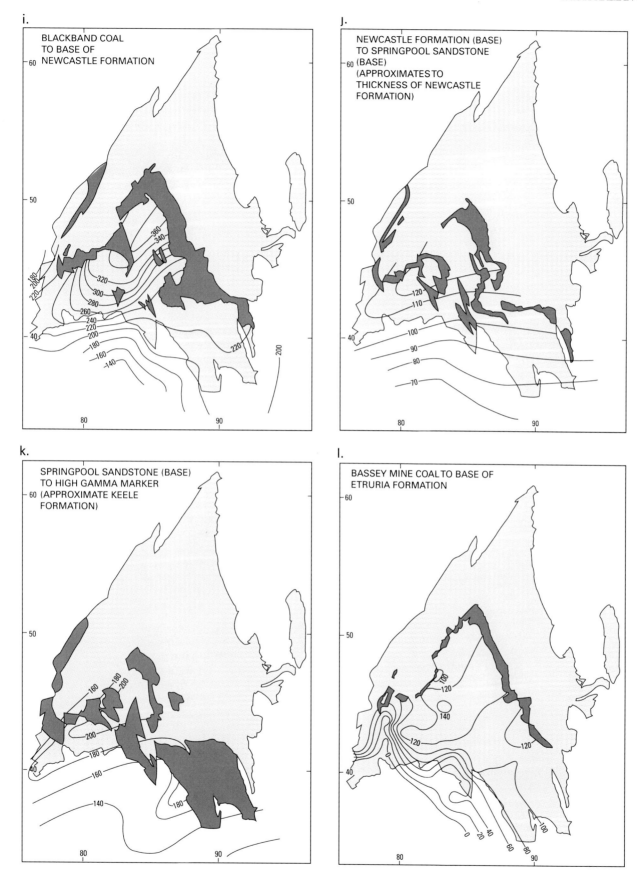

Figure 7 *(continued).*

Coalfield sequences by Guion and Fielding, who recognised the latter by the common occurrence of marine horizons. This distinction has been applied to the Potteries Coalfield (Figure 8), with some minor modifications.

Most sediments crossing the delta plains were transported by distributary channel systems. **Major channels**, between 2 and 10 kilometres wide, gave rise to the major sandstones of the coalfield, such as the Bullhurst Rock, Banbury Rock and Ten-Feet Rock. However, most were **minor channels**, in which were deposited thin, erosively based sandstones; they tend to occur in belts. These channel deposits commonly fine upwards and are epsilon cross-stratified, as may be seen in sandstones below the Yard Coal in the northern part of the district.

The channel systems were separated by extensive lakes. Where minor channels fed sediment into these, **lacustrine deltas** developed. These deltas formed progradational sequences, coarsening upwards from siltstone to sandstone. Many of the sandstones in the strata below the King Coal represent lacustrine deltas.

Sediments were also discharged from channels into lakes by flooding, or by breaching of the channel banks. **Overbank deposits**, caused by flooding, are commonly characterised by great variations in thickness. They are dominated by massive or poorly laminated siltstones, and thin ripple-laminated sandstone beds. Compressed plant debris is commonly well preserved in these sediments, which were actively aggrading. Such deposits form much of the lower part of the Middle Coal Measures. Following the breach of a channel bank, sediments were transported rapidly into the adjoining lake. The resulting **crevasse splay deposits** are characterised by upward-fining sharply bounded sandstones containing small cross-beds or ripple lamination. Near the breach the splay may be channelised and erosive, but elsewhere it is normally unconfined and non-erosive, forming sheet sandstones that thin distally.

The **lake deposits** that formed in the areas between distributary channels generally consist of pale grey to black mudstones, siltstones and claystones, with blackband or clayband ironstones, and cannel coals formed by the deposition of very fine-grained organic detritus. These deposits are commonly finely laminated. Lake deposits commonly coarsen upwards because of the progradation of lacustrine deltas; thus coarse siltstones and sandstones normally occur near the top of lacustrine sequences. Many lake deposits contain abundant roots, plant debris and, characteristically, a fauna of bivalves, ostracods and fish. A nonmarine setting is suggested by the bivalves, which have been compared with modern-day freshwater unionids (Trueman, 1946; Eagar, 1960; Calver, 1968). The lakes, at times, covered the entire district, as is illustrated by the widespread lateral extent of some of the blackband ironstones of the Upper Coal Measures. The largest lakes, for instance that in which the Accrington Mudstones of the Lower Coal Measures were deposited, covered most of the southern and central Pennine area, and were at times connected to the sea (Chisholm, 1990).

Where sediments were near or above the water surface, plants colonised them and, with time, soils formed. These siliciclastic palaeosols, more popularly known as **seatearths** or fireclays, resulted from the modification of sediments by weathering (leading to clay production), near-surface relocation of material, and the incorporation of organic matter. Seatearths commonly exhibit numerous rootlets, disturbed lamination, irregular sideritic concretions and crude horizonation. They commonly contain listric surfaces caused by compactional reorientation of clay minerals in the presence of organic debris (Schiller, 1980). Palaeosols formed in poorly drained conditions (hydromorphic or gley soils) are grey and contain sideritic concretions, whereas palaeosols formed in partially drained environments have a distinctive red, yellow, brown or mottled pigmentation and contain abundant sphaerosiderite. Partially drained palaeosols are only well developed near the top of the Coal Measures, especially above the Bassey Mine Coal (Figure 18). These were formed by temporary lowering of the water table, which allowed oxygenated water into the higher parts of the palaeosols (Besly and Fielding, 1989).

Coals are formed from peat, composed mainly of lycophyte remains which accumulated on the floor of equatorial rain forests in extensive sheet-like mires. Most coals represent rain-fed bogs which developed above the water table, as opposed to most seatearths and high-ash coals, which developed in swamps fed by groundwater (Fulton, 1987). An indication of the vertical variation in coal character in a single borehole, at Pie Rough, near Newcastle-under-Lyme, was given by Millot et al. (1946).

Occasionally, sea-level changes caused the delta plain to be inundated by marine waters. The **marine bands** which were deposited as a result are commonly less than a few metres thick, yet extend for many thousands of kilometres (Ramsbottom et al., 1978; Calver, 1968b). They consist of dark grey to black mudstones, many of them with distinctive faunas. These reflect water depth and substrate oxygenation; an ammonoid-pectinoid facies was formed farthest offshore, with a shoreward

Figure 8 *(opposite)* Classification and generalised sequence of the Westphalian and Namurian in the Potteries Coalfield showing the principal depositional environments of the intervals of the Coal Measures described in the text and their relationship to Westphalian nonmarine bivalve and marine band stratigraphy.

The diagram is based on Bittern's Wood Borehole, from which the gamma and sonic logs were derived. Depths are in metres below the rotary table of the drilling rig (139 m above OD). The position of coals and marine bands marked * in the borehole is uncertain; they are marked only in their approximate position. Namurian stages: P–A = Pendleian to Alportian, K–Y = Kinderscoutian to Yeadonian. Namurian groups: ES = Edale Shales, MG = Millstone Grit. Key to nonmarine bivalve subzones (modified after Ramsbottom et al. 1978): AD *Anthraconaia adamsi-hindi*, AT *Anthracosia atra*, CA *Anthracosia caledonica*, PH *Anthracosia phrygiana*, OV *Anthracosia ovum*, RE *Anthracosia regularis*, CR *Carbonicola crista-galli*, PS *Carbonicola pseudorobusta*, BI *Carbonicola bipennis*, TO *Carbonicola torus*, PR *Carbonicola proxima*, FP *Carbonicola fallax-protea*.

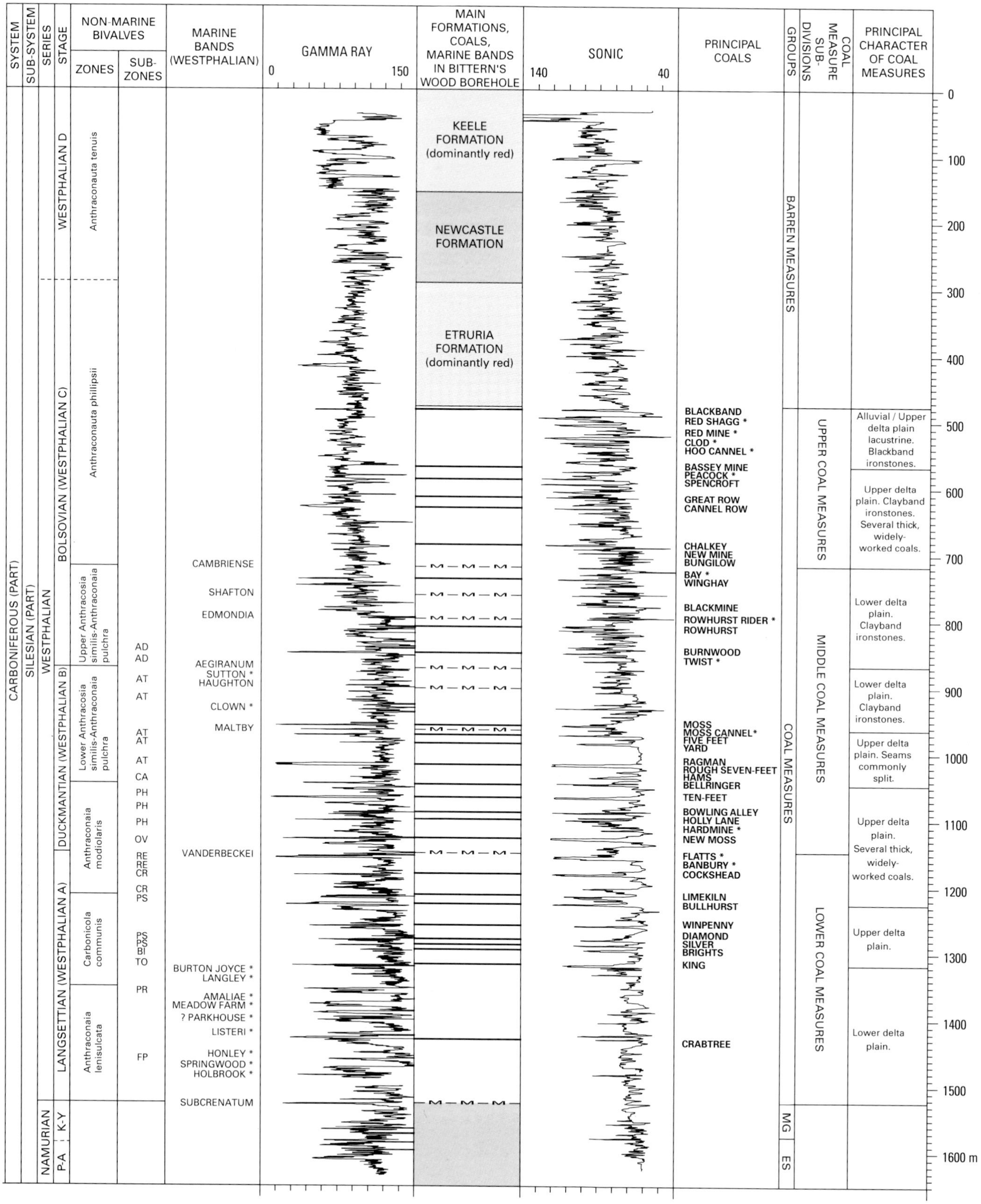
SYSTEM
SUB-SYSTEM
SERIES
STAGE
NON-MARINE BIVALVES
ZONES
SUB-ZONES
MARINE BANDS (WESTPHALIAN)
GAMMA RAY
0
150
MAIN FORMATIONS, COALS, MARINE BANDS IN BITTERN'S WOOD BOREHOLE
SONIC
140
40
PRINCIPAL COALS
GROUPS
COAL MEASURE SUB-DIVISIONS
PRINCIPAL CHARACTER OF COAL MEASURES
CARBONIFEROUS (PART)
SILESIAN (PART)
WESTPHALIAN
NAMURIAN
WESTPHALIAN D
BOLSOVIAN (WESTPHALIAN C)
DUCKMANTIAN (WESTPHALIAN B)
LANGSETTIAN (WESTPHALIAN A)
K-Y
P-A
Anthraconauta tenuis
Anthraconauta phillipsii
Upper Anthracosia similis-Anthraconaia pulchra
Lower Anthracosia similis-Anthraconaia pulchra
Anthraconaia modiolaris
Carbonicola communis
Anthraconaia lenisulcata
AD
AD
AT
AT
AT
AT
AT
CA
PH
PH
PH
OV
RE
RE
CR
CR
PS
PS
PS
BI
TO
PR
FP
CAMBRIENSE
SHAFTON
EDMONDIA
AEGIRANUM
SUTTON *
HAUGHTON
CLOWN *
MALTBY
VANDERBECKEI
BURTON JOYCE *
LANGLEY *
AMALIAE *
MEADOW FARM *
? PARKHOUSE *
LISTERI *
HONLEY *
SPRINGWOOD *
HOLBROOK *
SUBCRENATUM
KEELE FORMATION (dominantly red)
NEWCASTLE FORMATION
ETRURIA FORMATION (dominantly red)
BLACKBAND
RED SHAGG *
RED MINE *
CLOD *
HOO CANNEL *
BASSEY MINE
PEACOCK *
SPENCROFT
GREAT ROW
CANNEL ROW
CHALKEY
NEW MINE
BUNGILOW
BAY *
WINGHAY
BLACKMINE
ROWHURST RIDER *
ROWHURST
BURNWOOD
TWIST *
MOSS
MOSS CANNEL*
FIVE FEET
YARD
RAGMAN
ROUGH SEVEN-FEET
HAMS
BELLRINGER
TEN-FEET
BOWLING ALLEY
HOLLY LANE
HARDMINE *
NEW MOSS
FLATTS *
BANBURY *
COCKSHEAD
LIMEKILN
BULLHURST
WINPENNY
DIAMOND
SILVER
BRIGHTS
KING
CRABTREE
BARREN MEASURES
COAL MEASURES
MG
ES
UPPER COAL MEASURES
MIDDLE COAL MEASURES
LOWER COAL MEASURES
Alluvial / Upper delta plain lacustrine. Blackband ironstones.
Upper delta plain. Clayband ironstones. Several thick, widely-worked coals.
Lower delta plain. Clayband ironstones.
Lower delta plain. Clayband ironstones.
Upper delta plain. Seams commonly split.
Upper delta plain. Several thick, widely-worked coals.
Upper delta plain.
Lower delta plain.
0
100
200
300
400
500
600
700
800
900
1000
1100
1200
1300
1400
1500
1600 m

progression to productoid, myalinoid, *Lingula* and foraminiferal facies (Calver, 1968a,b). Estheriid bands, accompanied in the late Westphalian by *Leaia* bands, also provide local and regional correlation; these probably represent the most brackish environments, caused either by mixing of marine and lacustrine water or, particularly in the late Westphalian, by seasonal evaporation of ephemeral lakes. Individual marine bands may contain some or all of these facies in vertical section, reflecting the initial trangression, highstand, and basinward regression of the shoreline, as the sedimentary regime readjusted to the new sea level. Marine bands in which *Lingula* is the only macrofossil are known as *Lingula* bands; they are particularly common in the Lower Coal Measures below the King Coal.

Several sandstones and mudstones in the Coal Measures are reddened. In the Upper Coal Measures most of this reddening took place during or soon after deposition, but reddening in most other parts of the Coal Measures, especially east of the Potteries Syncline, occurred by near-surface oxidation during Permian or Triassic times (Crofts, 1953; Trotter, 1954; Earp and Calver, 1961). The extent of such reddening can be surprisingly large. For instance, the Coal Measures below the Hawksmoor Formation in Weston Coyney No. 1 Borehole have been completely reddened (and all coals oxidised) to 84 m, and selectively reddened to the base of the borehole, 766 m below the Hawksmoor Formation. Because of their higher porosity and permeability, some Coal Measure sandstones, despite lying within grey mudstone successions, have been reddened. Examples include the sandstone below the King Coal near Abbey Hulton [9108 4913], the Ten-Feet Rock between Milton and Norton in the Moors (Gibson, 1905) and the sandstone below the Twist Coal in the Berry Hill area [9000 4688] and the sandstone below the Bungilow Coal near Hanley [8877 4761].

BIOSTRATIGRAPHY OF THE COAL MEASURES

The greatest biostratigraphical resolution of the Westphalian Series in Britain is achieved using marine bands and nonmarine bivalve faunas, and these provide the main framework for recognition and subdivision of the Langsettian to Bolsovian stages within the Series. Some marine bands contain a unique biostratigraphical marker, whereas others are identified by their position relative to characteristic nonmarine faunas. Some of the marine and nonmarine assemblages can be traced from western North America across into eastern Europe.

The district has played a fundamental role in providing information on the fauna and flora of the British Coal Measures. Marine bands were recognised in the mid 19th century (Molyneux, 1864), and their more widespread stratigrapical use was pioneered by Stobbs (1902; 1905). Many nonmarine bivalve species were erected by Hind (1893, 1894, 1895, 1896), using material from this district. A detailed account of research up to the early part of this century was provided by Ward (in Gibson, 1905), who also documented the occurrence of fossils with respect to lithostratigraphical horizons. Particular attention was drawn by Ward to the abundant fish remains from the Potteries Coalfield, and Hind and Stobbs (1903) provided a strati-

graphical chart of the molluscan fauna. These studies contributed, along with parallel investigations in other coalfields (e.g. Eagar, 1956), to the subsequent development of nonmarine bivalve biozonation of the British Coal Measures, which culminated in a paper by Calver (1956). Hind's material was refigured and reinterpreted in a monograph by Trueman and Weir (1945–1968), which remains the key reference on nonmarine bivalves. The nonmarine bivalve zonation was applied to the North Staffordshire coalfields by Melville (1947). Knowledge of the Potteries Coalfield sequence recorded by Gibson et al. (1905) was significantly advanced by Magraw (1957) and by Earp and Calver (1961) in the light of detailed borehole information. This work provided the basic marine band sequence which was alluded to by Calver (1968b), and later correlated with the standard British nomenclature proposed by Ramsbottom et al. (1978). The standard marine band names are generally used in this account, but the older local names are indicated in the description of individual bands and in the figures which show borehole and shaft sections.

The Westphalian macroflora of the district has been described by Kidston (1892; 1905), Dix (1931), Crookall (1955–1976) and Cleal and Besly (1994). Some of the coals in the district yielded spores that were used in early palynological studies (Millot, 1939), and formed reference material for Smith and Butterworth's (1967) pioneering work on palynological zonation.

A synthesis of biostratigraphical and chronostratigraphical classification of the Westphalian Series is shown in Figure 8.

LITHOSTRATIGRAPHY OF THE COAL MEASURES

The evolution of a lithostratigraphy for the Westphalian rocks of the north Staffordshire coalfields has been complicated by attempts to establish a chronostratigraphical framework for correlation between different coalfields.

The first scheme erected for north Staffordshire was that of Smyth (1861):

4	Upper Measures
3	Pottery Coal and Ironstone Measures
2	Lower Thick Measures
1	Lowest Measures.

This classification reflected the economic importance of the ironstone measures (3), which were believed to be of latest Carboniferous age as, at this time, the Upper Measures (now Barren Measures) were thought to be of Permian age. In 1890 Ward proposed the following classification of the Coal Measures:

3	Upper Division	Blackband ironstones and red and variegated marls
2	Middle Division	Bassey Mine limestone to Rowhurst Coal
1	Lower Division	Upper group (Rowhurst to Winpenny Coals) Lower group (below Winpenny Coal)

This again emphasized the economic importance of the blackband ironstones, which he included in his Upper

Division, and also took account of the relative poverty of the measures below the Winpenny Coal.

Following the remapping of the district by the Geological Survey (Gibson et al., 1899a,b and 1900) the only part of the Coal Measures to be separately named was the 'Blackband Group', which was characterised by its important ironstones and ceramic clays. The position of the base of the group, taken by Gibson et al. at the Bassey Mine Coal, became a long-standing source of controversy, largely because the character of the group evolves upwards from the underlying succession without a clear break, and because chronostratigraphical importance was attributed to it (Kidston, 1905; Gibson, 1925, p.38). In the recent resurvey the 'Blackband Group' has not been mapped out as a separate stratigraphical unit.

The subdivision of the Coal Measures below the 'Blackband Group' only became standardised in 1957 when Stubblefield and Trotter used the Vanderbeckei and Cambriense marine bands as boundaries to redefine the Lower, Middle and Upper Coal Measures throughout England and Wales. These subdivisions continue to be widely recognised, and are used in this account, though with one modification, that the top of the Upper Coal Measures is now drawn at the base of the lowest primary red-bed formation (Etruria Formation). Correlation between the Potteries and Cheadle coalfields was established by Cope (1946).

LOWER COAL MEASURES

The Lower Coal Measures are of Langsettian (Westphalian A) age, being bounded by the base of the Subcrenatum Marine Band and the base of the Vanderbeckei Marine Band. They range between about 330 and 700 m in thickness and occur in both the Potteries and Shaffalong coalfields (Figure 7a–c). They are best known on the margins of the former, as only the lower part of the sequence (below the Pasture Coal) is preserved in the Shaffalong Coalfield.

The Lower Coal Measures are here divided into a lower part, comprising strata up to and including the King Coal, and an upper part comprising strata between the King Coal and the Vanderbeckei Marine Band. This subdivision is based on lithofacies, as the measures up to the King Coal are characterised by a strongly cyclical sequence (Eden, 1954) in which marine bands are common and coals are thin and laterally impersistent. These features are indicative of deposition in a lower delta plain environment, as opposed to the upper delta plain environment represented by the overlying part of the Lower Coal Measures (Figure 8).

STRATA BETWEEN THE SUBCRENATUM MARINE BAND AND THE KING COAL

The strata below the King Coal are exposed on the eastern side of the Potteries Coalfield and in the Shaffalong Coalfield, and are known at depth in the west of the Potteries Coalfield. In the Potteries Coalfield they are approximately 330 m thick in the east and thin notably westwards; in Apedale No. 2 Borehole they are about 117 m thick. Only about 180 m of the strata are preserved in the Shaffalong Coalfield, though these are thicker than

correlatives in the Potteries Coalfield to the west (Figure 7b), and in the Cheadle Coalfield to the east (Chisholm et al., 1988).

Strata between the Subcrenatum Marine Band and horizon of the Pasture Coal

The **Subcrenatum Marine Band** marks the base of the Coal Measures. The first record of the band in the Potteries Coalfield was in the Timbersbrook Borehole (Cope, 1948, p.14). The marine fauna includes the definitive ammonoid *Gastrioceras subcrenatum*. Its acme phase is in a pectinoid or ammonoid facies throughout the district. A typical section was encountered in the Ridgeway Borehole (549.63 to 550.75 m depth), where it consists of black mudstone. The basal 0.03 m lacks ammonoids, the fauna being limited to *Posidonia*, indeterminate mollusc spat and conodonts. The main acme phase of the marine band, 0.99 m thick, contains *Dunbarella* sp., *Posidonia gibsoni*, pyritic gastropod spat, *Anthracoceratites* sp., *Gastrioceras subcrenatum*, fish and conodont debris. The top 0.1 m of the marine band contains fish debris including *Elonichthys*. At surface the band is exposed at a locality near Ford Farm [9546 5160].

In the east of the district the Subcrenatum Marine Band is immediately overlain by commonly micaceous mudstones which contain rare plant and fish remains, and these are succeeded by a laterally persistent sandstone that correlates with the **Woodhead Hill Rock** of the adjacent Ashbourne (Chisholm et al., 1988), Buxton (Aitkenhead et al., 1985) and Macclesfield districts (Evans et al., 1968). The sandstone is well exposed in streams in the Werrington area, where it coarsens upwards (Howard, 1990). Petrographic samples of the sandstone from the Cheadle Coalfield (Chisholm et al., 1988, p.55) show that it is more feldspathic than the Namurian Chatsworth Grit and Rough Rock, verging on an arkosic composition. The Woodhead Hill Rock is between 25 and 47 m thick in the east of the district. Whether it occurs in the Apedale area is debatable; if present it is much thinner (Figure 9). In the eastern part of the Potteries Coalfield it is separated from the overlying Two Foot Coal by a thin development of micaceous siltstone with abundant rootlets, which thickens considerably into the Shaffalong Coalfield (Figure 9). The **Two-Foot Coal**, known as the Sweet Coal in the Shaffalong Coalfield, correlates with the Bassy Coal of Lancashire (Melville, 1947; Cope, 1948; Magraw, 1957). Though in most places it is only about 0.6 m thick, it was worked by numerous small pits in the Werrington and Shaffalong areas. The development of an ironstone, such as the Froghall Ironstone of the Cheadle Coalfield (Chisholm et al., 1988), at the horizon of the coal has not been recorded in this district. Nonmarine faunas occur above the Two Foot Coal. A stream section at Bank End Wood [5308 9068] yields *Carbonicola* aff. *fallax* and *C*. cf. *protea*, characteristic of the *Carbonicola fallax/protea* Subzone. These faunas correlate with those above the Bassy Coal of Lancashire (Evans et al., 1968, p.99). The sequence between the Two Foot and Crabtree coals is mostly mudstone, with thin marine horizons that represent one or more of the Holbrook, Springwood and Honley marine bands of the standard Westphalian sequence (Ramsbottom et al., 1978, pl. 2). The lowest of these, the **Holbrook Marine Band**, is possibly represented at an exposure [9604 5273] near Cheddleton, from which *Lingula*, *Posidonia* sp.? and spat have been obtained. nonmarine faunas occur above this horizon; assemblages typical of this interval (Eagar, 1956) were encountered in the Ridgeway Borehole (494.38 to 499.67 m), though the marine band itself was not recorded. The fauna includes *Carbonicola* cf. *haberghamensis, C.*

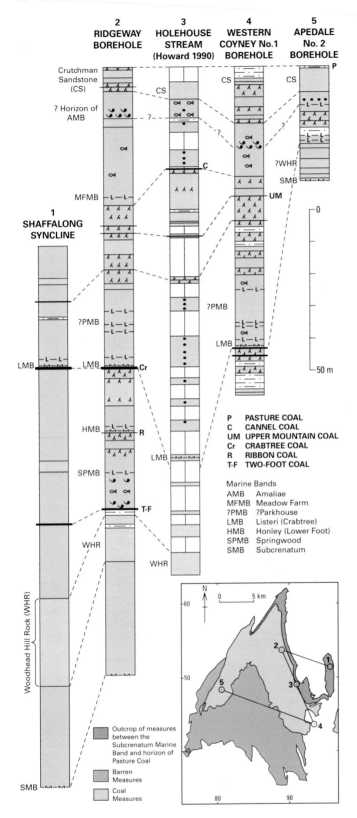

Figure 9 Correlation of the Lower Coal Measures between the Subcrenatum Marine Band and horizon of the Pasture Coal. For key see Figure 10.

rectilinearis, C. pilleolum, C. aff. *protea, C.* aff. *fallax, Naiadites* sp., *Geisina arcuata* and fish debris including *Rhabdoderma* sp. A higher band (488.7 to 490.95 m) contains C. aff. *protea, C.* cf. *artifex, C. protea* trans. *discus, Curvirimula* sp., *Naiadites* sp. and *Geisina arcuata*. This interval is exposed in a cliff [9595 5287] west of Deep Hayes Reservoir. **The Springwood Marine Band** is possibly represented by a *Lingula* band at 488.3 m in the Ridgeway Borehole, overlain by mudstones containing *Anthracosia* cf. *concinna, Carbonicola* cf. *limax, Curvirimula* sp., *Naiadites* sp, *Geisina arcuata*, acanthodian debris, *Elonichthys* sp., and *Rhabdoderma* sp. In the Werrington area these mudstones are locally overlain by laterally impersistent sandstones (Howard, 1990).

The **Ribbon Coal**, an impersistent seam which is less than 0.4 m thick in most places, is correlated with the Lower Foot Coal of Lancashire (Magraw, 1957). It lies approximately midway between the Two Foot and Crabtree coals. North of Longton it is overlain by a lenticular sandstone reaching 25 m in thickness near Moorside [9230 4865]. The Ribbon Coal is overlain by the **Honley (Lower Foot) Marine Band**, in an ammonoid or pectinoid phase. The best section is in the Ridgeway Borehole (473.61 to 475.11 m), where the main phase with mollusc spat, *Posidonia* cf. *gibsoni*, an orthocone nautiloid, *Anthracoceratites* sp., *Gastrioceras* sp., conodont debris and fish scales is overlain by 1.32 m of mudstone with *Lingula*, fish, and possible *Posidonia* sp. Surface exposure is known west of Deep Hayes Reservoir [9595 5287]. The marine band passes upwards into silty ganisteroid seatearths which may represent the position of the Lower Mountain Coal of Lancashire (Magraw, 1957).

The **Crabtree Coal** above the seatearths is equivalent to the Bullion Coal of Lancashire (Magraw, 1957). Otherwise known as the Stinking Coal because of its sulphurous nature, it has been widely worked, but only close to the outcrop. The coal normally forms a single leaf, up to 1.4 m thick. However, south of Werrington it is split (Howard, 1990) as in Weston Coyney No. 1 Borehole (Figure 9). The Crabtree Coal is overlain by a distinctive sequence of dark pyritic mudstones with several marine horizons. The lowest, and most faunally diverse, is the **Listeri (Crabtree) Marine Band**, which normally occurs within 2 m of the Crabtree Coal. The marine band is widespread and normally about 3 m thick. Because of the presence of easily identifiable ammonoids it is an important marker horizon. Debris from the bed can be easily recognised in old mine tips and helps to map the outcrop. It is the lowest horizon in the Westphalian to contain an arenaceous foraminiferal phase; this is typically present in the upper part. The Ridgeway Borehole (453.72 to 456.13 m) cored a representative section. The main faunal phase is in ammonoid or pectinoid facies (454.94 to 456.13 m) and contains *Lingula mytilloides, Caneyella multirugata, Dunbarella papyracea, Posidonia gibsoni*, gastropod spat, orthocones, *Metacoceras cornutum* var. *carinatum, Anthracoceratites* sp., *Gastrioceras listeri*, conodont debris and fish fragments including *Elonichthys* sp. Surface exposure occurs west of Deep Hayes Reservoir [9595 5287]. The marine band was recorded in the Shaffalong Coalfield by Scott (1927).

Dark mudstones overlying the Listeri Marine Band in the Ridgeway Borehole contain three marine beds and two nonmarine faunal horizons spread through a thickness of about 16 m. The correlation proposed by Magraw (1957) implies that all of these belong to the Listeri band, because he identified the Inch Coal as a 0.02 m seam at a higher level in that borehole. The **Parkhouse Marine Band,** which normally overlies the Inch Coal, was judged to be represented by fish-bearing mudstone at 425.6 m depth. However, the Parkhouse Marine Band could alternatively be represented by the upper two, or all three, of the marine bands that he considered to belong to the Listeri band (Figure 9). Similar spacings of these marine horizons are known in the East Midlands Coalfield (Eden, 1954). The lowest (at 445.13 to 445.39 m depth in the Ridgeway Borehole) contains *Curvirimula* sp. and *Geisina arcuata* in addition to *Lingula mytilloides* and abundant arenaceous foraminifera. The middle band (442.54 m) contains arenaceous foraminifera, *Lingula* and fish debris. The highest (437.26 to 437.36 m) contains arenaceous foraminifera, *Lingula mytilloides* and fish debris. This marine band is underlain by a thin nonmarine horizon with *Curvirimula* and *Geisina arcuata*. The only other record of what may be the Parkhouse Marine Band is in Weston Coyney No. 1 Borehole (829.36 m) where a similar fauna occurs, but apparently lacks arenaceous foraminifera.

The sequence above the dark marine mudstones is dominated by paler mudstones, siltstones, sandstones and several seatearths. On the eastern side of the coalfield three thin coals are known. The lowest of these reaches 0.6 m in thickness in the Werrington area, where it was locally exploited. In the Ridgeway Borehole this seam (at 425.7 m) is 0.02 m thick and was identified by Magraw (1957) as the Inch Coal of Lancashire, but this correlation is now in doubt (see above). A coal correlated with the **Upper Mountain Coal** of Lancashire, and which may be equivalent to the Split Coal of the Cheadle Coalfield (Chisholm et al., 1988), ranges locally up to 0.5 m thick, but in many places is represented only by a coaly mudstone. In the Holehouse stream section (Figure 9) it may be represented by a fusainous coal, 0.1 m thick, underlain and overlain by 0.4 m and 0.2 m respectively of coaly carbonaceous mudstone.

The **Cannel Coal** is made distinctive by the occurrence of marine mudstones above it. The coal appears to be thickest in the Werrington area, where it averages 1.1 m in thickness. It is exposed in Holehouse stream (Figure 9) and has been worked at surface and underground (Howard, 1990). Only in the Ridgeway Borehole (401.89 to 405.26 m) have samples from the overlying **Meadow Farm Marine Band** been retained. It contains arenaceous foraminifera, sponge spicules, *Lingula mytilloides* and fish debris.

No marine fauna corresponding to the Amaliae Marine Band has been identified with certainty in North Staffordshire. However, Magraw (1957) noted that nonmarine faunas in the Ridgeway and Weston Coyney No. 1 boreholes resemble that associated with the Amaliae band in Lancashire; the marine band is probably represented here by mudstones, at about about 385 m in the Ridgeway Borehole and 790 m in the Weston Coyney No. 1 Borehole.

The Knypersley Marine Band (Stobbs, 1905), exposed at the side of Knypersley Reservoir [8965 5518, 8969 5498], contains *Dunbarella papyracea*, *Posidonia* cf. *gibsoni*, *Hindeodella* sp., *Megalichthys hibberti* and Acanthodian spines (Evans et al., 1968, p.108). It appears to lie above the level of the Listeri band and was correlated tentatively with the Meadow Farm Marine Band by Evans et al. and with the Amaliae Marine Band by Ramsbottom et al. (1978). Neither correlation fits the sequence proved in the Ridgeway Borehole, which is sited only 1.2 km to the south-west, and the identity of the band remains in doubt. The fauna most closely resembles that of the Listeri Marine Band.

The mudstones containing the Meadow Farm Marine Band are overlain by a sequence of sandstones, siltstones, seatearths and thin coals which are equivalent to the **Crutchman** (or **Milnrow**) **Sandstone** of Lancashire (Magraw, 1957). The top of the sequence is normally marked by a sandy fireclay which probably marks the position of the **Pasture Coal** of Lancashire (Magraw, 1957). This coal has been rarely noted in the Potteries Coalfield, though a seam reaching 0.6 m in opencast trials south of Werrington [9302 4645] may represent it (Howard, 1990).

Strata between the horizon of the Pasture Coal and the King Coal

The Pasture Coal is overlain by a thick sequence of mudstones and siltstones which are correlated with the Accrington Mudstones of Lancashire (Magraw, 1957). These can be subdivided into upper and lower parts which correspond to the Shibden and Brighouse divisions respectively of Chisholm (1990) (Figure 10). These divisions, and the overlying Bradley Wood division (see below) are based on distinctive lithofacies, and are recognisable over most of the south Pennine area. The Shibden division is dominated by dark grey, commonly micaceous mudstones, though thin siltstones and sandstones occur near the top in the eastern part of the Potteries Coalfield and siltstones predominate in the west (Figure 10). The division is characterised by the occurrence of fish debris and common pyritic spherulites. Two marine bands, the Langley Marine Band and the Burton Joyce Marine Band, present in other parts of the Pennine Basin (Chisholm, 1990), have not been identified in north Staffordshire, but may be represented by pyritic horizons. Faunas containing the bivalves *Carbonicola torus*, *Curvirimula* sp. and *Geisina arcuata*, referable to the *Carbonicola torus* Subzone, occur at the junction between the Shibden and Brighouse divisions in the Ridgeway Borehole (333.3 to 333.45 m). The upper part of the Accrington Mudstones, ascribed to the Brighouse division, are generally grey-green and poor in mica, though rare beds rich in mica or fish debris, similar to those of the underlying division, do occur. The Accrington Mudstones pass up into a siltstone-dominated sequence equivalent to the **Old Lawrence Rock** of Lancashire (Magraw, 1957), or the Kingsley Sandstone of the Cheadle Coalfield (Chisholm et al., 1988). This part of the sequence, representing the upper part of the Brighouse division, is exposed at Knypersley Reservoir [893 553]. A tough siltstone bed worked in places as the Whetstone lies in this part of the sequence (Evans et al., 1968). The sequence between the Old Lawrence Rock and the King Coal comprises the Bradley Wood division of Chisholm (1990). This consists of micaceous lithologies similar to those of the Shibden division, but includes some green lithologies, like those of the Brighouse division. Mudstones dominate, though thin, interbedded siltstones and sandstones are common in the top part, particularly to the west. The mudstones contain the notable Daubhill Fauna of nonmarine bivalves (Magraw, 1957). In the Ridgeway Borehole (227.53 to 240.33 m) this includes the bivalves *Carbonicola* aff. *bipennis*, *C. torus*, *Curvirimula* sp., *Geisina arcuata*, fish debris and *Naiadites* sp. Seatearths occur below the King Coal in both Ridgeway and Weston Coyney No. 1 boreholes.

STRATA BETWEEN THE KING COAL AND VANDERBECKEI MARINE BAND

The Lower Coal Measures above the King Coal are devoid of marine faunas. They contain several laterally extensive coals, especially in the upper part where the Bullhurst, Cockshead and Banbury seams occur. These features are generally indicative of upper delta plain environments (Fielding, 1984b).

The isopachyte map for the Lower Coal Measures above the King Coal (Figure 7c) shows that the local depocentre had shifted only slightly westwards from that during deposition of the strata below the King Coal, and that the measures continued to thin towards the Western Anticline. Examination of interseam intervals (Corfield, 1991) suggests that there may have been a phase of rapid subsidence between deposition of the Bullhurst and Cockshead coals. However, most other intervals show relatively uniform thickness trends after the effects of differ-

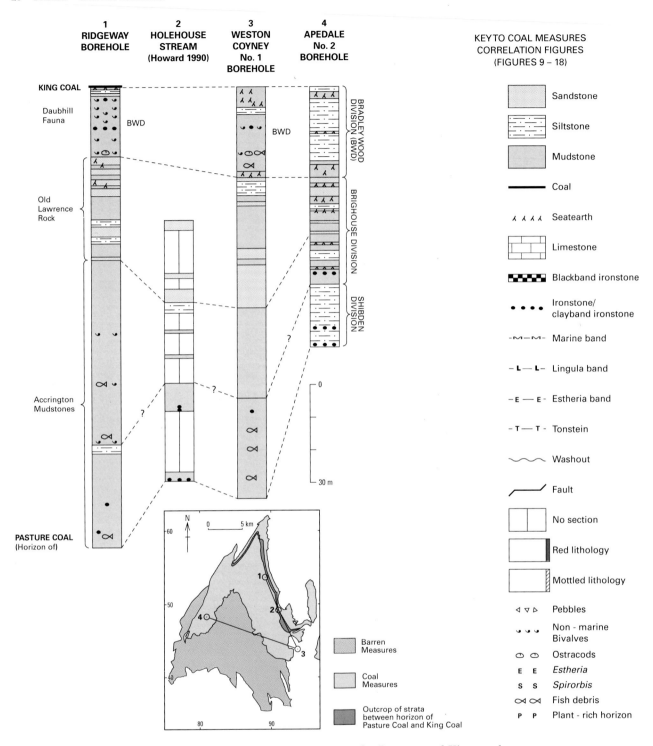

Figure 10 Correlation of the Lower Coal Measures between the Pasture and King coals.

ential compaction of varied lithologies have been taken into account.

For purposes of description the sequence between the King Coal and the Vanderbeckei Marine Band are informally divided into two parts. The strata below the Bullhurst Coal (Figure 11) are of minor economic importance and were generally worked on a small scale in northern and eastern parts of the coalfield, mostly in the Tunstall and Kidsgrove areas. In contrast, the seams above the Bullhurst Coal (Figure 12), especially the Bullhurst, Cockshead and Banbury seams, have been extensively worked and have been of major economic importance to the region. The strata between the King Coal and the Vanderbeckei Marine Band have generally been best exposed underground, particularly in 'cruts' (cross-cuts or cross-measure drivages); notable among

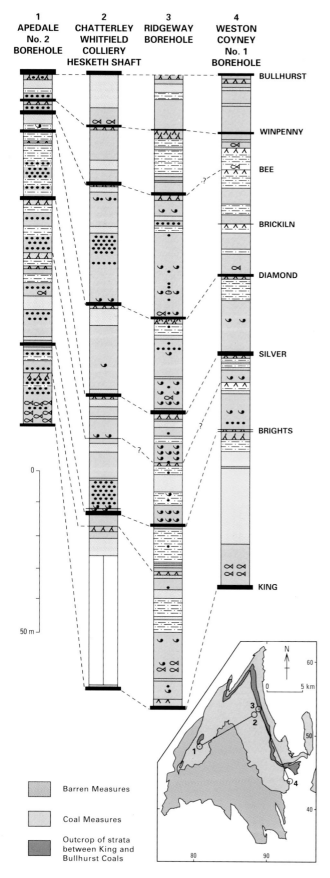

Figure 11 Correlation of the Lower Coal Measures between the King and Bullhurst Coals. For key see Figure 10.

these are the Bullhurst Main Crut at Norton Colliery, and the Hesketh Back Crut at Chatterley Whitfield Colliery (Figure 12).

Strata between the King Coal and the Bullhurst Coal

The **King Coal**, equivalent to the Woodhead Coal in the Cheadle Coalfield, is the lowest commonly worked seam in the Potteries Coalfield, though it has been mined almost exclusively close to surface. In most places it forms a single seam between 0.5 and 0.8 m thick. Erratic pebbles of quartzite were recorded in the coal in the Cheadle Coalfield (Lister and Stobbs, 1918; Stobbs, 1920, 1922) but have not been noted in the present district. Faunas immediately overlying the seam in the Ridgeway Borehole (213.21 to 222.58 m) are not diagnostic of a particular subzone; they comprise *Curvirimula trapeziforma*, *Naiadites* sp. and *Geisina arcuata*. The sequence between the King and Brights coals coarsens upwards overall from dark mudstones with notably rich fish faunas near the base to a siltstone and sandstone-dominated section near the top.

The **Brights Coal**, equivalent to the Rider Coal in the Cheadle Coalfield, has been worked mainly in the Tunstall and Kidsgrove areas where it is between 0.5 and 1.2 m thick. The sequence between the Brights and Silver coals consists of several thin cyclothems. Prominent mussel bands in the Ridgeway Borehole (between 144.47 and 167.49 m) yield assemblages characteristic of the *Carbonicola bipennis* Subzone. These include the eponymous species, along with *Anthraconaia* sp., *Carbonicola* ex gr. *communis*, *C.* cf. *obliqua*, *C.* cf. *polmontensis*, *C.* cf. *subconstricta*, *Curvirimula subovata*, *C. trapeziforma*, *Naiadites* sp., *Geisina arcuata* and *Spirorbis* sp.

The **Silver Coal**, which is mostly less than 0.8 m thick, has been little worked, and then only where it has a competent sandstone roof. The overlying mudstones contain nonmarine faunas, which in the Ridgeway Borehole (117.96 to 133.98 m) include *Curvirimula trapeziforma*, *Naiadites quadratus* and *Geisina arcuata*. Mudstones above the Silver Coal in Chatterley Whitfield No. 2 Underground Borehole (87.32 to 87.78 m) contain the lowest faunas indicative of the *Carbonicola pseudorobusta* Subzone of the *C. communis* Zone and are overlain by siltstones and sandstones.

The **Diamond Coal**, which is equivalent to the Cobble Coal in the Cheadle Coalfield, is less than 0.8 m thick. It is overlain by a mudstone sequence rich in ironstones, fish, ostracods and nonmarine bivalves. In the Ridgeway Borehole (73.23 to 99.44 m) the faunas include *Carbonicola aldamae*, *C. communis*, *C.* cf. *polmontensis*, *C. robusta*, *Curvirimula candela*, *C. subovata*, *C. trapeziforma* and *Geisina arcuata*. A thin coal occurs above the Diamond Coal in the Werrington area. It may correspond with the Wasp Coal in the Biddulph district (Howard, 1990).

The **Brickiln Coal** is equivalent to the Mans Coal in the Cheadle Coalfield. The seam commonly forms a single leaf between 0.5 to 0.8 m thick, and is overlain by measures rich in sandstones and siltstones. These are notably variable in thickness (Evans et al., 1968; Howard, 1990). The overlying **Bee Coal** has often been confused with the Brickiln Coal, and correlations of both seams remain tentative (Figure 11). The seam is commonly less than 0.7 m thick. The next higher seam, the **Winpenny Coal**, is equivalent to the Foxfield Coal in the Cheadle Coalfield and normally forms a single seam between 0.3 and 1.1 m thick. It was usually mined for household usage, particularly in the Tunstall and Werrington areas. It was more extensively worked than the underlying seams because of its proximity to the Bullhurst Coal and because in these areas it generally has a sandstone roof. The immediate roof of the

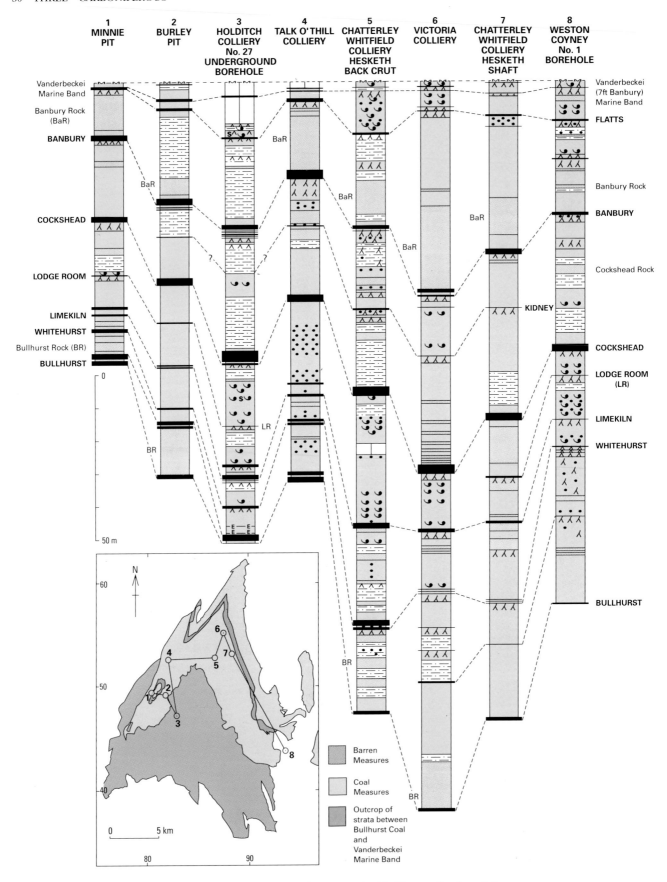

Figure 12 Correlation of the Lower Coal Measures between the Bullhurst Coal and the Vanderbeckei Marine Band. For key see Figure 10.

Winpenny Coal contains an impoverished fauna, locally with fish remains, as in Holts Barn Borehole (1194.28 m), but at Adderley Green Colliery, *Carbonicola* cf. *communis* also occurs. Between Kidsgrove and Apedale, the sequence between the Winpenny and Bullhurst coals is dominated by sandstone (Crofts, 1990b; Wilson, 1990).

Strata between the Bullhurst Coal and the Vanderbeckei Marine Band

The **Bullhurst Coal**, known as the Two Yard Coal at Norton Colliery, is equivalent to the Alecs Coal in the Cheadle Coalfield. It has been worked extensively in the Potteries Coalfield and commonly forms a single seam between 1.3 and 2.0 m thick, though near Madeley it approaches 5.5 m thick. Between Chesterton and Hanley it is commonly split; in the Burley Pit, Apedale, the parting is generally less than 0.2 m thick, but locally reaches 18.4 m and is almost entirely composed of sandstone. The lower leaf is known locally as the Little Coal, while the upper is known as the Tops and Middles (Wilson, 1990). The coal from the eastern part of the Potteries Syncline was extensively used by manufacturing industries and as a domestic fuel (Homer, 1875). In the western part the seam was worked as a bituminous coking coal. It tends to be somewhat sulphurous; its roof ('the hussle') often caused mine fires because it was liable to spontaneous combustion due to oxidation of contained pyrite (Stobbs, 1915). The Bullhurst Coal is in most areas overlain by mudstones; locally these contain notable fish and *Estheria* faunas. In places (Figure 12) the mudstones are thin or absent and the coal is closely overlain by a sandstone, the **Bullhurst Rock**. Locally this reaches over 25 m in thickness (Howard, 1990), but commonly it is split into several leaves by siltstones (Figure 12). The base is locally erosional, and the Bullhurst Coal has been removed in several areas, such as north-west of Longton (Cope, 1954; Howard, 1990) and near Biddulph (Evans et al., 1968). The Bullhurst Rock is well exposed in Dallows Wood [8904 5464] near Knypersley Reservoir. The sandstone fines upwards into siltstones and mudstones containing mussels and, towards the depocentre (Figure 12), thin coals.

The **Whitehurst Coal**, locally known as the Bottom Rider Coal, has been little worked. It commonly forms a single seam, 0.7 to 1.0 m thick, though in some parts of the coalfield it is split into several leaves. The highest fauna of the *C. pseudorobusta* Subzone occurs in mudstones and siltstones above the coal, for example in Holts Barn Borehole (1139.49 to 1152.91 m), where it includes the bivalves *Carbonicola* cf. *pseudorobusta*, *Curvirimula candela*, *C. subovata*, *Naiadites* sp., the ostracod *Geisina arcuata*, and fish remains.

A variable thickness of mudstones, siltstones and sandstones separates the Whitehurst Coal from the overlying **Limekiln Coal**, known also as the Sudden, Whitehurst Rider, Double Rider, Middle Rider or Top Rider Coal. In the past it has probably also been confused with the Whitehurst Coal. It was most important economically in the Biddulph valley (Evans et al., 1968). It ranges up to about 1 m in thickness, though is often split in the western part of the Potteries Coalfield. It is overlain by a sequence notable for numerous ironstones, and for faunas characteristic of the *Carbonicola crista-galli* Subzone. The faunas are well represented in the Holts Barn Borehole (1106.12 to 1127.61 m) and include *Anthraconaia* sp. nov., *Carbonicola* cf. *crista-galli*, *C. rhomboidalis*, *C.* cf. *robusta*, *Naiadites* sp., *Carbonita humilis*, *Geisina arcuata* and fish remains. A thin coal, the **Lodge Room Coal** or Top Rider Coal, is commonly associated with siltstones and sandstones in this interval, particularly in the south, west and central parts of the coalfield.

The **Cockshead Coal**, known as the Eight-Foot Banbury at Madeley Colliery, or the Eight-Feet Nabbs on the Western Anticline, is equivalent to the Dilhorne Coal of the Cheadle Coalfield. It normally forms a single seam between 2.3 and 2.6 m thick, making it easy to mine. On the east of the Potteries Syncline the Cockshead Coal was worked as a steam coal, for use largely in the iron and steel industry, but on the west side it was worked as a coking coal. The coal is low in ash (Millot, 1941), and commonly has clean, curved, slickensided joint surfaces, making it easily recognisable (Cope, 1954). The purity of the coal and lack of partings make puzzling the discovery of a quartzite cobble in the middle of the seam at Deep Pit, Hanley (Lister and Stobbs, 1917). At the top of the seam is a cannel coal, grading into a black coaly mudstone (Stobbs, 1915). In some areas this is immediately overlain by ironstone-bearing mudstones containing the highest faunas in the *Carbonicola crista-galli* Subzone. A typical assemblage is present in Holts Barn Borehole (1092.1 to 1095.6 m) and includes the bivalves *Carbonicola crista-galli*, *C. os-lancis*, *C.* cf. *rhomboidalis*, *C.* cf. *subconstricta*, *Carbonita* sp., *Geisina arcuata*, *Spirorbis* sp. and fish remains. Other bivalve taxa recorded from this horizon in the district include *Anthracosphaerium* cf. *boltoni* and *Naiadites* cf. *flexuosus*. The holotype of *C. crista-galli* and the lectotype of *C. rhomboidalis* (designated Trueman and Weir, 1947, pl. 7, figs. 1–3) come from the roof of the Cockshead Coal at Adderley Green Colliery. Well-preserved fish (Molyneux, 1864) and *Spirorbis* at this horizon are notable near Longton. These faunas are restricted to the lowermost part of the sequence between the Cockshead and Banbury coals. Higher strata in this interval are dominated by siltstones and thin sandstones, and include a laterally persistent sandstone known as the Cockshead Rock, from which the lectotype of *C. obtusa* (Trueman and Weir, 1947, pl. 7, figs. 12–13) was derived at Chatterley Whitfield Colliery. Over most of the coalfield a thin coal, the **Kidney Coal**, occurs towards the top of this sequence (Figure 12). A kaolin tonstein, 7 mm thick, occurs in the seatearth immediately below the Banbury Coal at Florence Colliery (Barnsley et al., 1966).

The **Banbury Coal**, known also as the Seven-Foot Banbury at Victoria, Sneyd and Madeley collieries, the Seven-Feet Nabbs on the Western Anticline, the Froggery at Norton Colliery, and the Bambury or Frogrow at Chatterley Whitfield Colliery, is equivalent to the Little Dilhorne Coal in the Cheadle Coalfield. It forms a single seam, commonly between 1.3 and 2.3 m thick (though this varies considerably: Gibson, 1925; Cope, 1954), making it easy to work. It has been worked from most of the major collieries in the district, in the western part of the Potteries Coalfield as a household or gas coal, and as a steam coal in the east. The immediate roof of the Banbury Coal marks the entry of assemblages belonging to the *Anthracosia regularis* Subzone, although definite examples of the eponymous taxon enter higher in the local sequence. There are few records, but *Anthracosia* sp. is recorded from Chatterley Whitfield Colliery, and *Naiadites* sp., *Carbonita* sp., *Spirorbis* sp. and fish debris are known from Sneyd Colliery. The interval between the Banbury and Flatts coals is made notable by the presence of a major sandstone, the **Banbury Rock**. This is over 30 m thick at its maximum (Figure 12) and represents a major channel. It fines upwards from the base, which contains abundant rip-up clasts, to generally fine-grained sandstones, as seen at an exposure at Miry Wood [812 494]. The siltstone-dominated facies that occur laterally and above the Banbury Rock represent crevasse-splay, overbank, or channel fill deposits. Locally these deposits supported vegetation, forming seatearths and thin coals. The channel and its associated facies occur throughout the Potteries Coalfield. The top of the interval between the Banbury and Flatts coals is marked locally by ironstone-bearing mudstones.

The **Flatts Coal**, known as the Littlemine Coal at Sneyd and Adderley Green collieries, has been of little economic importance, being less than 1 m thick. It is separated from the Vanderbeckei Marine Band by a sequence containing ironstone-rich mudstones, and minor sandstones and siltstones associated with seatearths and thin coals (Figure 12). An abundant fauna is present in the immediate roof of the coal throughout the district. This includes *Anthraconaia* aff. *insignis*, *A.* aff. *modiolaris*, *A.* cf. *williamsoni*, *Anthracosia regularis*, *Anthracosphaerium* cf. *cycloquadratum*, *Anthracosia aquilina* inter *lateralis*, *Carbonicola oslancis*, *C. rhomboidalis*, *C. obtusa*, *Carbonita humilis*, *Geisina arcuata*, *Naiadites* aff. *carinata* inter *triangularis*, *N.* cf. *productus*, *N. productus* inter *quadratus*, *N.* cf. *tumidus*, *Spirorbis* sp. and fish debris. Strata immediately below the Vanderbeckei Marine Band in Bowsey Wood Borehole (853.95 to 854.51 m) show a typical assemblage, with *Anthraconaia* cf. *modiolaris*, *Anthracosia regularis*, *A.* aff. *phrygiana*, *Naiadites* cf. *quadratus*, *G. arcuata*, an estheriid, and fish debris.

MIDDLE COAL MEASURES

The Middle Coal Measures comprise the strata between the base of the Vanderbeckei Marine Band and the top of the Cambriense Marine Band. The interval includes the whole of the Duckmantian Stage (Westphalian B) and early part of the Bolsovian Stage (Westphalian C). The Middle Coal Measures range between about 250 and 680 m in thickness in the Potteries Coalfield.

The Middle Coal Measures are here divided into lower and upper parts, which are further subdivided. The lower part, between the Vanderbeckei and Maltby marine bands, contains many thick, extensively worked coals and is devoid of marine bands (apart from the one at the base). These strata were probably deposited in an upper delta plain setting, like the strata immediately underlying them. The upper part, above the base of the Maltby Marine Band, contains fewer coals, which tend to be patchily or thinly developed, and includes several marine bands. These strata may have been deposited in a lower delta plain setting (Figure 8; Fielding, 1984b).

Strata between the Vanderbeckei and Maltby marine bands

The strata between the Vanderbeckei and Maltby marine bands thicken towards Hanley, and towards the northern part of the coalfield (Figure 7d). Where isopachytes have been generated for individual interseam thicknesses (Corfield, 1991) they generally mirror the pattern recognised for the entire interval between the Vanderbeckei and Maltby marine bands.

For purposes of description the strata are further subdivided here (Figure 8) into the sequence between the Vanderbeckei Marine Band and the Bellringer Coal, in which coals are laterally persistent and may be traced with ease around the Potteries Coalfield (Figure 13), and the sequence above the Bellringer Coal, in which coals vary considerably in thickness, as do intervening strata (Figure 14), making correlation more difficult.

Strata between the Vanderbeckei Marine Band and Bellringer Coal

The **Vanderbeckei Marine Band**, which marks the base of the Middle Coal Measures, is locally known as the Seven-Feet Banbury Marine Band (after a local name for the Banbury Coal), though it is commonly separated from that seam by over 25 m of strata (Figure 12). The marine band is widespread, and is around 3 to 4 m thick. Its acme fauna is in an ammonoid or pectinoid phase, but a close interbedding with nonmarine faunas is typical in the lower and upper parts of the band. In Bowsey Wood Borehole (853.67 to 854.05 m depth) the base of the band contains arenaceous foraminifera (including *Glomospira* sp.), *Lingula* sp., *Anthracosia* sp. and *Hollinella*. The acme fauna (853.21 to 853.67 m) includes *Lingula mytilloides*, *Dunbarella* sp., *Soleniscus* sp., *Anthracoceratites vanderbeckei*, conodont and fish debris. The waning phase (850.01 to 853.21 m) comprises arenaceous foraminifera, sponge spicules, *Paraconularia* cf. *crustula*, *Lingula mytilloides*, *Hollinella* sp. and fish debris. Other taxa recorded from the district include: *Agathammina* sp., *Ammonema* sp., *Hyperammina* sp., *Dunbarella papyracea*, *Myalina* sp., *Posidonia* sp., *Euphemites* sp., *Dithyrocaris* sp. and *Geisina arcuata*. The marine band was previously exposed in Kidsgrove [8396 5455] (Wilson, 1990) but is now only easily accessible at Miry Wood, Apedale [8118 4940], where the fauna includes *Lingula mytilloides*, *Hollinella clay-crossensis*, *Dunbarella papyracea* mut., ammonoid ghosts and fish debris (Wilson, 1990; Rees, 1993). The marine band is normally overlain immediately by mudstones rich in nonmarine bivalves typical of the *Anthracosia ovum* Subzone, with fish and ostracods. In Weston Coyney No. 3 Borehole (295.35 to 300.84 m) *Anthracosia* cf. *ovum*, *A. phrygiana*, *A.* aff. *phrygiana*, *Naiadites* cf. *quadratus* and *Spirorbis* sp. occur. *Anthraconaia* cf. *williamsoni* is also recorded from this interval at Fenton. Above the Vanderbeckei Marine Band at Miry Wood, *Naiadites flexuosus*, *N.* cf. *productus*, *Anthracosia ovum*, *Anthraconaia* and *Spirorbis* are found (Wilson, 1990). These mudstones grade upwards into sandstones and siltstones which dominate the upper part of the interval between the marine band and the New Moss Coal. Seatearths, and locally coals, are commonly developed towards the top of these (Figure 13).

The **New Moss Coal**, known as the Stinkers at Sneyd and Mossfield collieries, and the Newmine at Parkhall Colliery, is equivalent to the Four Foot Coal in the Cheadle Coalfield. The seam mostly occurs as a single leaf, though on the eastern and western margins of the Potteries Coalfield it is commonly split. In general it thickens from the western side of the coalfield, where it is about 0.5 m thick, to Hanley, where it is about 0.8 m thick, to the Longton area in the east, where it is about 1.7 m thick and has been most extensively worked. On the western side of the coalfield in Bittern's Wood Borehole, and possibly in Bowsey Wood Borehole, the New Moss Coal appears to be combined with the Hardmine Coal (information from J O'Dell of British Coal). The New Moss Coal at Chatterley Whitfield Colliery appears to thin over an unusually sandstone-rich sequence, proved in Hesketh Shaft (Figure 13), suggesting compactional control of sedimentation. The coal is commonly overlain by mudstones containing faunas that include *Anthracosia* sp., *Anthracosphaerium* cf. *exiguum*, as in Weston Coyney No. 3 Borehole (267.46 to 280.32 m). These mudstones are overlain by a sequence dominated by siltstones and sandstones with abundant plants (Gibson, 1925), previously well exposed near Kidsgrove (Crofts, 1990b). However, mudstones predominate in the south-eastern part of the coalfield, and in the north under the Hardmine Coal there is a white pottery clay (Evans et al., 1968).

The overlying **Hardmine Coal**, which takes its name from its extreme hardness, was known as the Top Muckrow at Madeley Colliery and is equivalent to the Littley Coal in the Cheadle Coalfield. It normally forms a single seam, 1.0 to 1.5 m thick. In the south-western part of the coalfield the seam is split into a lower thin leaf and an upper thicker leaf by up to 4.0 m of rooted siltstone, and locally appears to be combined with the New Moss Coal (see above). It was principally worked for use in blast furnaces as, on the eastern side of the Potteries Syncline where it was chiefly mined, it is bright, with a low ash content, and virtually sulphur-free. The seam contains lithologies made up largely of spores, including the 'Fretters' — a coal preferred by glaze manufacturers in the potteries (Stobbs, 1915). The coal is overlain by mudstones bearing nonmarine bivalves that have been referred to the *Anthracosia phrygiana* Subzone (Ramsbottom et al. 1978). Faunas which mark the onset of this subzone occur in roof mudstones reported to be above the Hardmine Coal in Florence Colliery Underground Borehole No. 4 (102.87 to 110.79 m). These include *Anthraconaia modiolaris*, *A. williamsoni*, *Anthracosia aquilina*, *A. beaniana*, *A. beaniana* inter *ovum*, *A. disjuncta*, *A. ovum*, *A. phrygiana*, *Anthracosphaerium affine*, *A. exiguum*, *Naiadites quadratus*, *N.* cf. *triangularis*, *Spirorbis* sp., *Carbonita humilis* and fish debris. *Anthracosia* cf. *aquilinoides*, *A. ovum* inter *phrygiana*, *A.* cf. *phrygiana* and *Anthraconaia* cf. *curtata* have also been recorded from this horizon north-east of Longton, as have the holotypes of *Carbonicola venusta*, *Anthraconaia obovata* and *Anthracosphaerium exiguum*. The mussel-rich mudstones grade upwards into a siltstone-dominated sequence. On the eastern margin of the coalfield siltstone dominates the entire sequence between the Hardmine and Holly Lane coals. The latter is commonly underlain by a seam less than 0.5 m thick, which has been worked in a few places only.

The **Holly Lane Coal**, known as the Bottom Two Row at Madeley Colliery, the Under Two Row in the Potteries Syncline and the Two Row on the Western Anticline, is equivalent to the Yard Coal of the Cheadle Coalfield. It is normally 1.1 to 1.3 m in thickness, was easily and cheaply mined, and was a popular household coal. Towards the local depocentre (Figure 13) the seam is commonly split by up to 5 m of mudstone. The coal is generally overlain by a canneloid mudstone, or 'cat' (Stobbs, 1915), which generally contains an impoverished fauna including *Anthraconaia wardi*, the type specimen of which possibly comes from this horizon at Adderley Green Colliery (Trueman and Weir, 1967, pl. 53, fig. 25), and *Anthracosia concinna* as at Park Hall Colliery. However, faunas overlying the Holly Lane Coal are much more abundant in Florence Colliery Underground Borehole No. 4 (81.84 to 87.48 m), where they include the bivalves *Anthraconaia lanceolata*, *A.* cf. *beaniana*, *A.* cf. *concinna* and *Naiadites quadratus*. A further faunal band, a little higher in the sequence (70.71 to 73.71 m) contains *Anthracosia aquilina* inter *phrygiana*, *A.* cf. *ovum*, *A.* cf. *phrygiana*, *Naiadites* sp., *Spirorbis* sp. and fish debris. Locally the roof of the Holly Lane Coal is rich in plants (Gibson, 1925).

The **Bowling Alley Coal**, known as the Top Two Row at Madeley and Holditch collieries, Tatchen End or Little Row on the Western Anticline, or Magpie in the Biddulph area, is equivalent to the Half Yard Coal of the Cheadle Coalfield. It commonly forms a single leaf about 1.0 m thick which has been extensively mined, particularly in the Longton area, and was popular in ironmaking (Homer, 1875). In many places it is overlain by canneloid mudstones or calcareous mudstones containing *Spirorbis* (Stobbs, 1902). Faunas overlying the coal in Florence Colliery Underground Borehole No. 4 (56.08 to 67.03 m) include cf. *Anthraconaia pulchella*, *Anthracosia* cf. *aquilina*, *A.* cf. *beaniana*, *A.* cf. *disjuncta*, *A. nitida*, *A. ovum* inter *aquilina*, *A. phrygiana*, *A. phrygiana* inter *disjuncta*, *A.* cf. *planitumida*, *A.* cf. *subrecta*, *Anthracosphaerium* cf. *exiguum*, *Naiadites quadratus*, *Spirorbis* sp., *Carbonita humilis*, cf. *Hiboldtina* sp. and fish debris. Other additional taxa from this level in the district include *Anthraconaia salteri*, *Anthracosphaerium affine* and *A. turgidum*. These mudstones are overlain by a sequence dominated by siltstones and sandstones, with mudstone intervals locally containing thin ironstones. The sandstones, which are locally rich in plants (Gibson, 1925), are known as the **Bowling Alley Rock**. Where the sandstones are poorly developed, in the south-western part of the coalfield, a minor

coal is present between 2 and 10 m below the overlying Ten-Feet Coal (Figure 13).

The **Ten-Feet Coal**, also known as the Ten Foot, or the Main Coal at Sneyd Colliery, is equivalent to the Two Yard Coal of the Cheadle Coalfield. It is normally 1.8 to 2 m thick over the Potteries Coalfield, though between Chatterley, Holditch and Madeley collieries it may be up to 3.5 m thick. It commonly forms a single seam, but is split at Victoria Colliery and in the Longton area. It was mined in the east of the Potteries Coalfield principally as a manufacturing or domestic coal, and in the west as a bituminous coking coal. The seam generally has a low ash content. Faunas in roof-mudstones to the coal occur locally, as in the Longton area. They are well represented in Bowsey Wood Borehole (785.54 to 786.97 m) and include *Anthracosia aquilina* inter *phrygiana*, *A.* cf. *aquilina*, *A. beaniana*, *A.* cf. *ovum*, *A. ovum* inter *aquilina*, *A. ovum* inter *beaniana*, *A. phrygiana*, *Anthracosphaerium* cf. *exiguum*, *Naiadites* sp. and fish debris. However, the sequence above the Ten-Feet Coal is generally dominated by a widely developed sandstone, the **Ten-Feet Rock**, with associated siltstone facies. Locally this may be over 30 m thick, as at Talk o' t' Hill Colliery [822 527]. It forms a prominent, persistent topographic feature, like the ridge at Norton-in-the-Moors [894 515]. In places the sandstone directly overlies the Ten-Feet Coal, and may erode into it, for instance south of Knutton (Wilson, 1990). Evidence for the erosion, associated with complex channel bank collapse, was clear at Holditch Colliery (information from J O'Dell of British Coal). The sandstone is mostly fine to medium grained, and micaceous. It usually appears to be massive, though exposures such as that in a cutting on the Leek road [8980 5048] may show lateral accretion surfaces if examined in detail. Isopachytes for the Ten-Feet Rock have been generated by Corfield (1991), showing that the area of maximum thickness broadly coincides with the depocentre for the Ten-Feet to Bellringer interval as well as with that for the sequence between the Vanderbeckei and Maltby marine bands (Figure 7d).

The Ten-Feet Rock has been reddened over much of the eastern limb of the Potteries Syncline between Tunstall and Berry Hill (Gibson, 1925; Rees, 1990a; Howard, 1990). This widespread reddening is likely to be of Permo-Triassic age.

Strata between the Bellringer Coal and Maltby Marine Band

These strata are notable for seam splitting and thickness changes, and for great lateral variation in interseam sequences. Few coals have diagnostic characteristics and seams are locally washed out by overlying sandstones, a feature clearly seen in Brown Lees opencast site (Figure 7a).

Variability is most marked between the Hams and Yard coals (Figure 14). The difficulties of seam correlation have led to confusion of seam names, and several seams have been misidentified in the past. Even now seam correlation in these strata is imprecise and the correlation of seams shown in Figure 14 is one of several that could be proposed. Large changes of interseam thickness occur over small areas. For instance, at Silverdale Colliery the interval between the Hams and Ragman/ Rough Seven-Feet coals varies between 0 and 15 m over a lateral distance of less than 200 m. These variations probably result from differential subsidence during deposition, induced by a phase of normal faulting (Corfield, 1991). Mapping the location of seam splits is made difficult by the problems of seam correlation. Corfield (1991) suggested that most seams split along lines trending approximately west-north-west. However, in some areas, as at Silverdale, the split lines appear to trend more nearly north–south (information from J O'Dell of British Coal). The Hams and Rough Seven-Feet coals form a single multiple seam in the area south of Keele (see Highway

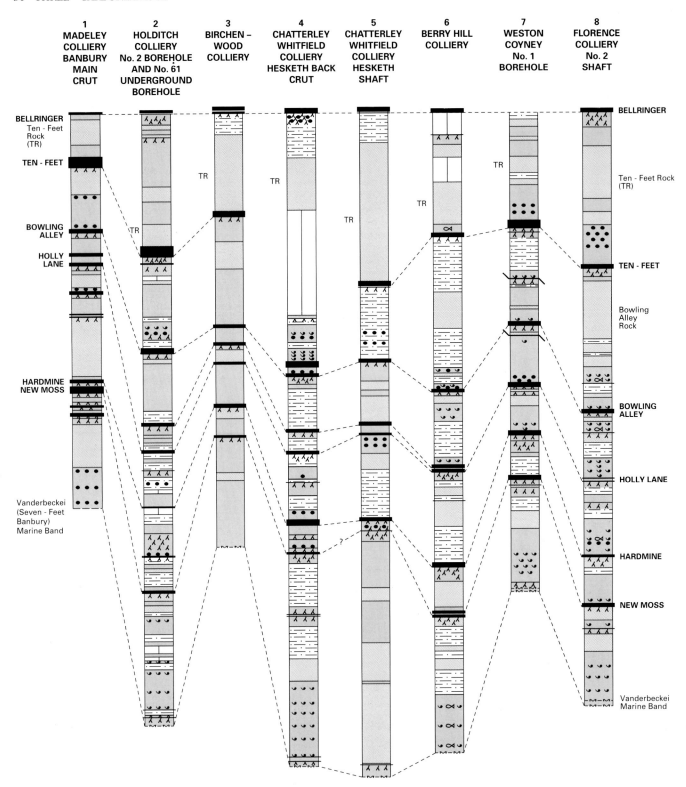

Figure 13 Correlation of the Middle Coal Measures between the Vanderbeckei Marine Band and the Bellringer Coal. For key see Figure 10.

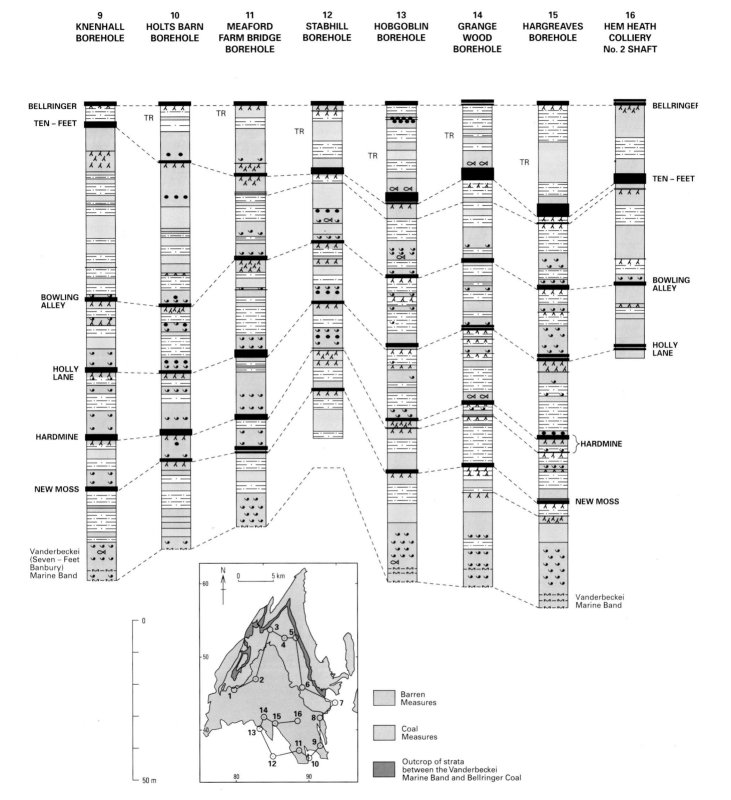

Figure 13 (*continued*).

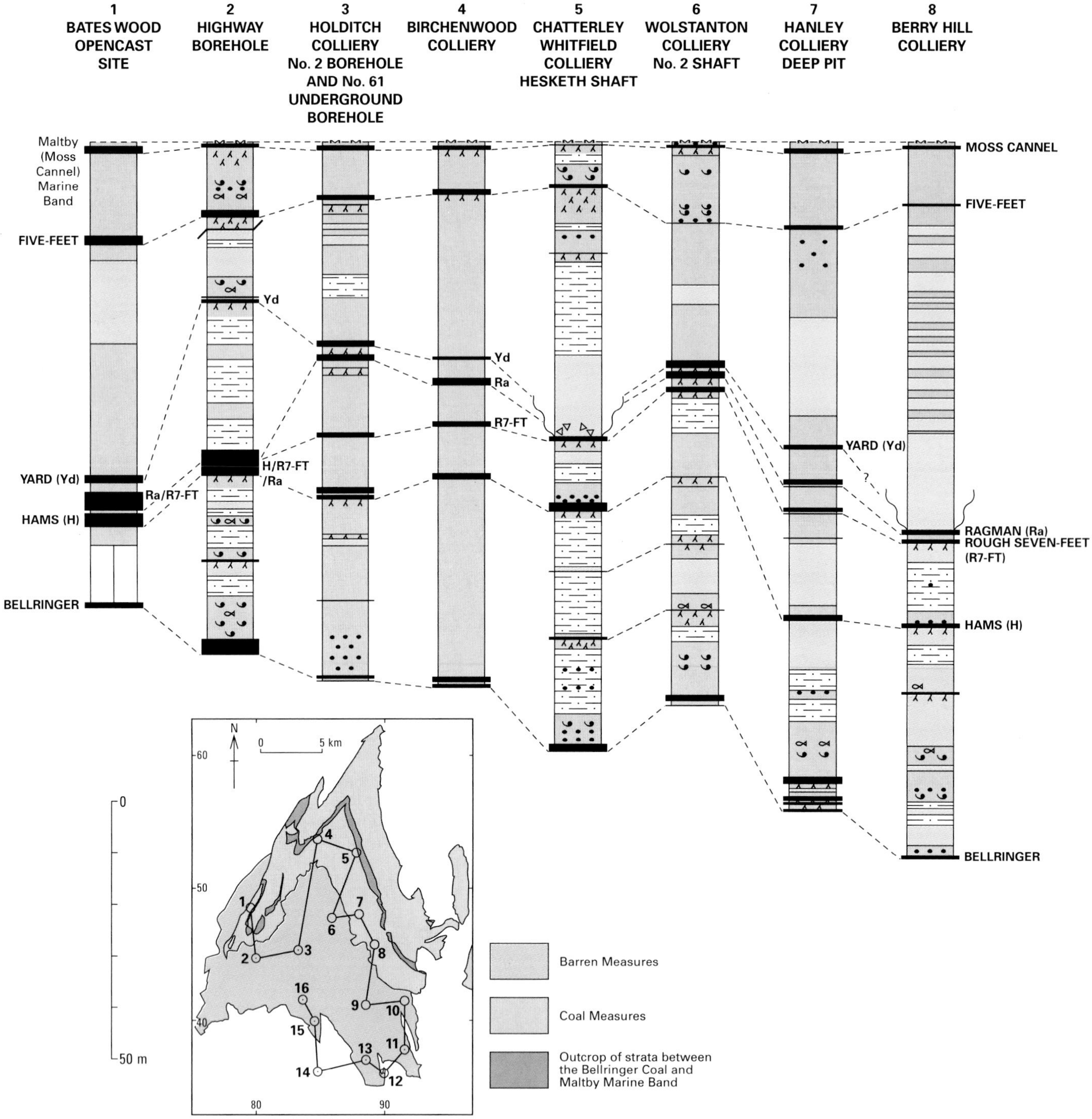

Figure 14 Correlation of the Middle Coal Measures between the Bellringer Coal and the Maltby Marine Band. For key see Figure 10.

Borehole, Figure 14). Most of these seams are low in sulphur and ash, which made them popular for use in blast furnaces.

The **Bellringer Coal**, known as the Stoney Eight Feet at Hanley Deep Pit and the Ten Feet Rider on the Western Anticline, derives its name from its hardness, which causes it to ring when struck (Homer, 1875). Although formerly sought for use in blast furnaces, the extent of its working was limited because of its poor roof and because it is normally less than

0.5 m thick. It was widely worked only in the northern part of the coalfield, where it is commonly thicker than 1 m. In places the coal is split into two leaves by up to 6 m of siltstones and mudstones containing root traces, particularly in the south-eastern and south-western parts of the coalfield and in the Hanley area. The roof of the coal normally consists of mudstones containing fish debris, as in Holts Barn Borehole (884.07 to 885.44 m). However, in Norton Colliery Under-

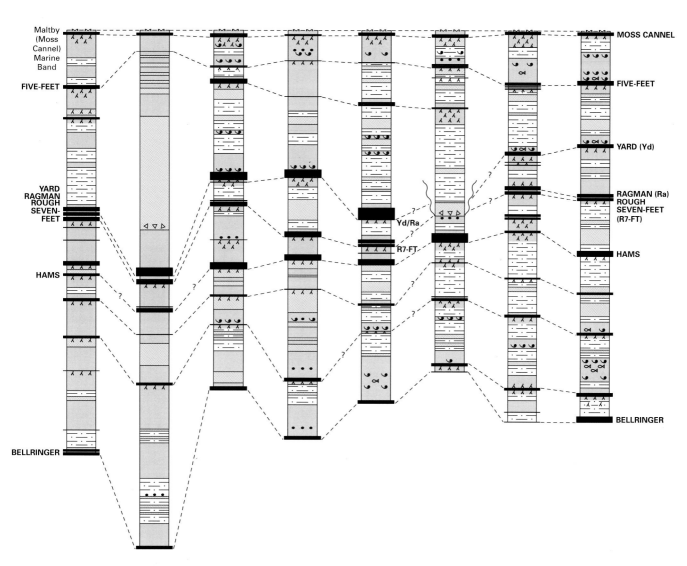

Figure 14 *(continued).*

ground Borehole No. 3 a diverse nonmarine bivalve assemblage referable to the *Anthracosia caledonica* Subzone occurs above the Bellringer Coal (34.44 to 37.79 m). It includes *Anthracosia aquilina* inter *phrygiana*, *A.* cf. *caledonica*, *A. caledonica* inter *lateralis*, *A.* cf. *lateralis*, *A. lateralis* inter *planitumida*, *A. phrygiana* and *Naiadites* sp. *Spirorbis* sp., *Carbonita* sp. and fish debris are also present. These mudstones form the basal part of the lowest of at least three cyclothems that occur between the Bellringer and Hams coals. Mudstones containing ironstone nodules or nonmarine bivalves overlie all but a few of the coals and seatearths in this interval. Above one of the coals in Norton Colliery Underground Borehole No. 3 (47.04 to 47.37 m), an estheriid band with poorly preserved anthracosiids, estheriids and fish debris occurs. This is the correlative of the **Lowton Estheria Band** of the Lancashire Coalfield, an important marker which extends throughout much of northern Britain and the southern North Sea. Sandstones occur at several

horizons between the Bellringer and Hams coals (Figure 14).

The **Hams Coal** was also known as the Old Whitfield Coal at Chatterley Colliery, Seven-Feet Coal at Victoria Colliery, Five-Feet Coal on the Western Anticline, Rough Seven-Feet Coal at Hanley Deep Pit, and Birches Coal in the Fenton and Longton areas. It varies considerably in thickness, ranging commonly up to about 2.4 m, and is split in places between Wolstanton and Fenton. It was mined largely for use in blast furnaces. Although locally combined with the Rough Seven-Feet and Ragman coals, the Hams is separated in most areas from these seams by at least 5 m of variable strata. The roof of the Hams Coal marks the entry of assemblages referable to the *Anthracosia atra* Subzone; *Anthracosia acutella* and *A. atra* are known from this horizon at Florence Colliery, Longton.

The **Rough Seven-Feet Coal** was previously confused with the Hams Coal at collieries in the Fenton and Longton areas, except at Fenton Colliery, where it was included with the combined Yard/Ragman Coal. Although rarely greater than 1.0 m thick, it varies considerably in thickness, and near Madeley it is over 2.4 m. The roof of the coal is commonly rich in plant fossils. The **Ragman Coal** was known as the Two Feet at Sneyd Colliery, the Rough Seven-Feet at Holditch and part of the Yard Coal in the Fenton and Longton areas. It commonly forms a single seam between 0.7 and 1.4 m in thickness. A polished, slab-like argillite cobble was found in the seam at Sneyd Colliery. Lister and Stobbs (1917) believed this to be an erratic, perhaps transported by entanglement in the roots of a floating tree.

The **Yard Coal** was known as the Five-Feet Coal at Madeley Colliery and the Little Row Coal at Chatterley Colliery. In the Fenton and Longton areas what is known as the Yard Coal includes the Ragman Coal, and at Fenton Colliery also included the Rough Seven-Feet Coal (Figure 14), but in most areas it forms a single distinct seam. Although generally low in sulphur, the coal locally contains an irregular pyritous coal known as 'Grisley' (Stobbs, 1915). The Yard Coal commonly has a cannel roof, which is overlain by mudstones bearing fish and nonmarine bivalves; the latter in Holts Barn Borehole (804.06 m) include *Anthracosia* cf. *atra*. These mudstones normally coarsen up into a sequence dominated by sandstone and siltstone, some of the sandstones having channelised bases. Locally, along the eastern margin of the Potteries Coalfield, the channels have eroded into the underlying coals, for instance at Chatterley Whitfield Colliery (Figure 14). In the Longton area, some of the channel sandstones are over 40 m thick (Rees, 1990b). The sequence contains notably varied faunas. Trueman and Weir (1956, pl.30, fig.14) recorded *Naiadites* cf. *angustus* from between the Yard and Moss coals at Florence Colliery, Longton. A slightly higher horizon, 15.24 m below the Five Feet Coal, is represented in an assemblage obtained from an old railway cutting [8015 4719] near Leycett, which contains *Anthracosia* cf. *atra*, *A. planitumida*, *Anthracosphaerium propinquum*, *Naiadites obliquus*, *Carbonita humilis* and *C. pungens*. Higher still [8009 4713], 6.1 m below the Five Feet Coal, *Anthracosia* cf. *atra*, *A.* cf. *lateralis*, *Anthracosphaerium exiguum*, *A. turgidum* and *Naiadites obliquus* are present. Approximately 10 m below the Five-Feet Coal, a laterally persistent seam, occasionally confused with the Yard or Five-Feet Coal, occurs in the southern part of the coalfield (Figure 14). This may be a leaf split from the Five-Feet Coal.

Although the **Five-Feet Coal** rarely meets up to its name, normally being less than 1.0 m in thickness, it generally forms a single seam. The roof of the coal has a widespread mussel band. Faunas from Silverdale Colliery comprise *Anthraconaia* cf. *curtata*, *A.* aff. *ellipsoides*, *A.* cf. *librata*, *A. salteri*, *Anthracosia* cf. *atra*, *A.* cf. *concinna*, *Anthracosphaerium affinis* inter *propinquum*, *A. propinquum* (including the holotype, Trueman and Weir, 1953, pl. 25, fig. 1) and *A. turgidum*. Additional components from this district include *Anthracosia aquilina*, *A.* cf. *aquilinoides*, *A.* cf. *faba*, *A.* cf. *fulva*, *Anthracosphaerium exiguum*, *A.* cf. *radiatum*, *Naiadites* cf. *obliquus*, *Carbonita fabulina*, *C. humilis* and fish debris. In Holts Barn Borehole (770.53 to 770.99 m) an estheriid band with *Euestheria* and fish is present at this horizon. Unlike most of the interseam sequences between the Bellringer Coal and Maltby Marine Band, that between the Five-Feet and Moss Cannel coals is dominated by mudstones. These commonly contain ironstone nodules, fish, *Spirorbis* and nonmarine bivalves.

The **Moss Cannel Coal**, also known as the Single Five-Feet, the Single Two-Feet on the Western Anticline, the Little at Florence Colliery, or the Little Row in the northern part of the coalfield is, as its name suggests, a canneloid coal. It normally forms a single seam less than 1.0 m thick. A thin mudstone, locally containing ironstone nodules, forms the roof of the coal, separating it from the overlying Maltby Marine Band.

STRATA BETWEEN THE MALTBY AND CAMBRIENSE MARINE BANDS

The strata between the Maltby and Cambriense marine bands include five further, regionally developed, marine horizons. The coals in this interval, with some exceptions such as the Moss, Rowhurst, and Winghay seams, are thinner than those in the strata below and above, and the strata contain many sandstones, the petrography of which was described by Malkin (1961). The combination of these features suggests deposition in a lower delta plain setting (Figure 8). The isopachyte maps for these strata (Figure 7e–f) show that their thickness distribution is more complex than for underlying sequences (Figures 7b–e), but that there is still a general thickening on the east crop of the Potteries Syncline.

The sequence is described in two parts. The lower part, below the Aegiranum Marine Band, contains very few workable coals (Figure 15) and is of Duckmantian (Westphalian B) age. Above this the sequence contains thicker coals (Figure 16), many of them with ironstones in the roof measures. These are clayband ironstones, lacking the carbonaceous laminae present in blackband ironstones. They were commonly mined in the past, in particular the more carbonaceous ironstones such as the Burnwood, which have been described as 'semiblackband'. The Aegiranum Marine Band marks the approximate base of these 'Clayband Measures', and also defines the base of the Bolsovian Stage (Westphalian C).

Strata between the Maltby and Aegiranum marine bands

These strata contain the Maltby, Clown, Haughton and Sutton marine bands. Isopachytes for the interval are shown in Figure 7e. The base of the sequence is marked by the **Maltby Marine Band**, locally known as the Moss Cannel Marine Band. In its acme phase this is in a myalinid facies. In Holts Barn Borehole (763.22 to 763.45 m depth) it is 0.23 m thick with *Lingula* and fish remains at the base, overlain by pyritic mudstone with *Lingula* sp., *Anthracosia* sp. (stunted), *Naiadites* sp., cf. *Geisina* sp., estheriids and fish debris. This horizon yielded the holotype of *Naiadites obliquus* at Madeley Colliery (Trueman and Weir, 1956, pl.31, fig. 3). In Hem Heath Colliery Underground Borehole No.8 (17.68 to 20.57 m) the marine band is in leaves. The basal 0.76 m contains arenaceous foraminifera, sponge spicules and *Lingula mytilloides*. A further band, 0.31 m above this and 0.35 m thick, contains the same fauna but with additional *Myalina* cf. *perlata* and *Geisina subarcuata*. Overlying this is an interval 0.25 m thick with *Anthracosia* sp. (stunted) and *Naiadites* sp. A band 0.02 m thick with *Lingula* sp. lies 0.06 m above and is followed by 0.84 m of mudstone with *Anthracosia* sp. and *Naiadites* cf. *triangularis*. A further marine horizon, with arenaceous foraminifera and myalinids, occurs 0.3 m higher. *Myalina* sp. is also recorded from Chatterley Whitfield, Sneyd and Victoria collieries. Immediately overlying the Maltby Marine Band in Holts Barn Borehole (757.73 to 762.84 m), there is a well-developed mussel band with *Anthracosia acutella*, *A.* cf. *atra*, *A.* cf. *caledonica*, *A.* cf. *concinna*, *A. planitumida*, *Anthracosphaerium* cf. *radiatum*, *Naiadites alatus* and fish debris. From elsewhere in the district, *Anthracosia aquilinoides*, *A.* cf. *carissima*, *A.* cf. *similis*, *Naiadites* cf. *daviesi* and *N.* aff. *productus* are also recorded at this level.

The **Moss Coal** is otherwise known as the Mossfield Coal at Sneyd and Fenton collieries, and the Four-Feet at Madeley and Holditch collieries. It thickens towards a local depocentre near Hanley, where the upper part is a thick cannel (National Coal

Board, 1960). The coal normally varies in thickness between 1.0 and 2.0 m. Whether active faults influenced its deposition is uncertain, though Homer (1875) suggested that it thickens southwards across the Shelton Fault. The coal arguably was the most consistent high-quality coal in the district. It was one of the most extensively worked seams in the Potteries Coalfield, largely for domestic use, even though it has incompetent roof and floor mudstones. These commonly contain ironstones and in many places are fossiliferous, with fish, ostracods and nonmarine bivalves. In Holts Barn Borehole (750.65 to 753.77 m) the mudstones contain *Anthracosia* cf. *barkeri*, *A.* cf. *caledonica*, *Naiadites* cf. *obliquus* and fish debris, and form the type horizon of *Anthracosphaerium gibbosum*, the lectotype of which comes from Fenton (Trueman and Weir, 1953, pl. 24, fig. 45). This bivalve horizon is overlain at 745.39 m in Holts Barn Borehole by an estheriid band, the **Birchenwood Estheria Band**, which contains *Anthracosia* aff. *fulva* and *Euestheria* sp. This band is also recorded from Clayton Borehole (715.82 to 717.12 m), where it is associated with *Anthraconaia* cf. *sagittata*, *Anthracosia* cf. *aquilinoides*, *A. atra*, *A.* cf. *barkeri*, *Carbonita humilis*, *Euestheria* sp. and fish debris. Overlying faunas in Clayton Borehole (712.77 to 715.52 m) include *Anthracosia atra*, *A.* cf. *barkeri*, *A. barkeri* inter *atra*, *A.* cf. *phrygiana*, *Spirorbis* sp. and *Carbonita humilis*, but this fauna is absent in places, for example in the Holts Barn Borehole.

Above these mudstones are two coals, together known as the **Birchenwood Coal**, or Granville Coal near Hanley, though they are commonly separated by over 5 m and rarely form a combined seam (Figure 15). The Birchenwood Coal is unfortunately named, as the seam initially given this name at Birchenwood is the Moss Coal (Earp and Calver, 1961, p.167). The Birchenwood Coal is absent, or poorly developed, on the western side of the coalfield. Elsewhere it contains inferior cannels and was only worked to any extent in the Kidsgrove and Hanley areas. The coal has a poor mudstone roof with nonmarine bivalves, including *Anthacosphaerium turgidum*, underlying the Clown Marine Band. Mudstones between the Birchenwood Coal and the Clown Marine Band in Clayton Borehole (696.7 to 697.0 m) have also yielded *Anthracosia atra*, *Naiadites* sp., *Spirorbis* sp. and fish debris.

The **Clown** (**Longton Hall**) **Marine Band** (Ward, in Gibson, 1905, p.49; Earp and Calver, 1961) normally occurs within a metre of the top of the Birchenwood Coal. It is widespread, but very thin and restricted to a *Lingula* phase. A typical section is present in Clayton Borehole (696.32 to 696.57 m), where *Planolites ophthalmoides*, *Lingula mytilloides* and fish debris occur. It is commonly overlain by ironstone-bearing mudstones containing fish and nonmarine bivalves, which include in the Holts Barn Borehole (716.58 to 730 m) a fauna comprising *Anthraconaia* cf. *librata*, *Anthracosia* aff. *aquilina*, *A. atra*, *A.* cf. *barkeri*, *A.* cf. *fulva*, *A.* cf. *lateralis*, *A.* cf. *simulans*, *Anthracosphaerium* sp. nov. aff. *radiatum*, *Anthracosphaerium* sp. nov. aff. *turgidum*, *Naiadites* sp., *Carbonita humilis* and fish debris. These mudstones are overlain by siltstones and sandstones, which are commonly rooted and contain coals (Figure 15).

The **Doctor's Coal**, also known as the Doctor's Mine Coal on the Western Anticline, is developed over most of the coalfield. However, because it is normally less than 0.5 m thick, it has been little worked. In most areas the coal forms a single seam. It is overlain by the **Haughton** (**Doctor's Mine**) **Marine Band**. This band is widespread and in its acme phase contains some elements of a shelly benthonic fauna. The best sections are in the Clayton (647.17 to 653.19 m) and Holts Barn (687.48 to 691.82 m) boreholes, where it consists of several alternations of *Lingula* and foraminiferal phases (Earp and Calver, 1961, fig. 6). The composite fauna includes *Ammonema* sp., sponge spicules, *Lingula mytilloides*, *Orbiculoidea* sp., *Planolites ophthalmoides*, *Sphe-*

nothallus stubblefieldi, cf. *Tomaculum* sp., crinoid and fish debris. From Berryhill Colliery, *Euphemites* cf. *anthracinus*, *Nuculopsis* sp. and a coiled nautiloid are also recorded. The marine band is in places immediately overlain by mudstones containing a nonmarine fauna, such as those containing *Anthracosia* cf. *atra* in Clayton Borehole (645.41 to 645.49 m).

The **Bee Coal**, normally about 10 m above the Haughton Marine Band, occurs in the western and central parts of the coalfield. It is locally overlain by nonmarine mudstones which in the Clayton Borehole (642.90 to 643.66 m) contain *Anthracosia atra*, *A.* cf. *planitumida*, *A.* cf. *simulans*, *Naiadites* cf. *alatus*, *Carbonita humilis*, *C. scalpellus* and fish debris. In many places, however, the Bee Coal is immediately overlain by the **Sutton** (**Clayton**) **Marine Band** (Trotter, 1955; Earp and Calver, 1961). This marine band, previously exposed in Apedale (Wilson, 1990), is laterally impersistent. The best sections are in the Clayton (637.95 to 641.58 m) and Sutton (127.4 to 127.71 m) boreholes where *Agathammina* sp., *Ammonema* sp., *Planolites ophthalmoides* and *Lingula mytilloides* occur.

The sequence between the Sutton and Aegiranum marine bands consists of a varied sequence of siltstones, sandstones and nonmarine mudstones. Some sandstones form persistent topographic features in the Kidsgrove area, such as that which forms the high ground around Talke Pits [828 526]. The sequence contains fish, but very impoverished bivalve faunas. In the Clayton Borehole, overlying the Sutton Marine Band (630.94 to 632.61 m) *Naiadites* sp. and fish debris occur. The same fauna recurs at 615.70 to 616.03 m and immediately beneath (603 to 603.2 m) the Aegiranum Marine Band. This highest fauna contains *Naiadites* cf. *obliquus*, *N.* cf. *productus*, *Spirorbis* sp., *Carbonita humilis* and fish debris. *Anthraconaia* sp. and *Anthraconauta* cf. *phillipsii* have been found in this band in a disused railway cutting [8205 4917] near Chesterton. Coals between the Sutton and Aegiranum marine bands are laterally impersistent, though three are notable. The first occurs within 15 to 20 m of the Sutton Marine Band and appears to be developed over the central and southern parts of the coalfield. The second occurs within 15 m of the Aegiranum Marine Band only in the south-western part of the coalfield. The third, the **Gin Mine Coal**, commonly occurs within 5 m of the Aegiranum Marine Band, and locally reaches 1.0 m in thickness. It is normally separated from the marine band by thin nonmarine mudstones. In the past it has been confused with the Twist Coal, which lies above the Aegiranum Marine Band.

Strata between the Aegiranum and Cambriense marine bands

These strata, like those below the Aegiranum Marine Band, were probably deposited in a lower delta-plain setting; the occurrence of the Aegiranum, Edmondia, Shafton and Cambriense marine bands suggests that the delta plain was near to sea-level and was subject to frequent marine incursions. A selection of sections is shown in Figure 16. The sequence between the Aegiranum and Cambriense marine bands is distinguished from that between the Maltby and Aegiranum marine bands by the presence of well-developed clayband ironstones, which commonly form sheet-like or nodular bodies in the mudstones above coals. Unlike the blackband ironstones, they contain no carbonaceous laminae (Boardman, 1978; 1989). They have iron-rich cores and magnesium and calcium-rich margins, and are interpreted by Curtis et al. (1975) as being diagenetic segregations produced during burial, at levels several hundred metres below the sediment–water interface. The lateral variation in character of the clayband ironstones was described by Dewey (1920), who noted that many are absent in the Fenton and Silverdale areas. The ironstone overlying the first coal above the Aegiranum Marine Band is

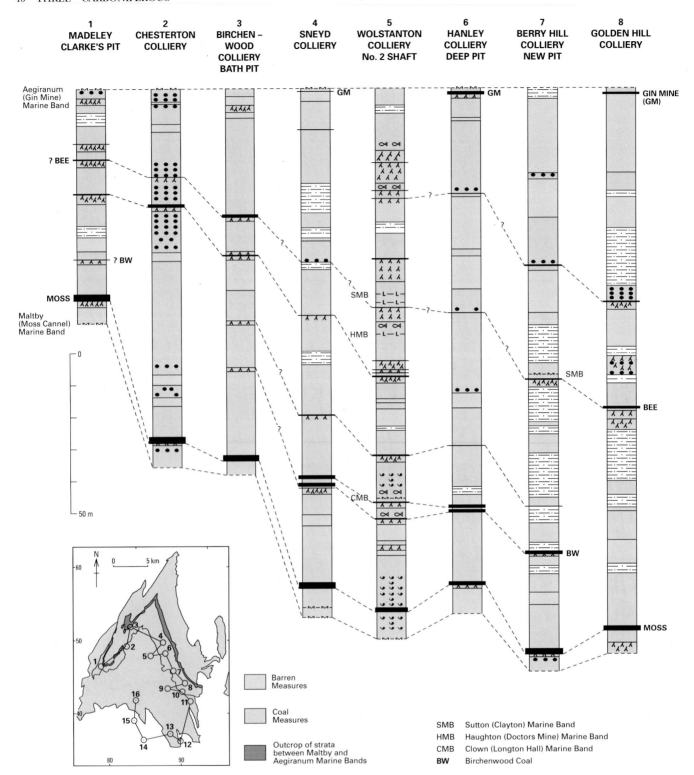

Figure 15   Correlation of the Middle Coal Measures between the Maltby and Aegiranum marine bands. For key see Figure 10.

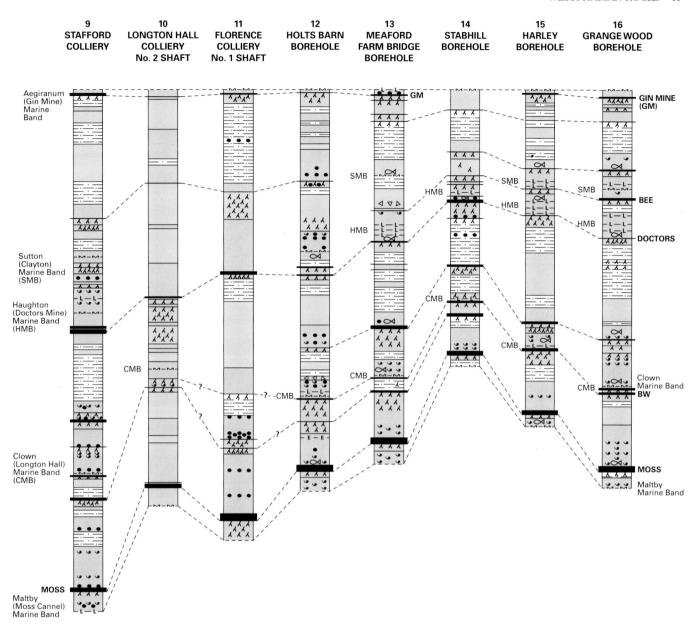

Figure 15 *(continued).*

the lowest of the economically important ironstones of the Potteries Coalfield (Homer, 1875), so the base of the informally named 'Clayband Measures' was drawn approximately at the marine band.

Excavations for ironstone or mudstone have provided several excellent exposures in the strata between the Aegiranum and Cambriense marine bands. Of note are Birchenwood Brickpit [8538 5418 to 8545 5403], which exhibits most of this sequence, and Acreswood Brickpit [8785 5139 to 8770 5132] which formerly exposed most of the sequence between the Stafford and Rowhurst Rider coals; these are detailed by Rees and Clark (1992).

The strata between the Aegiranum and Cambriense marine bands contain several well-developed tonsteins, which are devitrified volcanic ashes. These form excellent time markers and may be used widely in correlation (Barnsley et al., 1966; Wilson et al., 1966).

Isopachytes for the strata between the Aegiranum and Cambriense marine bands are shown in Figure 7f. The thickest development is between Hanley and Madeley, and the strata thin towards the south and also more rapidly towards the Red Rock Fault. The latter effect is accompanied, in the vicinity of the Western Anticline, by a decrease in the amount of sandstone in the sequence (Trotter, 1955; Earp and Calver, 1961). The last-named authors noted that the marine bands in the area also thicken, perhaps because the area was closer to the open sea than the rest of the coalfield.

The **Aegiranum Marine Band** is also known in the Potteries Coalfield as the Gin Mine Marine Band, after the coal it overlies, or less commonly the Speedwell, Nettlebank or Florence Colliery Marine Band (Stobbs, 1905). It consists of dark grey to black mudstones, with notable calcareous laminae or nodules in places (Cope, 1954). The marine band is widespread and contains the most diverse marine fauna in the

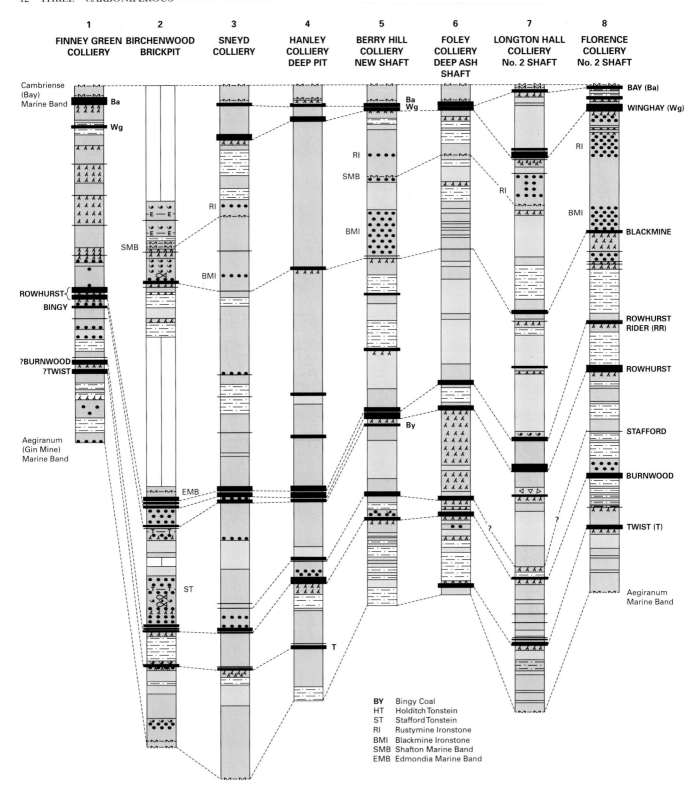

Figure 16 Correlation of the Middle Coal Measures between the Aegiranum and Cambriense marine bands. For key see Figure 10.

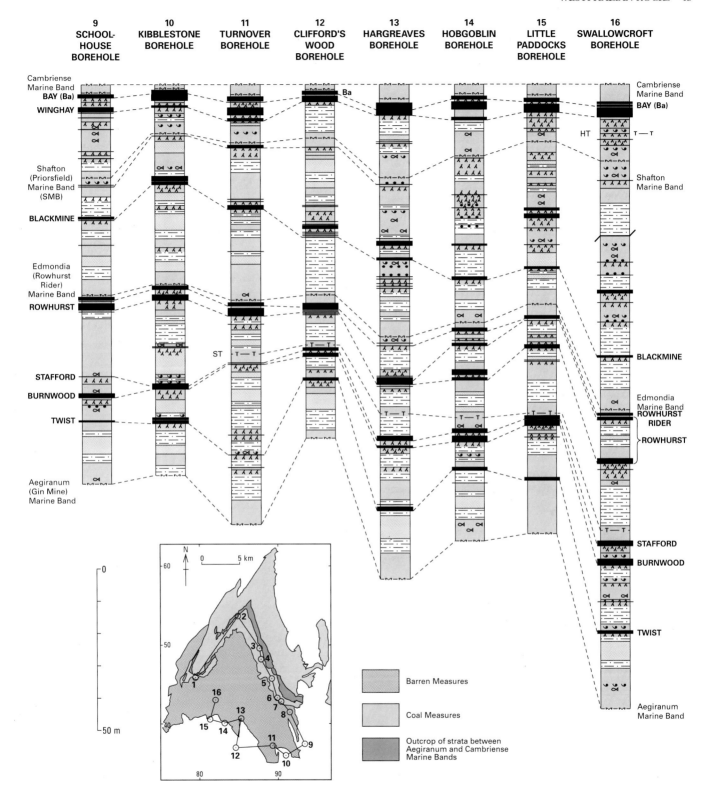

Figure 16 *(continued).*

district. Its acme phase is in an ammonoid/pectinoid facies, which is associated with a relatively diverse shelly benthonic fauna. An ankeritic limestone, known as the cank, is commonly present towards the base. Although it was previously exposed in Apedale (Wilson, 1990) and Birchenwood Brickpit (Rees and Clark, 1992) the best sections are in boreholes: Bowsey Wood (705.81 to 713.21 m), Clayton (599.92 to 602.97 m) and Holts Barn (639.77 to 640.46 m).

In Holts Barn Borehole the marine band is thin and dominated by an ammonoid/pectinoid phase. The lower part of the band (640.23 to 640.46 m) contains *Lingula mytilloides*, *Orbiculoidea* sp., *Dunbarella* sp., *Pernopecten carboniferus*, cf. *Schizodus* sp., *Donetzoceras* sp., *Politoceras politum*, *Sphenothallus stubblefieldi*, *Hollinella* sp., crinoid, conodont and fish debris. Above this (639.77 to 640.0 m) the fauna is limited to *Agathammina* sp., *Ammonema* sp., *Lingula mytilloides* and fish debris. In Bowsey Wood the band forms three leaves, separated by nonmarine mudstones with fish debris. The basal part (712.77 to 713.21 m) contains *Lingula mytilloides*, *L. pringlei*, *Donetzoceras aegiranum*, turreted gastropod, conodont and fish debris. The ankeritic horizon (712.20 to 712.62 m) contains *Lingula* sp., a chonetoid, a cephalopod and fish debris and is overlain by conulariids, *Lingula* sp., cf. *Phestia acuta*, ammonoid, ostracod, conodont and fish debris (711.4 to 712.14 m). The middle leaf (708.96 to 710.41 m) contains *Lingula mytilloides* and *Sphenothallus stubblefieldi*. The upper leaf (705.81 to 708.66 m) is dominated by *Lingula mytilloides* and fish debris, except between 707.59 to 708.2 m where this fauna is accompanied by arenaceous foraminifera, including *Ammonema* sp. The richest assemblages have been obtained from Nettlebank Colliery [886 503], but their precise distribution within the marine band is approximately not known. The fauna is as follows; solitary coral indet., *Lingula mytilloides*, *Crurithyris carbonaria*, *Dictyoclostus craigmarkensis*, *Linoproductus* sp., *Neochonetes granulifer*, *Rugosochonetes skipseyi*, *Tornquistia* cf. *gibbosa*, *Anthraconeilo taffiana*, *Aviculopecten* sp., *Dunbarella macgregori*, *Nuculopsis gibbosa*, *Pernopecten carboniferus*, *Posidonia sulcata*, *Pseudamussium fibrillosus*, *Euphemites anthracinus*, *E. urei*, *Platyconcha hindi*, *Trepospira radians*, *Domatoceras* sp., *Ephippioceras costatum*, *Huanghoceras costatum*, *Metacoceras* cf. *cornutum*, *M.* cf. *perelegans*, '*Orthoceras*' *sulcatum*, cf. *Pseudorthoceras* sp., *Donetzoceras aegiranum*, *Politoceras politum*, *Hollinella* sp., *Archaeocidaris* sp. and fish debris. Other taxa recorded from this district include *Orbiculoidea* cf. *nitida*, *Lissochonetes minutus*, *Tornquistia diminuta*, cf. *Cypricardella* sp., *Phestia attenuata*, *Pleurophorella* sp., *Posidonia* cf. *gibsoni*, *Straparollus* sp., *Trepospira radians*, cf. *Coelogasteroceras* sp., cf. *Cycloceras* sp., *Solencheilus* sp., *Gastrioceras* sp., and *Hollinella* cf. *bassleri*. The marine band is locally overlain by mudstones rich in fish and nonmarine bivalves, which in Clayton Borehole (597.41 to 599.47 m) include *Naiadites* cf. *productus*. The mudstones tend to grade upwards into a sequence of siltstones interbedded with sandstones. The sandstones generally form good topographic features in the Hanley, Sneyd Green and Silverdale areas. They are capped by a mudstone containing the **Twist Coal**. This seam, also known as the Pottery Coal or Gin Coal, occurs over most of the coalfield, but has been commonly worked only on the eastern limb of the Potteries Syncline. It tends to be less than 1.0 m in thickness and was mined because of its low sulphur and ash content, which made it popular for iron or glass working (Homer, 1875). *Naiadites* cf. *productus* and fish debris accompany *Euestheria* in the roof of the coal in Bowsey Wood Borehole (700.89 to 701.27 m). The same estheriid band overlies the seam in Clayton Borehole (580.19 to 581.1 m), where it contains *Anthraconaia hindi*, *Carbonita humilis*, *Euestheria vinti* and fish debris. The nonmarine bivalve assemblage is characteristic of the *Anthraconaia adamsi-hindi* Subzone. Higher in the sequence, in Clayton Borehole (567.92 to 568.07 m and

574.40 to 575.31 m), are horizons with *Naiadites* sp., ostracod and fish debris. The upper bed contains *Naiadites* cf. *productus*. Although locally, as in the Chatterley Whitfield area, the Twist Coal is associated with a worked ironstone, over most of the coalfield it is the overlying Burnwood Coal that has the better-developed ironstone.

The **Burnwood Coal**, known as the Littlemine Coal in the Longton area, normally forms a single seam between 0.8 and 1.2 m thick, though is locally split. It is generally overlain by dark mudstones containing laterally impersistent nonmarine faunas. At Great Fenton Colliery, for example, *Anthraconaia* aff. *pulchra* and *Naiadites* cf. *elongatus* are recorded. The ironstones associated with the Burnwood Coal are described as 'semi-blackband', having poorly developed carbonaceous laminae. The ironstones mostly overlie the coal though locally occur below, or within it. At Silverdale Colliery lenses of ironstone within the seam are up to 0.9 m thick and 15 m wide, and occur more than 0.1 m from the roof of the seam; they have a distinctive spherulitic texture (information from J O'Dell of British Coal). Where a packet of ironstone is well developed above the coal, particularly east of the Potteries Syncline, it is known as the Burnwood Ironstone, Littlemine Ironstone or Newmine Ironstone, and was frequently worked with the coal. Locally, in the area of the Western Anticline, the coal contains a tonstein (Appendix 2). The Burnwood Ironstone yields a fauna characteristic of the *adamsi-hindi* Subzone, including *Anthraconaia adamsi*, *A. dolobrata*, *A. expansa*, *A. hindi*, *A. pulchra*, *A.* aff. *stobbsi*, *A. wardi*, *A.* cf. *warei*, *Naiadites* cf. *elongatus*, *N.* cf. *melvillei*, *Spirorbis* sp. and arthropod debris. Some of the best material is from Newchapel Colliery [865 547], which is the type locality for *Anthraconaia pulchra* (Trueman and Weir, 1965, pl.37, figs. 16, 17). The lectotype of *Anthraconaia expansa* comes from this horizon at Great Fenton [882 433] (Trueman and Weir, 1965, pl. 36, figs. 1, 2), and the holotype of *Anthraconaia stobbsi* from Fenton (Trueman and Weir, 1967, pl.53, fig. 30).

Away from the local depocentre (Figures 7f and 16), the **Stafford Coal** mainly occurs within 5 m of the Burnwood Coal, though near the depocentre the seams are commonly separated by over 10 m of mudstones and, locally, siltstones and sandstones. The Stafford Coal is generally less than 1.2 m thick and in places is absent, though its position is commonly marked by a seatearth (Figure 16). The coal is overlain by fish-bearing mudstones and siltstones which contain the kaolinitic **Stafford Tonstein** (Earp and Calver, 1961, p.162; Wilson, 1962, p.40; Barnsley et al., 1966; Wilson et al., 1966; Appendix 2). This bed, up to 0.17 m thick, was formerly well exposed in Birchenwood Brickpit [854 541] and in Acreswood Brickpit [878 513] (Rees and Clark, 1992). The sequence between the Tonstein and the Rowhurst Coal is dominated by siltstones and sandstones.

The **Rowhurst Coal**, known as the Ash Coal in the Longton area and the Main Coal at Florence Colliery, was widely worked, largely as household or steam coal (Homer, 1875), though over most of the district it tends to be sulphurous. The coal is generally in several leaves separated by mudstone partings. In the Hanley area the lowest leaf was known as the Bingy, Binghy or Bingay Coal. The thickness of the Rowhurst Coal varies considerably, and appears to be greatest, locally exceeding 2 m, in the Silverdale and Longton areas, where it has been most extensively worked. Locally in the Trentham area it, and the overlying Rowhurst Rider Coal, are composed largely of cannel. In the Tunstall area two named tonsteins are associated with the seam. These are the **Rowhurst No. 2 Tonstein**, below the lowermost leaf of the seam (Barnsley et al., 1966) and the crandallite-bearing **Rowhurst No. 1 Tonstein**, between the top and middle leaves (Barnsley et al., 1966; Wilson et al., 1966). Elsewhere, as at Florence, only a single tonstein has been recorded within the seam (Appendix 2). The roof of the coal is variable in character, and locally is rich in plants or in ironstones with fish debris (Gibson, 1925; Cope, 1954). In the Knowl Wall Borehole, an estheriid band with *Naiadites* sp. and *Euestheria* sp. is present at 604.06 m and *Naiadites* cf. *productus*, *Carbonita humilis* and *C. inflata* occur above the coal at 597.41 to 597.48 m. The overlying **Rowhurst Rider Coal** commonly occurs within 2 m of the Rowhurst seam in the central and northern parts of the coalfield, though is separated by up to 20 m of mudstone and siltstone in parts of the Trentham and Longton areas (Figure 16). The seam has been the prime target for recent licensed adit mines (Chapter 9). *Naiadites* sp. and *Geisina subarcuata* commonly occur in the roof mudstones of the Rowhurst Rider Coal, as in Bowsey Wood Borehole (672.85 to 673.3 m). These mudstones and siltstones are overlain by the **Edmondia Marine Band**, locally known as the Rowhurst Rider Marine Band (Earp and Calver, 1961). The marine band normally occurs immediately above the Rowhurst Rider Coal, though locally north of Tunstall it is separated from it by up to 15 m of mudstone. The marine band is widespread and characterised by numerous arenaceous foraminifera. *Lingula* is relatively rare, but is the dominant element in Blacklake Borehole (Earp and Calver, 1961). The best sections have been in High Lane opencast site (Figure 7a), the brickpits at Acreswood and Birchenwood (Rees and Clark, 1992) and in the boreholes at Bowsey Wood (669.42 to 672.85 m), Clayton (527.28 to 530.96 m) and Knowl Hall (593.77 to 596.34 m). The composite fauna comprises *Planolites ophthalmoides*, *Agathammina* sp., *Ammonema* sp., *Glomospira* sp., *Lingula mytilloides*, *Curvirimula* sp.?, *Geisina subarcuata*, *Hollinella* sp. and fish debris.

The interval between the Edmondia and Shafton marine bands consists of a mixed sequence of lithologies containing several thin coals. These tend to be laterally discontinous and thin, causing difficulties in correlation. The correlation shown in Figure 16 is thus tentative. The sequence contains several ironstone horizons in seatearths or mudstones bearing nonmarine bivalves (Earp and Calver, 1961) and fish (Ward, *in* Gibson, 1905, p.323). One of the best sections is in Bowsey Wood Borehole, where several horizons with cf. *Anthraconauta phillipsii*, *Naiadites* sp., *Carbonita* sp. and fish debris overlie the Edmondia Marine Band. The uppermost of these (659.0 to 659.36 m) contains *Euestheria* sp. The **Blackmine Coal** was known as the Sulphur Coal at Berry Hill Colliery, the Big Mine on the Western Anticline, the Brownmine in the Tunstall area and the Billey Coal at Sneyd Colliery. It is commonly split into several leaves and rarely attains a metre in thickness. An ironstone, the Blackmine Ironstone, locally overlies the coal and was mined to mix with richer blackband ores. Nonmarine faunas are associated with the ironstone, and have been correlated with those of the Main Estheria Band of the East Pennine Coalfield (Edwards, 1951, p.69; Earp and Calver, 1961). In Bowsey Wood Borehole (643.33 to 647.24 m) they include *Naiadites* sp., *Geisina subarcuata* and fish debris; an estheriid band with *Euestheria* sp. occurs at 644.19 to 644.32 m. Faunas from other localities at this horizon include *Naiadites melvillei* (holotype, Trueman and Weir, 1957, pl. 30, fig. 16), *N. productus*, *Spirorbis* sp., *Carbonita bairdioides*, *C. fabulina*, *Geisina subarcuata* and fish debris. A further faunal band in Bowsey Wood Borehole (636.5 to 636.73 m), with *Geisina subarcuata* and fish debris, immediately underlies the Shafton Marine Band.

The **Shafton Marine Band** was known locally as the Priorsfield Marine Band. The bed is widespread and in its acme phase is developed in a pectinoid facies. The best sections are in the boreholes at Bowsey Wood (630.88 to 636.27 m), Crowcrofts (550.42 to 550.62 m) and Holts Barn (529.36 to 530.43 m), where the composite fauna comprises *Planolites ophthalmoides*, *Ammodiscus* sp., *Glomospira* sp., sponge spicules, *Lingula mytilloides*, *Orbiculoidea* cf. *nitida*, *Dunbarella* sp., gastropod and fish debris. The marine band was exposed at Weston Sprink [930 432], where a detailed section was logged by Stobbs (1905); at Birchenwood Brickpit [854 541]; and on the A500 road [8299 5181]. The marine band is commonly overlain by mudstones with ironstones, the **Rustymine Ironstone**, also known as the Priorsfield Bass Ironstone, which locally contains rich fish faunas and nonmarine bivalves (Ward, in Gibson, 1905, p.323). *Naiadites melvillei* is recorded from this interval in Clayton Borehole (525.02 m). In Crowcrofts (544.07 to 549.86 m) and Holts Barn (525.63 to 526.39 m) boreholes several nonmarine faunal bands occur, with a composite assemblage comprising *Anthraconaia varians*, *A.* cf. *stobbsi*, *Naiadites* sp., *Geisina subarcuata*, *Euestheria vinti* and fish debris. The estheriids are present in two bands, one in the roof of the Shafton Marine Band and the other below the Winghay Coal. Several species of *Naiadites* are recorded from the district including *N. daviesi*, *N. hindi* and *N. triangularis*; a possible *Anthraconauta* sp. is also present. These ironstones and mudstones pass upwards into a sequence dominated by siltstones and sandstones with several thin coals, and towards the base of the Winghay Coal, thick seatearths. A tonstein, the **Holditch Tonstein**, has been noted at this level towards the centre of the coalfield (Appendix 2; Figure 16).

The **Winghay Coal**, otherwise known as the Knowles Coal in the Longton area and the Brownmine Coal at Silverdale Colliery, has not been widely worked in the northern part of the coalfield, largely because it commonly contains dirt partings or is split. However, it is generally over 2 m thick, and where thickest, in the southern part of the coalfield, it has in recent times been one of the most economically important seams. In places it forms a composite seam with the overlying Bay Coal, but over most of the coalfield the two seams are separated by seatearths. The interval between the seams varies dramatically in the Fenton area, where it decreases from over 14 m to zero in a distance of less than 1.5 km. This seam split, associated with slumped beds at Hem Heath Colliery (information from J O'Dell, of British Coal), was also noted by Earp and Calver (1961) in the south of the coalfield. Locally, ironstones in the mudstones above the Winghay Coal were worked, and yielded good fish faunas (Molyneux, 1864). Records of bivalves are rare, but *Naiadites* sp. has been found in a railway cutting [8383 5243] near Kidsgrove.

The **Bay Coal**, otherwise known as the Lady Coal, is rarely more than 1.0 m thick and has only been worked where it forms a composite seam with the Winghay Coal. It is overlain by the **Cambriense Marine Band**, formerly known in this district as the Bay Marine Band. This is the highest marine band in the Carboniferous sequence and is widespread in the coalfield. Its acme development is in a productoid facies. The best sections are in the boreholes at Holts Barn (510.23 to 511.61 m) and Bowsey Wood (613.87 to 616.53 m), where the composite fauna includes *Planolites ophthalmoides*, *Tomaculum* sp., *Lingula mytilloides*, indet. crustacean fragments., *Hollinella* sp., conodont and fish debris. Other faunal elements recorded from the district include arenaceous foraminifera, *Orbiculoidea* sp., *Productus* sp., *Phestia* sp., *Myalina* sp?, turreted gastropods, cf. *Jordanites cristinae* and *Euestheria* sp. The marine band was also well exposed near Bathpool, Kidsgrove [8384 5240] (Crofts, 1990b). It was first recorded by Molyneux (1864) and is notable for the presence of *Dunbarella papyracea*, which commonly occurs in marine faunas at lower levels in the Coal Measures. The top of the Cambriense Marine Band marks the top of the Middle Coal Measures.

UPPER COAL MEASURES

The Upper Coal Measures are all of Bolsovian (Westphalian C) age. They have been of major economic importance to the district, as they contain many extensively worked coals, ironstones and clays. The mudstones above the coals, particularly in the sequence above the Great Row, tend to have sparser invertebrate faunas, and better developed floras, than those in the Middle Coal Measures. These features probably reflect their environment of deposition in an upper delta plain or alluvial plain setting (Fielding, 1984b). The Upper Coal Measures pass laterally towards the south and west into red beds of the Etruria Formation (Figure 18 inset). This transition, which reflects the existence of better-drained depositional environments in the south and west, is demonstrated by facies variations across the district. For instance the black, mussel- and ostracod-bearing lacustrine mudstones that form the roof of the Great Row Coal at Silverdale pass southwards into pale mudstones, seatearths and thin coals at Whitmore (information from J O'Dell, British Coal). The major sandstones of the Upper Coal Measures tend to be thinner than those of the Lower and Middle Coal Measures; their petrography was described by Malkin (1961). The Upper Coal Measures range up to about 400 m in thickness.

During deposition of the Upper Coal Measures, marine transgressions failed to reach Britain and the climate became more seasonal and drier. Lacustrine mudstones and limestones took over the cyclical position previously occupied by marine bands. Faunas in these younger Westphalian strata tend to show low diversity but large population size. Consequently they are of less value in correlation between coalfields. The roof mudstones of the Cambriense Marine Band effectively form the base of the *Anthraconauta phillipsii* Zone. Typical faunas of the Upper Coal Measures include *Anthraconaia saravana, Anthraconauta calcifera, A. phillipsii, Anthracopupa brittanica, Carbonita humilis, C. inflata, C. fabulina, C. pungens, C. salteriana, C. scapellus, Cypridina radiata, Eocypridina radiata, Hiboldtina wardiana, Spirorbis* sp., *Euestheria* cf. *mathieui, E. simoni* and fish.

The Upper Coal Measures are described in two parts. The lower part (Figure 17), between the top of the Cambriense Marine Band and the Bassey Mine Coal, constitutes the upper part of the informally named 'Clayband Measures'. Of these measures those above the base of the Cannel Row Coal were called the 'Great Row Measures' by Earp and Calver (1961), because they were regarded as transitional in nature between the 'Clayband Measures' and the 'Blackband Group'.

The upper part of the Upper Coal Measures contains distinctive blackband ironstones. These strata (Figure 18), between the base of the Bassey Mine Coal and the top of the Blackband Coal, were referred to as the 'Blackband Group' in previous surveys.

STRATA BETWEEN THE CAMBRIENSE MARINE BAND AND THE BASSEY MINE COAL

These strata range up to 250 m in thickness (Figure 7g), and show thickness patterns similar to those of the strata between the Aegiranum and Cambriense marine bands (Figure 7f). They are locally absent in the southern part of the district, as in Sidway Mill Borehole, where the Upper Coal Measures have passed entirely into the Etruria Formation. The strata have been extensively documented, as they were formerly frequently worked for coal, clay and ironstone. Detailed sections are recorded from the Tunstall area [8580 5233, 8593 5303] (Rees and Clark, 1992), and in the Fenton area [900 441, 903 443] (Rees, 1990b). A detailed section of parts of these measures, and the underlying Cambriense Marine Band, was exposed in the Bathpool area [8384 5240] (Crofts, 1990b).

Between the Cambriense Marine Band and the Chalkey Coal is a sequence of very varied thickness containing several coals that cannot easily be correlated. These strata are notable in containing thick seatearths, numerous ironstones and brittle dark grey mudstones rich in fish detritus. Where channel-related sandstones and siltstones form a substantial proportion of the sequence, as in the Longton, Fenton and Silverdale areas, seatearths are not so well developed. Such sandstones below the Great Row and Spencroft coals at Silverdale are associated with several metres of conglomerate, bearing well-rounded quartz, quartzite and igneous pebbles derived perhaps from a southern source. The roof of the Cambriense Marine Band marks the base of the *Anthraconauta phillipsii* Zone. The highest occurrence of *Naiadites* and *Geisina*, including *G. subarcuata*, is recorded at this level, as in the Bowsey Wood Borehole (609.45 to 612.65 m). Several other mudstones, containing nonmarine bivalves, ostracods and fish, occur in the overlying strata. Large numbers of fish have been collected, locally from several horizons (Ward, 1890), as at Fenton, and are preserved in the Ward Collection at Hanley Museum. Three named seams are present in the beds below the Chalkey Coal; they are the **Ragmine Coal**, the **Bungilow Coal** (known as the Hanbury Coal, or Hanbury Mine Coal in the Fenton and Longton areas) and the **New Mine Coal**, and appear to be well developed only in the Hanley and Fenton areas. Individually the seams rarely exceed 1.0 m in thickness, are commonly split, and were infrequently worked. Ironstones are developed above the seams over much of the coalfield. Sandstones between the New Mine and Chalkey coals and below the Bungilow Coal form topographic features in the area between Longton and Newchapel. The sandstone below the Bungilow is characteristically reddened in Hanley.

The **Chalkey Coal** was known near Longton as the Chalkey Mine, and at Silverdale Colliery as the Yard. Like the underlying seams of the Upper Coal Measures it is generally less than 1.0 m thick, though is widely correlatable and commonly has a cannel roof. In places this is overlain by a formerly muchworked ironstone, the Chalkey Mine Ironstone, which was used as a mix for blackband ironstones. Similar ironstones, known as Flowery Mine Ironstones (Homer, 1875) also occur locally at higher levels between the Chalkey Mine Ironstone and the Deep Mine Coal.

The **Deep Mine Coal**, known as the Blackstone Coal at Silverdale Colliery, is thicker than the underlying seams of the Upper Coal Measures, commonly exceeding 1.0 m in thickness. The lower part of the sequence between the Deep Mine and Wood Mine Coals consists of mudstones bearing fish and nonmarine bivalves. Ironstones are present, including the Deepmine Ironstone. The overlying **Wood Mine Coal**, otherwise known as the Sheath Mine Coal at Silverdale Colliery, tends to be thinner than the Deep Mine Coal and is absent or thin in the Hanley area. In parts of the south and west of the coalfield it is split into two leaves. The coal is generally overlain by mudstones; oil has been extracted from these in the past (Homer, 1875). Locally the mudstones are associated with a nodular ironstone, the Woodmine Ironstone, more commonly known in the Hanley area as the **Pennystone Ironstone**. This was known as a 'semi-blackband' and was often smelted with blackband ironstones.

The **Cannel Row**, also known as the Little Mine Coal on the Western Anticline and the Five Feet Coal in the Trentham area, has been one of the most extensively worked seams in the coalfield. It is generally between 1.5 and 2.0 m thick, was easy to work and was popular for kiln firing. The coal derives its name from its cannel roof, which commonly grades up into an ironstone. This is normally overlain by fissile mudstones, the 'Cannel Row Half Yards' which have in the past been used for oil production (Homer, 1875). Locally between Trentham and Silverdale the roof consists of listric seatearth mudstones which in places form linear roof intrusions, with irregular margins, into the Cannel Row (information from J O' Dell, British Coal). A tonstein has been recorded in the lower part of the interval between the Cannel Row and Great Row towards the south-western part of the coalfield. In the same area an ironstone, normally lying within 5 m of the Cannel Row, becomes increasingly calcareous and grades locally into a limestone bearing ostracods, plant debris and *Spirorbis* (Stobbs, 1905). Although a seatearth is commonly developed beneath the Great Row, locally the seam is directly underlain by a dark competent mudstone which was often mistaken for coal (Stobbs, 1915).

The **Great Row**, which was the most extensively worked coal in the Upper Coal Measures, is between 1.5 and 4.0 m thick over most of the Potteries Coalfield, is low in ash and has few partings. It is commonly split, though mainly into a few, thick leaves which could be individually worked. These properties made it cheap to work, even though its roof is generally poor; a cannel within the top metre of the coal was often used to make a working roof (Cope, 1954). The coal was worked for household purposes, iron and steel making and general manufacturing, and was particularly popular in the pottery industry as it burns with a long flame. The **Gubbin Ironstone**, one of the most worked 'semi-blackband' ironstones of the coalfield, commonly contains abundant bivalves. It occurs in many areas in the interval between the Great Row and Spencroft seams, and is best developed in sequences containing few sandstones. The thickness between the Great Row and Spencroft seams decreases towards the south-east, where they are separated by less than 10.0 m of seatearths and thin coals.

The **Spencroft Coal** was extensively worked for the firing of pottery. Associated seatearths were also widely worked. The coal is commonly split, though this has not affected its popularity as at least one leaf is normally over 1.2 m thick. Over most of the Western Anticline it forms three leaves (see frontispiece), the middle one containing a tonstein (Figure 17). The strata between the Spencroft and Peacock coals are notable in containing laterally persistent sandstones, which have been mapped over much of the northern, central and western parts of the coalfield. They form particularly clear topographical features in the Fenton area. In the southern part of the coalfield the sequence is dominated instead by seatearths, associated with the plant-rich 'Peacock Marl', and thin coals (Figure 17).

The **Peacock Coal** gained its name from its iridescent appearance when fractured. It was frequently worked, along with its associated marls, for used by the pottery industry. It generally forms a single seam between 1.0 and 2.0 m thick, though locally is split. Like the sequence below the Peacock Coal, the sequence between the Peacock Coal and the Little

Row is dominated by seatearths. Fossil trees have been recorded south of Tunstall (Floyd, 1964).

The **Little Row**, also known as the Yard Coal, tends to be less than 1.0 m thick, though in the Hanley and Silverdale areas it is locally thicker. It has a higher ash content than most underlying seams of the Upper Coal Measures. Seatearths are prominent in the strata between the Little Row Coal and the Bassey Mine Coal; sandstones and siltstones are common only on the southern margin of the coalfield.

BASSEY MINE COAL AND OVERLYING STRATA

These strata include the 'Blackband Group' of previous surveys, defined by Malkin (1961) and Boardman (1978) as the measures between the Bassey Mine and Blackband coals, plus the overlying beds up to the base of the Etruria Formation. The sequence contains four blackband ironstones, the Bassey Mine, Red Mine, Red Shagg, and Blackband Ironstones. Seatearths, coals and mudstones otherwise dominate the sequence. Lithological variations within these are discussed by Malkin (1961). Reddened mudstones present at specific horizons within these strata increase both in number and thickness to the south and west, where they eventually pass laterally into the Etruria Formation.

The 'Blackband Group' was of immense value in the past. In addition to the ironstones, the associated coals were mined, chiefly for domestic use, the thick seatearths were worked for the manufacture of bricks, tiles and kiln furniture, and the mudstones were used to produce oil (Homer, 1875; Cadman, 1901). A coal was rarely worked underground without its associated ironstone; since the time when working of the ironstones became uneconomical the seams have only been extracted in opencast operations.

It is no coincidence that five of the six towns of the Potteries were originally sited on the outcrop of the 'Blackband Group'. Numerous pits were dug to provide resources for local industry; it is the more recent of these that have provided much of our stratigraphical knowledge of the strata, such as Berry Hill Marl Pit [896 458], Cobridge Marl Pit [880 487] (Rees, 1990a), Scotia Bank Brickworks [8685 5070], and Newfield Marl Pit [858 523] (Rees and Clark, 1992).

The blackband ironstones are sideritic carbonaceous mudstones, and range up to 4.3 m thick (Boardman, 1978). They exhibit an alternation of conspicuous carbonaceous (often coaly) and sideritic laminae up to 15 mm thick. They commonly contain rich nonmarine bivalve and ostracod faunas, together with *Estheria*, *Spirorbis* and fish remains, and preserve primary structures such as root traces and microscopic details of plant remains (Boardman, 1978; 1989). They commonly overlie coal seams, or are separated from them by cannel coals, and are mostly overlain by varved mudstones or oil shales. Inferior ironstones that overlie the blackband ironstones were often given colourful names (Stobbs, 1915) and used for building, or packing, in mines. Noting the association of the ironstones with lacustrine rocks, and the presence of features suggesting development of siderite soon after deposition, Boardman (1989) proposed that the ironstones are fossil bog iron ores. He suggested that they were formed by mixing of iron bicar-

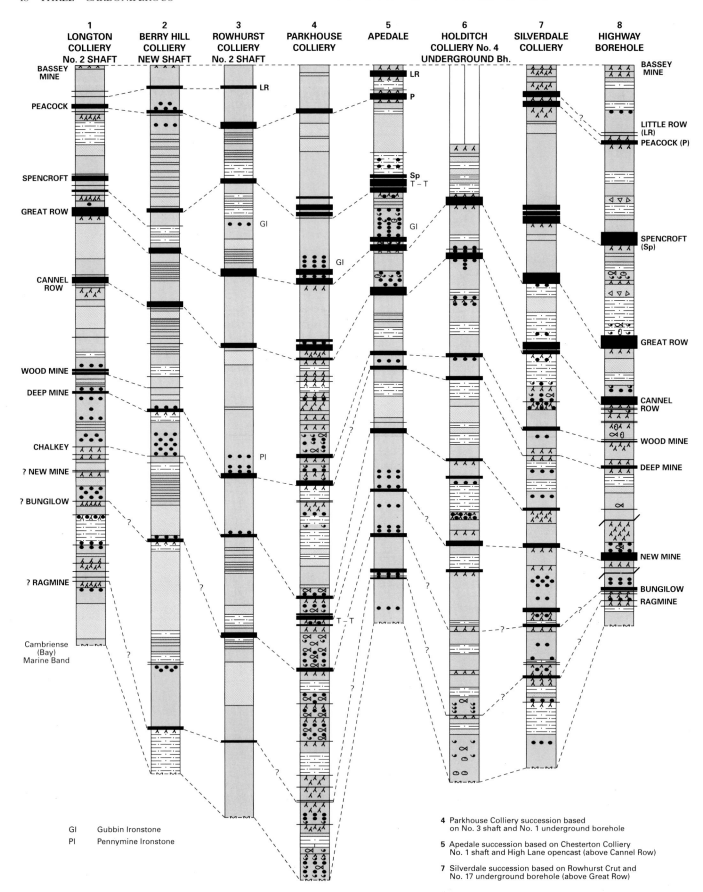

Figure 17 Correlation of the Upper Coal Measures between the Cambriense Marine Band and the Bassey Mine Coal. For key see Figure 10.

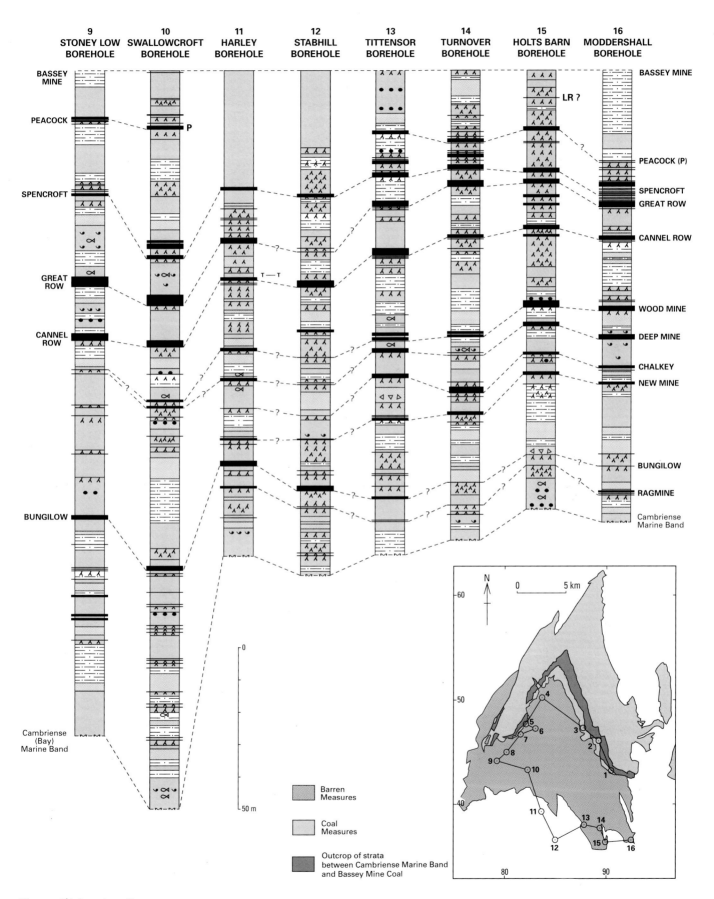

Figure 17 *(continued).*

bonate-saturated water (from acidic carbon-dioxide rich peat) with oxygenated lake surface water, causing precipitation of ferric oxyhydroxides; these were converted by biological reduction processes into siderite at shallow depth. The total iron content commonly decreases upwards through the ironstone from about 40 per cent near the base to between 20 and 30 per cent near the top. Boardman suggested that this results from a progressive dilution of the iron precipitates in the lakes by an increasing argillaceous influx. Analyses of ironstones also prove the presence of other metallic elements, derived in part from minute crystals of galena and sphalerite (Cadman, 1901).

The blackband ironstones pass laterally into limestones or calcareous mudstones towards the south and west of the Potteries Coalfield (Malkin, 1961; Earp and Calver, 1961), where the sequence is thinner, and where the Red Mine, Red Shagg and Blackband ironstones are poorly developed or absent. In the Apedale area, to the west, limestones occur instead of the Bassey Mine Ironstone, above the Bassey Mine Coal (Boardman, 1978).

The reddened mudstones between coals, which represent partially drained palaeosols, have proved to be of stratigraphical significance (Figure 18). The principal red horizons occur between the Bassey Mine and Hoo Cannel coals, between the Clod and Red Mine coals, and between the Red Shagg and Blackband coals, though they also occur at other levels. The interseam intervals that are least commonly reddened are between the Hoo Cannel and Clod coals and between the Red Mine and Red Shagg coals. These intervals also have the greatest proportion of sandstones and siltstones associated with distributary channels. This suggests that the reddening present in other interseam intervals was caused by better drainage, away from distributary channel systems. The reddened horizons generally thicken and become more common towards the south (Figure 18). In the northern part of the coalfield, as at Parkhouse Colliery (Boardman, 1978) and at New High Carr Colliery [839 512] (Figure 18), the strata between the Bassey Mine Coal and the Etruria Formation are almost entirely grey. The areal relationship between reddened and thinner measures above the Blackband has been noted by Boardman (1978). The reason why the blackband ironstones can be traced laterally into limestones in these areas has yet to be satisfactorily explained.

The **Bassey Mine Coal** was also known as the Pottery Bassey Mine Coal on the Western Anticline. It was worked mainly on the eastern limb of the Potteries Syncline, along with the Bassey Mine Ironstone, and was used principally in brick manufacture (Homer, 1875). In places it contains a sphaerosideritic ironstone in the middle of the seam, known as 'Peel' or 'Shotty' (Stobbs, 1915). Normally the coal is overlain directly by the **Bassey Mine Ironstone**, which ranges up to 1.5 m thick in the Tunstall area (Homer, 1875), though is conspicuously absent in the southern part of the coalfield. The ironstone commonly contains *Anthraconauta* cf. *phillipsii*, *Carbonita humilis*, *Spirorbis* sp. and fish debris. Inferior ironstones occur in the sequence beteen the Bassey Mine and Hoo Cannel coals. This sequence is characteristically red-mottled over much of the coalfield (Figure 18). It also contains up to three limestones (generally

two), which individually may attain 0.5 m in thickness. The lower limestone is locally known as Wards limestone as it was taken by Ward (1890; Ward, in Gibson, 1905, p.331) to mark the base of his 'Upper Division' which equated with Gibson's 'Blackband Group' (Dix and Trueman, 1931). It contains *Spirorbis* sp., *Anthraconauta* sp., *Carbonita humilis*, *C. pungens*, *C. salteriana* and fish debris (Earp and Calver, 1961). The upper limestone also contains *A. phillipsii* and *Euestheria minuta*, and contains more insoluble material than the lower limestone (Malkin, 1961). It was originally taken as the base of the 'Blackband Group' by Gibson (1899a, pp.123–124). An oil shale occurs locally below the Hoo Cannel Coal (Stobbs, 1915) and was worked in the Tunstall area.

The **Hoo Cannel Coal**, as its name suggests, consists largely of cannel and was thus only worked on a minor scale, commonly with associated seatearths. It is normally overlain by mudstone with abundant *Estheria* and is generally separated by less than 10 m of overlying strata from another coal, sometimes referred to as the upper leaf of the Hoo Cannel Coal. Whether this is accurately named is unclear, as although the seams appear to converge in the south-east of the coalfield (Figure 18) they are nowhere known to form a single seam. In the central part of the coalfield between Kidsgrove and Trentham a sandstone is developed between them. The sequence above these seams in the northern part of the coalfield appears to consist of ironstone-bearing seatearths and thin coals, but sandstones and siltstones predominate in the south. The Hoo Cannel Coal appears to have been eroded locally by channels in the Madeley and Hanley areas.

The **Clod Coal**, known as the Black Bass Coal in the south-west of the coalfield, includes several impure coals and in all but a few areas forms several leaves, separated by canneloid mudstones. Misidentification of the latter as coals has led in the past to incorrect reports of the Clod Coal as a single seam over 4.0 m thick. Only at High Lane (Figure 18) does the combined thickness of the leaves reach this figure, and rarely are any of the leaves thicker than 1.0 m. The sequence between the Clod and Red Mine coals is dominated by seatearths and mudstones, and contains no notable ironstones.

The **Red Mine Coal** has been principally worked along with the more economically important ironstone and oil shales associated with it. The coal is commonly between 0.5 and 0.9 m thick, in a single seam. Locally, however, the seam is split by a thin mudstone which has been worked for oil (Stobbs, 1915). Boardman (1989) described a 0.2 m thick cannel containing nodular siderite within the seam at Mitchell's Wood opencast site [830 510]. The **Red Mine Ironstone** normally lies directly on the Red Mine Coal, though in some places, as at High Lane (Figure 18), other lithologies intervene. The ironstone is generally between 0.6 and 1.2 m thick. It is absent from most of the southern part of the coalfield as, locally, is the coal, though its position in boreholes may be inferred from the presence of calcareous mudstones (Figure 18). The ironstone is commonly overlain by a cannelloid mudstone which contains abundant *Anthraconauta phillipsiii* (Stobbs, 1915; Dewey, 1920). This in turn is overlain by notable high-yielding oil shales (Homer, 1875; Cadman, 1901; Stobbs, 1915). The sequence between these and the Red Shagg Coal is sandstone-rich, when compared with most of the sequence above the Bassey Mine Coal. Reddening is localised (e.g. at High Lane opencast; Figure 7a). The seatearths under the Red Shagg Coal caused problems in working the seam because of their swelling properties, but were utilised for brickmaking (Homer, 1875).

The **Red Shagg Coal** ranges in thickness between 0.5 and 0.7 m, and was rarely worked without the overlying **Red Shagg Ironstone**, normally between 0.3 and 1.0 m thick. Unlike some of the other blackband ironstones it rarely contains nonmarine bivalves, though plant roots, *Stigmaria ficoides*, are common. The ironstone is overlain by an oil shale which commonly passes up into reddened or mottled mudstones; these are characteristic of the lower part of the sequence between the Red Shagg and Blackband coals in several parts of the coalfield (Figure 18). The upper part of the sequence consists largely of siltstones and sandstones and contains a widely developed coal, up to 0.5 m thick. The seatearth of the Blackband Coal has been worked for brickmaking.

The **Blackband Coal** is known as the Half Yards Coal in the Tunstall area and the Ethel or Ethyl, Gutter or Stokesley Coal in the Fenton area. Although the thickness is variable, generally less than 0.8 m, it was mined for domestic use. It normally forms a single leaf; where split, commonly only the lower leaf was workable. The **Blackband Ironstone**, where present, almost invariably rests directly on the coal, and is generally between 0.4 and 1.2 m thick. Homer (1875) noted that where the Blackband Coal is thick, the overlying ironstone is thin, and vice versa. The ironstone is commonly associated with faunas including *Anthraconauta phillipsii*, *Carbonita* sp., *Euestheria* sp. and *Spirorbis* sp. The ironstone is locally overlain by pyritous mudstones with thin ironstones, some of blackband character. The pyrite was the source of many mine fires (Stobbs, 1915). The mudstones overlying the ironstone are also noted for their abundant *Calamites* stems. The sequence between the Blackband Ironstone and the base of the Etruria Formation is very variable because the base of the latter is diachronous; for instance, near Hanley the boundary varies locally over a stratigraphical thickness of about 20 m. The strata are largely composed of seatearths, though siltstones, sandstones and coals, some of which may be traced laterally over large areas of the coalfield, also occur. In Bowsey Wood Borehole the sequence is over 100 m thick, and consists of mudstones, variably red and grey (Earp and Calver, 1961). The latter contain *Spirorbis* sp., *Anthraconauta phillipsii*, *A. wrighti*, *Carbonita humulis*, *C. inflata*, *C. salteriana* and *Euestheria* sp.

Barren Measures

Carboniferous strata overlying the Coal Measures are referred to here as Barren Measures, an informal name derived from the lack of exploitable coal within them. The earliest use of a similar term (Barren Measures, or Barren Coal Measures) appears to be by Vernon (1912) who defined it (p.593) in the Warwickshire Coalfield for predominantly red formations above the productive Coal Measures, up to and including the Keele Beds. In the present district the Barren Measures comprise the Etruria, Newcastle and Keele formations, of Westphalian age, and the Radwood Formation, which is probably of Stephanian age.

The names of these units are altered somewhat from those used in previous classifications (e.g. Trotter, 1960): the formal name Etruria Formation is used instead of Etruria Marl Series; the informal term Newcastle formation replaces the named Newcastle-under-Lyme Series (or Newcastle Beds, or Newcastle Group), the informal term Keele formation is used for the lower part of the former Keele Sandstone Series (or Keele Beds, or Keele Group), and the formal name Radwood Formation is applied to the upper part of the old Keele Sandstone Series.

The thick sequence overlying the Etruria Formation is sandstone-rich at its base and broadly fines upwards. Its lower part is dominantly grey and contains several good

lithological and biostratigraphical markers; its upper part consists mainly of red beds that contain few clear markers. When the sequence was mapped late in the last century, it was subdivided by Gibson (1899) on grounds of colour into a lower, grey, Newcastle-under-Lyme Series and an upper, red, Keele Sandstone Series (after the Keele and Penkhull Beds of De Rance, 1898). As this subdivision was easy to apply, its usage persisted, with minor changes of name, and extended over most of the English Midlands. Most recently (Wilson et al., 1991), the units were referred to as formations, though without formal definition.

Recent research has uncovered many problems with the traditional subdivision based on colour. Not only do grey strata occur within the Keele formation, but the base of the latter is notably diachronous and many beds can be traced laterally from the Newcastle formation, where they are grey, into the Keele formation, where they are red. Recent detailed work on the sequence suggests that it can be subdivided much more effectively on the basis of characteristics other than colour. This work shows that the red beds can be split into two parts; an upper, mud-dominated division, and a lower, sandstone-rich division. The upper part, containing sublitharenite sandstones, has great similarity with the Enville Beds of south Staffordshire and Shropshire, and is now known as the Radwood Formation. The lower part, which is dominated by lithic arenites, is difficult to distinguish on the basis of petrography, geochemistry, sedimentology, or geophysical log character from the Newcastle formation, and can be regarded as the upper, oxidised part of the same primary stratigraphical unit. The recent work of Besly and Cleal (1997) suggests that this combined unit is equivalent to the Halesowen Formation of the Warwickshire, South Staffordshire and Shropshire coalfields, and it is likely that the unit will eventually be referred to by that name. For the present, it seems best to consider the Newcastle and lower part of the Keele formations as separate, but informal, units; they are referred to here as the Newcastle formation and the Keele formation.

Etruria Formation

The Etruria Formation consists of distinctive reddish brown and multicoloured, largely structureless mudstones with subordinate beds of green sandstone. It generally forms a wide vale, as at Etruria, with the higher beds cropping out on the face of the Newcastle formation escarpment. Because of the soft and easily weathered nature of the formation, there are no significant natural exposures. The best surface sections have been in brick quarries and although most of these have been backfilled, records exist for many that were open during the 1960s and 1970s.

Continuous sections through the formation have been recorded in a few colliery shafts and cored boreholes. South of its outcrop the formation has been penetrated in numerous coal boreholes; most of these were not cored, so stratigraphical details originate from geophysical logs and cuttings descriptions.

The formation was originally named by Gibson (1899) as the Etruria Marl Series, from exposures in quarries

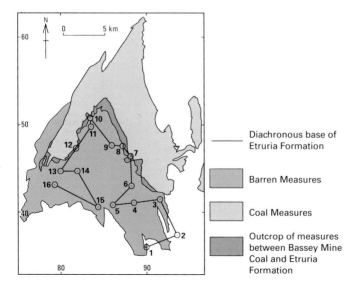

1 HOLTS BARN BOREHOLE
2 SCHOOLHOUSE BOREHOLE
3 FLORENCE COLLIERY No. 2 SHAFT
4 HEM HEATH COLLIERY No. 2 SHAFT
5 TRENTHAM BOREHOLE
6 FENTON MANOR COLLIERY HOME SHAFT
7 HANLEY (Rees 1990)
8 RACECOURSE COLLIERY No. 3 SHAFT

BLACKBAND
RED SHAGG
RED MINE
CLOD
HOO CANNEL
BASSEY MINE

Etruria Formation

CLOD
HOO CANNEL
BASSEY MINE

Diachronous base of Etruria Formation

Barren Measures

Coal Measures

Outcrop of measures between Bassey Mine Coal and Etruria Formation

Figure 18 Correlation of the Upper Coal Measures between the Bassey Mine Coal and the base of the Etruria Formation. For key see Figure 10. Inset box shows the lateral passage between the Etruria Formation and Coal Measures in the area south-west of Keele.

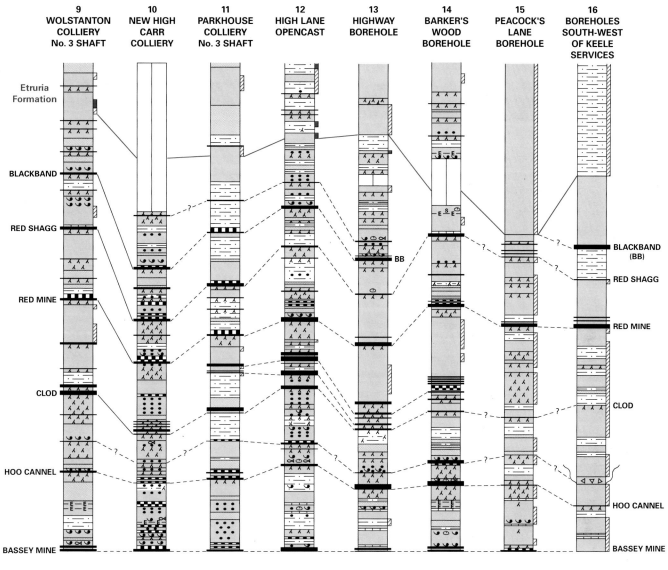

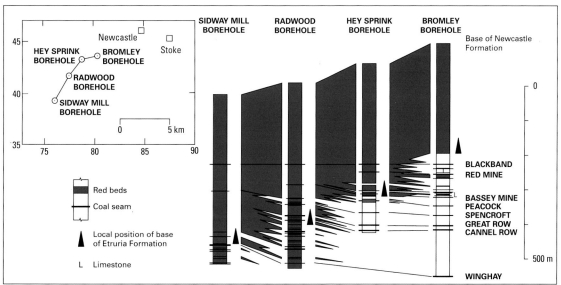

Figure 18 *(continued).*

near Etruria. The term 'Marl' has been dropped from the formal formation name, as it is not an accurate description of the mudstones; in informal usage, however, ceramic clays from this formation are still referred to as 'Etruria Marl'.

The base of the formation is defined at the incoming of more or less persistent red beds at the top of the Coal Measures. Exact definition is difficult, as there is an upward transition from grey Coal Measures into red beds through a complex interdigitation over an interval of up to 150 m (Boardman, 1978). As a result, the base is strongly diachronous, occurring at lower stratigraphical horizons in the south and involving the complete lateral passage of the Upper Coal Measures of the exposed coalfield into red beds (Figure 18 inset). In the present survey the transitional zone was classed as Coal Measures. The base of the formation has no distinctive log character. An estimate of its position is provided by coals in the Upper Coal Measures which persist as thin grey mudstone intercalations within the lateral facies transition into red beds; these can be recognised in dominantly red-bed sequences from the gamma ray, density and neutron logs in some boreholes. Haslam (1993) attempted to distinguish between the Coal Measures and Etruria Formation in the Sidway Mill Borehole on the basis of geochemistry. He found that although values for the mobile elements Mg, K, Ca, and Sr are generally higher in the Coal Measures than the Etruria Formation, levels of immobile elements are very much the same.

The top of the formation is defined by the base of the Newcastle formation. This occurs at a pronounced change of facies, between massive structureless mudstones of the Etruria Formation and the grey laminated basal mudstones of the Newcastle formation. The boundary commonly approximates to the colour change, although locally the topmost metre of the Etruria Formation is grey or mottled. In boreholes the boundary is clearly defined by a shoulder in the gamma ray log (Figure 19). This corresponds to the upward change from the kaolinite-dominated mineralogy of the Etruria Formation mudstones (see below) to the illitic mudstones of the Newcastle formation.

The Etruria Formation has been referred to the late Bolsovian (Westphalian C) stage (Ramsbottom et al., 1978), on the basis of the sporadic occurrence of *Anthraconauta phillipsii* and *Anthraconaia* cf. *saravana* (Gibson, 1901). An early Westphalian D microflora was obtained from coals at the supposed base of the Etruria Formation in the Pie Rough (Keele No. 1) Borehole (reported by Butterworth and Smith, 1976), but recent geophysical evidence suggests that these coals in fact lie in the Newcastle formation. Butterworth and Smith reported similar floras in a coal high in the Etruria Formation at Chesterton. This is probably at the same horizon as a coal and ironstone referred to by Gibson (1901) as containing *Anthraconauta phillipsii*.

The formation is dominated by mudstone of massive appearance, with crude bedding defined only by colour variations between dark brownish red, variegated shades of purple and ochre, and pale grey. Dark grey, locally carbonaceous mudstones form a small proportion of the formation, especially towards the base, and they occa-

sionally contain thin coal seams. Thin limestones are sporadically present in the upper part of the formation.

The mudstones consist of a mixture of three principal mineralogical components: silt grade detrital quartz; kaolinite, in a highly disordered form; and haematite (Holdridge, 1956; Keeling and Holdridge, 1958; Biggs, 1987). Goethite and calcite are present as major components in places, and illite, chlorite and anatase form minor components. The distribution of calcite and illite is stratigraphically controlled. The predominance of kaolinite over illite in the formation is borne out by low values for K_2O and K_2O/Al_2O_3, illustrated by Haslam (1993). His geochemical profiles for the Sidway Mill Borehole show a progressive upward decrease in K_2O, K_2O/Al_2O_3, Rb, Sr and, especially, Ba, leading to extreme depletion near the top of the formation.

Sandstones are greenish grey lithic arenites, forming the distinctive 'Espley Grits' (Williamson, 1946). These are fine- to coarse-grained sandstones which locally contain lenticular conglomerates, especially near the base. Identifiable lithic components include polycrystalline quartz of metamorphic provenance, quartzite, jasper, and weathered igneous fragments (Malkin, 1961). The greenish coloration is due to the presence of abundant chlorite-rich matrix, largely derived from the alteration of detrital basic igneous grains (Besly, 1983). The sandstones are impersistent and generally do not exceed 6 m in thickness. However, where they are most extensively developed, in the middle of the formation in the south and east of the Potteries Coalfield, they reach thicknesses of over 15 m. Palaeocurrents measured in the sandstones suggest that the main sediment transport direction was to the north or north-west.

Ironstones in the formation are rare, compared with the underlying Coal Measures, though there are notable examples locally. One about 45 m above the Blackband Coal, for instance, is over 2 m thick in the Keele area.

Numerous mudstones containing in-situ roots and rootlets were recognised as palaeosols during the 1963 survey and feature in many of the sections and boreholes measured at that time. Subsequently, Besly and Fielding (1989) have shown that pedogenic modification is even more extensive in the formation; pedogenic features dominate the sequences characterised by distinctive patterns of variegation in shades of brownish yellow (goethite pigment) and grey and purple (haematite pigment). Such sequences are interpreted as ferruginous and ferralitic palaeosols, which locally exhibit characteristic palaeosol profiles. Palaeosols often lie at the tops of upward-fining sequences of silty mudstone and mudstone above sandstone. Besly (1988) has interpreted the Etruria Formation in terms of alluvial floodplain deposition. Thin sandstones are interpreted as crevasse splay deposits, whilst thicker sandstones, within which upward-fining and lateral accretion units are present, are interpreted as deposits of meandering fluvial channels. The geochemical characteristics of mudstones and sandstones (low values for mobile elements, high values for immobile elements) demonstrate deposition under conditions of intense weathering, combined with generally good drainage (Haslam, 1993).

Figure 19 Bromley Borehole: geophysical logs of the type borehole section through the Etruria and Newcastle formations. For key see Figure 21.

The thickness of the Etruria Formation varies from 210 m to 430 m, with the greatest recorded thickness at Sidway Mill Borehole. The thickening here is due to the diachronous nature of the base of the formation (Figures 7h, 18), in which large thicknesses of strata containing important coal seams such as the Great Row and Cannel Row pass southwards into red beds almost devoid of coal seams. The pattern of subsidence is best shown on an isopach map that ignores the facies change by showing data for the interval between the Blackband Coal and the base of the Newcastle formation (Figure 7i). The map has been derived from borehole penetrations that are apparently unfaulted. It shows the persistence of minor depositional axes trending approximately along the axis of the present Potteries Syncline, and in a north–south direction in the area to the south of Longton. These are separated by a zone of reduced subsidence in the south-west part of the explored coalfield. The Western Anticline also appears to have been a zone of reduced subsidence, although evidence for this is tentative, being restricted to two poorly recorded boreholes at the southern end of the structure.

The formation is informally split into three divisions (Wilson et al., 1992; Figure 20) on the basis of the detailed records from Wolstanton No. 3 colliery shaft (Malkin, 1961; Keeling, 1961) and the Silverdale area, especially Barker's Wood Borehole. Details given here relate to areas in which continuing quarrying operations, or the infilling of existing large excavations, will ensure the availability of exposure for the foreseeable future. The pattern of quarrying is such that individual features are not commonly exposed for more than a few years. Details of backfilled quarries that were open in the 1970s were recorded by Besly (1983). Backfilled quarries are mentioned here only where unique features have been observed.

The lower division, between 60 and 80 m thick, typically comprises reddish brown and variegated mudstones, greenish grey laminated mudstones and up to five thin coals. Its most characteristic feature is the widespread occurrence of greenish grey sandstones, particularly in the north and west. These are up to 14 m thick in Wolstanton No. 3 Shaft, and form escarpment features in the Tunstall area, to the north of Parkhouse, and to the west of Silverdale. Gamma ray logs of closely spaced boreholes in the Keele area show that the sandstones are laterally variable, implying that the topographic features seen at outcrop may not be formed by individual beds. Locally sandstones are completely lacking in this division. The detailed log of a large exposure in Fenton Manor Quarry north of the Longton Fault [885 451] is shown in Figure 21. About 34 m of variegated and reddish brown mudstones with at least one complex palaeosol horizon (figured by Besly and Turner, 1983) and several seatearth horizons were recorded. The topmost of the seatearths comprises 0.6 m of slickensided carbonaceous mudstone. Minor sandstones up to 0.3 m in thickness are also present in the mudstone sequence. The top of the quarry face was formed by over 6 m of medium- to coarse-grained sandstone, which locally erodes down into the topmost seatearth. Sandstones and interbedded sandstones and mudstones close to the base of the formation were also exposed

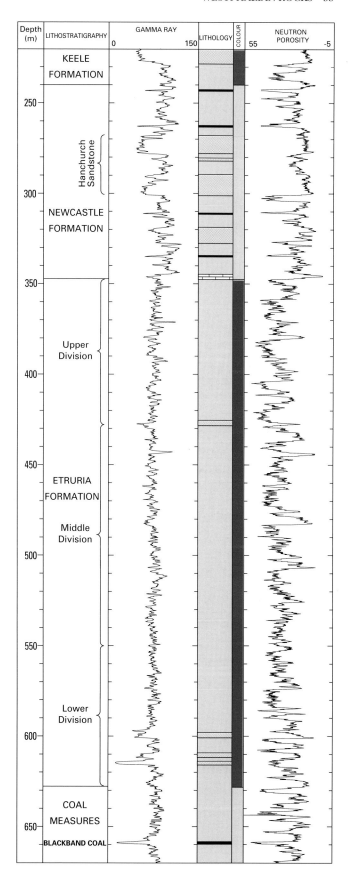

in the margin of the colliery yard at Silverdale [8158 4704], and in the partly backfilled quarry of Silverdale Tilery [804 462]. A typical sequence of colour-banded and mottled mudstones was exposed in the quarry at Bankeyfields [849 519].

The middle division is 105 to 170 m thick at outcrop, being thickest in the north. It consists of structureless mudstones, both reddish brown and variegated, with impersistent greenish grey sandstones. A more laterally persistent group of sandstones occurs between Fenton and Hollybush. The top of the division is marked between Chesterton and Wolstanton by a thin coal, which may be correlated with reasonable confidence with carbonaceous mudstone horizons recorded in the Barker's Wood and Newstead boreholes (Figure 20). The division has been extensively worked for clay, as it generally lacks the deleterious calcareous bands present in the upper division, and contains fewer sandstones than the lower division. Areas of former working at Chesterton and Tunstall are now exhausted or sterilised by development. Quarries are still operating on the east side of Bradwell Wood [around 848 509] and in the Knutton and Silverdale area. The best sections in the middle division are currently at Knutton [828 468] and Walleys [832 460] quarries near Silverdale. The two quarries are at approximately the same stratigraphical level, and expose sequences of varicoloured, yellowish brown, grey and reddish brown mudstones, in part silty and interbedded with fine- and medium-grained sandstone. In Knutton Quarry silty and interbedded units pass laterally into mudstone. In Walleys Quarry (Figure 21b) cross-bedded upward-fining sandstones are exposed at both the top and base of the section. In both quarries the outcrop data are supplemented by records of exploration boreholes which show the downward continuation of the sequence in a succession of seatearth mudstones and structureless mudstones containing grey carbonaceous horizons; these probably belong to the lower division of the formation. A significant proportion of the middle division, totalling 115 m (with gaps), was observed in 1963 in three quarries at Apedale Road, Chesterton [826 493]. This important section is now infilled, although it was partly recorded by Besly (1983). At the top of the section there was a rare exposure of 0.28 m of coal capped by 0.42 m of blackband ironstone with scattered mussels. The coal was encountered near the top of Parkhouse Colliery No. 3 Shaft, and can also be recognised in Wolstanton No. 3 Shaft where it is 0.05 m thick.

The upper division occurs above the coal developed in the Chesterton area and at equivalent horizons farther south. It varies in thickness at outcrop from 115 m in the north to 80 m in the south, and consists of reddish brown structureless mudstones with some variegated bands. The division is characterised by extensive beds of calcareous mudstone, which in most cases contain calcite concretions, 0.5 to 2 mm in diameter, known locally as 'shot'. Beds of this lithology, which are extremely rare in the lower and middle divisions, form up to 38 per cent of the thickness of the upper division (in Wolstanton No. 3 Shaft), and dominate the top 20 m of the formation. The upper division is also characterised by a lower content of illite than is found in the other divisions. Sandstones are generally impersistent, fine grained and greenish grey. In places a well-bedded pale grey ostracod-bearing limestone occurs about 40 m above the base of the division; it was 0.32 m thick in Holditch No. 2 Shaft, and 0.09 m thick in a quarry near Chatterley [8432 5075]. The division outcrops on the face of the escarpment of the Newcastle formation and is affected by

Figure 20 Sections of the Etruria Formation encountered in colliery shafts, borehole cores and exposures in and near the outcrop.

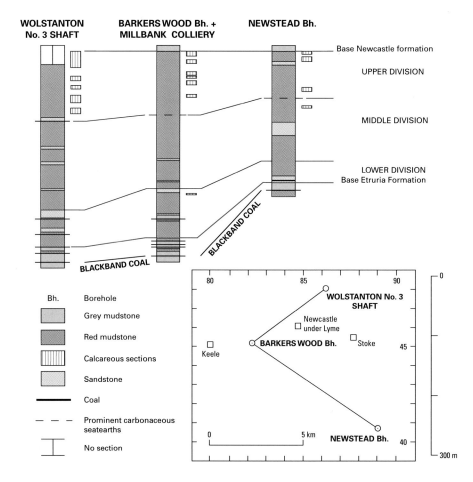

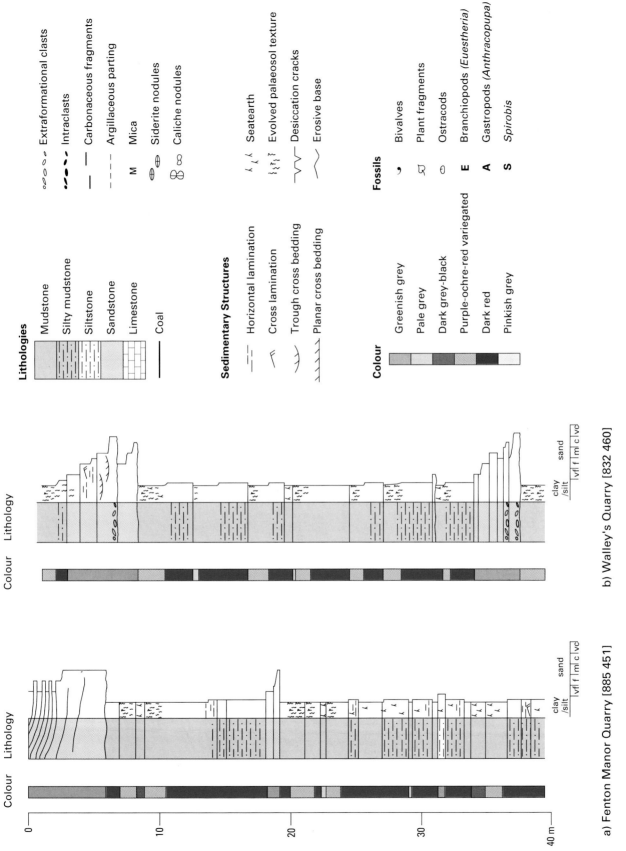

Figure 21 Representative sedimentological logs of the Etruria Formation: a. shows part of the lower division of the formation exposed in July 1984 at Fenton Manor Quarry, b. is a composite section of part of the middle division of the formation exposed between 1977 and 1990 at Walleys Quarry.

landslips and solifluction lobes. Currently the lower beds are seen in the abandoned Springfields Quarry [862 443] in 14 m of reddish brown and variegated mudstones containing a greenish yellow sandstone 2.1 m in thickness. The top of the division is locally marked by irregular grey blotches or 'pipes' descending from the base of the Newcastle formation into reddish brown mudstones. The top is exposed in disused quarries around Parkhouse and Chatterley. At Metallic Tilery quarry [8402 4976] 15 m of reddish brown and variegated mudstone, partially 'shot'-bearing, and a 0.6 m thick sandstone are exposed. A similar 39 m-thick sequence of strata is poorly exposed in the nearby disused Bradwell Wood Quarry [843 504]. The Madeley Quarry [789 451] exposes 26 m of reddish brown and variegated 'shot'-bearing mudstones, including a seatearth, and a greenish grey coarse-grained sandstone between 0.5 and 1.2 m thick. South of Longton, at Lightwood Quarry [2949 9085], about 9 m of reddish brown and variegated mudstones, with a sandstone, are exposed some 30 m below the top of the division (Rees, 1990b).

Not all elements of the tripartite division of the formation can be recognised in the south of the area, away from the outcrop, where the formation thins and the base occurs at lower stratigraphical horizons. The correlation datum afforded through the diachronous base of the formation by coals in the Upper Coal Measures does, however, allow recognition around Tittensor and Moddershall of the lower, sandstone-rich division of the outcrop area. Higher sandstones in this southern area can probably be correlated with the sandstone-rich interval mapped in the middle division of the formation in the Fenton and Hollybush areas, and recognisable in the Newstead Borehole. As in the case of the sandstones in the lower division, borehole geophysical logs show a complex pattern of lateral variation in this sandstone-rich unit. Of borehole penetrations to the south and west of the outcrop, the most detailed record is of the fully cored Sidway Mill Borehole, where the sequence is also thickest (430 m). Significant thicknesses of the formation were also cored in Hobbergate, Little Paddocks and Kibblestone boreholes. Lithologies are similar to those seen at outcrop. Seatearth mudstones and grey beds are common, especially in the part of the sequence which is laterally equivalent to the Upper Coal Measures in the exposed coalfield. Coals are developed locally, such as the seam 0.21 m thick at about 202 m below the top of the formation at Sidway Mill, which can be correlated (by means of wireline logs in adjoining boreholes) with the Blackband Coal in the Silverdale area. Similarly, in the cored section in Little Paddocks Borehole, the Red Shagg and Red Mine coals of the exposed coalfield area can be identified as thin coal seams developed in otherwise entirely red-bed facies. The seams can be correlated with the outcrop area by means of geophysical logs in the intervening boreholes, the Blackband Coal forming a prominent log marker in all but the extreme south of the area.

NEWCASTLE FORMATION

The Newcastle formation consists of grey mudstone, siltstone and sandstone interbedded with thin coals. The formation, which has an informal status, was first named the Newcastle-under-Lyme Series by Gibson (1899) after a former exposure in a railway cutting in the town (Figure 22). The basal sandstones commonly form an escarpment with lower slopes in the softer Etruria Formation, well seen at Bradwell Wood, Wolstanton, Trentham and Keele. Exposure is generally limited to the lowest beds, seen in the tops of quarries of the Etruria Formation, and rare stone pits in higher sand-

stones. In the 1980s the formation was temporarily exposed in road cuttings on the M6, A34, A500 and A525. Continuous sections through the formation have been recorded in a few colliery shafts, shallow cored boreholes sunk in the Keele and Butterton areas, and in deep boreholes at Barker's Wood, Hobbergate, Meaford Hall No. 1, Newstead and Sidway Mill (Figure 22). Further stratigraphical information is derived from borehole geophysical logs and cuttings descriptions.

The base of the formation is drawn at the sharp contact between the laminated grey silty mudstones of the Newcastle formation and the structureless, massive, dominantly red mudstones of the Etruria Formation (Pollard and Wiseman, 1971). This boundary does not always coincide with the colour change, as the top metre of the Etruria Formation is variably grey-mottled. The Metallic Tileries Quarry [8402 4976], a designated SSSI, is the reference section. In geophysical logs the boundary is marked by a distinct shift in the gamma ray log (Figures 8, 25). A group of limestones and lime-cemented silty mudstones in the basal few metres of the formation are commonly recognisable on the neutron porosity log. Geochemical profiles show a pronounced discontinuity at the base of the Newcastle formation (Haslam, 1993), with marked increases in the levels of several elements, notably K, Rb and Sr. The discontinuity is also illustrated by the relationship between the immobile elements Ti, V and Cr, the beds of the Newcastle formation having proportionately more Cr.

The top of the Newcastle formation is defined by the diachronous base of the lowest red or purplish grey beds of the dominantly reddened sequence of the Keele formation. It occurs below the Hanchurch Sandstone in the south-east of the area, and near Keele (Figure 23), and above the Springpool Sandstone in Sidway Mill Borehole (Figure 22). In the adjoining Stafford district, there is a complete lateral facies change into the Keele formation, and the Newcastle formation is not recognisable. The top of the formation does not have a distinctive wireline log signature. However, the base of the Springpool Sandstone, in the Keele formation, can usually be recognised on wireline logs and forms a correlation datum near the grey-to-red colour transition over most of the district.

Mudstones at the base of the formation at Metallic Tileries Quarry yielded the bivalve *Anthraconauta tenuis* (Myers, 1954) and a microflora indicating an early Westphalian D age (Turner, 1991). A Westphalian D spore flora was recoverd from coals higher in the formation by Butterworth and Smith (1976).

The formation is dominated by olive-grey and pale to medium grey silty mudstones. The principal clay minerals present are kaolinite and illite, in approximately equal amounts. The mudstones contain abundant plant debris, and are commonly rooted. Thin coals are present throughout. In the upper part of the formation, where red and varicoloured mudstones occur, coals are partially altered to limestone, similar to those described by Mykura (1960). Thin limestones of primary origin are also present, especially near the base of the formation, where they contain algae and a restricted fauna. Sandstones in the formation are grey, fine- to medium-grained micaceous lithic arenites, which weather to a distinctive brown colour. Locally they are red (see above), and are indistinguishable from those in the Keele formation. The sandstones are dominated by lithic grains of low-grade metasedimentary origin; the substantial clay content of these grains gives the sandstones a distinctive high gamma ray response. Tonsteins have locally been recorded in the formation (Appendix 2).

The lenses of algal limestone near the base of the formation at Metallic Tileries Quarry are interpreted as the deposits of a hypersaline lagoon (Pollard and Wiseman, 1971). The overlying beds include widespread thin micritic limestones with some ostracods, interbedded with, and overlain by, shaly mudstones with nonmarine bivalves, *Euestheria*, ostracods and fish remains. These beds are interpreted by Besly (1988) as being lacustrine in origin. Overlying beds mark the infilling of the lake and the establishment of a freshwater deltaic environment, similar to that which existed during sedimentation of the Coal Measures. Overbank deposits of siltstones, mudstones and thin sandstones are interbedded with channel and crevasse splay sandstones. Periods of swamp development led to the formation of thin coal seams. The thicker, laterally persistent Hanchurch Sandstone probably represents the fill of major fluvial or distributary channels. Palaeocurrent directions measured in the formation have a wide distribution, with a significant component of flow towards the west and south-west.

The isopach map shown in Figure 7j approximates the thickness of the formation and is based on boreholes in which a lack of faulting can be confidently inferred, except in the western part of the area, where modern data are largely lacking. The pattern of thickness change in the Newcastle formation is similar to that in the Etruria Formation, showing a general thinning southwards. In the west of the district the thickness of the formation, as penetrated in Bowsey Wood Borehole (based only on cuttings examination) and calculated from mapping in the Craddocks Moss area (Figure 22, column 3), suggest that the pattern of thinning associated with the Western Anticline earlier in the Westphalian persisted during deposition of the Newcastle formation.

Beds below the Hanchurch Sandstone

The basal 3 to 15 m of the formation consist of calcareous shaly silty mudstones containing up to three limestones, each varying between about 0.1 and 1.8 m in thickness. Locally the limestones, the uppermost of which contains ostracods and fish, pass laterally into the mudstones. The mudstones are grey, locally with chocolate coloured blotches, and in places contain ostracods and fragmentary bivalves. These beds are exposed at Metallic Tileries Quarry (figured by Pollard and Wiseman, 1971, and by Besly, 1988), where two limestones, each about 0.25 m thick, are underlain by mudstones containing lenses of algal limestone. At Keele Tileworks Quarry, Madeley Heath [789 451], the upper limestone, 0.18 m thick, which locally splits, and the lower limestone, 0.1 m thick, are tough grey ostracod-bearing calcisiltites. Two limestones were also logged

at this level at Highfields Tileries [859 488] and Lightwood Lodge [918 405] by Gibson (1905). The limestones are particularly well developed in Redheath Plantation No. 3 and Springpool No. 1 boreholes, where thicknesses of about 2 m were proved (Figure 22). The limestone-bearing unit is overlain by 2 to 3 m of grey shaly mudstones containing mussels, ostracods, *Euestheria* and fish scales. In Sidway Mill Borehole the mudstones beneath the limestone are some 50 mm thick and also fossiliferous. *Anthraconauta phillipsii*, *Euestheria* and ostracods occur above the limestone and *Anthraconauta calcifera*, ostracods and fish scales below it.

These mudstones pass up into a sequence of grey mudstones, siltstones and sandstones, containing between two and five coals, which is over 70 m thick near Stoke-on-Trent City General Hospital (Figure 22). Two of the coals are of sufficient thickness to be recognised on gamma ray, density and neutron porosity logs , and can be widely correlated in boreholes (Figure 22); they lie respectively about 12 m above the base of the formation, and about 10 m below the base of the Hanchurch Sandstone. An intermediate seam can be correlated locally in the Keele area. Locally, sandstones form a strong topographic feature which characterises the base of the formation (see above). The lowest 18 m of this sequence is best exposed in the Keele Tileworks Quarry, Madeley Heath [789 451], where the lacustrine sequence overlying the basal limestones is overlain by an upward-coarsening sequence of siltstone and sandstone, capped by a seatearth and thin coal. This sequence is overlain by rooted mudstone, a thin coal, and cross-bedded flaggy fine-grained sandstone. The sequence up to the first sandstone is seen at the Metallic Tilery [840 497] and was formerly more extensively exposed in the widening of the nearby A34 road. At the time of survey the lowest sandstone was also exposed on the crest of the escarpment at Bradwell Wood [8440 5041] in thinly bedded fine-grained facies overlying laminated greenish grey siltstone. The best exposures at this level are at Apedale South Quarry [827 489] (Wilson, 1990; Rees, 1993), where the lowest sandstone is absent. The overlying mudstones are rich in plants, those above the lowest coal containing whole pteridosperm fronds. Three coals, between 0.05 and 0.3 m thick, are also exposed in mudstones near Butterton Grange Farm [8391 4221]. Mussels occur locally above the second and third coals. In Meaford Hall No. 1 Borehole the topmost coal seam is overlain by a thick bed of fossiliferous mudstone containing mussels, ostracods, fish and *Spirorbis* (Figure 22).

Hanchurch Sandstone

The beds just described pass up into a sequence 22 to 50 m thick without coals. It is commonly dominated by sandstones, collectively known as the Hanchurch Sandstone, which locally approaches 30 m in thickness but shows marked variation associated with lateral transitions from channel to overbank facies. This is well illustrated at Hem Heath Colliery, where the lowest 11 m of the Hanchurch Sandstone in the colliery drift pass laterally into mudstones with thin sandstone beds in the No. 2 Shaft (Figure 22). Job's Wood Quarry [823 460], a recognised Regionally Important Geological Site near Keele, exposes 10 m of greenish grey fine-grained sandstone in the lower leaf of the Hanchurch Sandstone, which forms a marked topographic feature at Hanchurch Plantation. A nearby quarry [8408 4213] is in 5.5 m of thinly to thickly bedded fine-grained sandstone, with some cross- bedding. Associated mudstones with *Euestheria*, *Spirobis* and rare nonmarine bivalves were temporarily exposed in cuttings on the A500 road and M6 motorway. Boreholes near Keele revealed a westward and northward transition in the colour of the Hanchurch Sandstone from grey to purplish grey

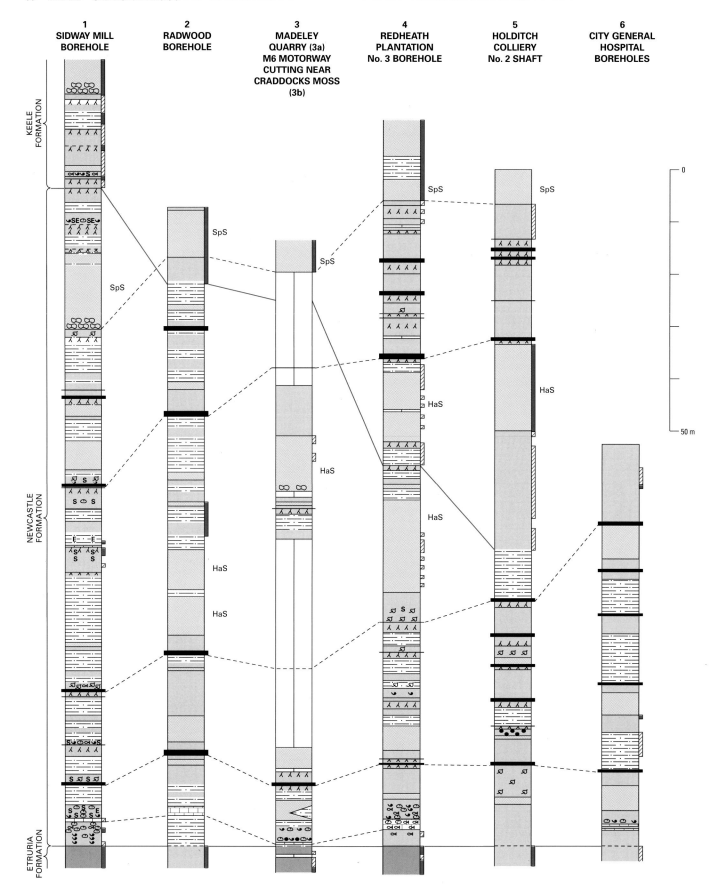

Figure 22   Correlation of sections of the Newcastle formation across the district.

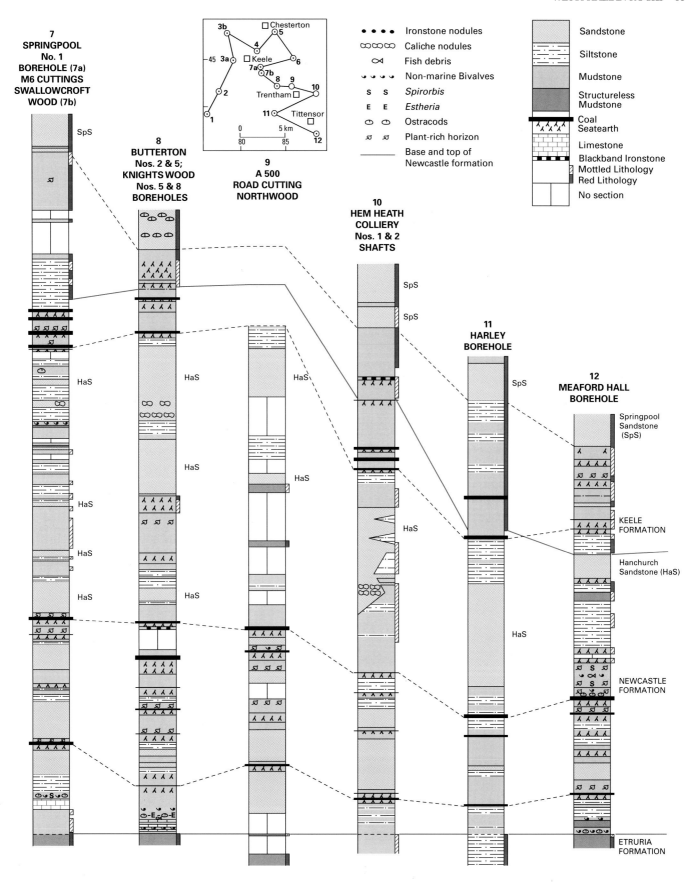

Figure 22 (continued).

(Figure 23). Where affected by this reddening the Hanchurch Sandstone is indistinguishable from sandstones in the Keele formation, whose base has therefore been mapped locally to include this unit. Sporadic reddish brown coloration affects some mudstones at the approximate level of the Hanchurch Sandstone over a wide area where the sandstone is itself grey. This is particularly the case at Hem Heath Colliery (Figure 22). Temporary exposures of the Hanchurch Sandstone at Trentham [869 419] proved yellowish brown sandstone, but with indications of interbedded red mudstone. Similarly red-mottled structureless mudstones were seen just below the lowest leaf of the Hanchurch Sandstone in the Northwood road cutting (Figure 22).

Beds above the Hanchurch Sandstone

The beds overlying the Hanchurch Sandstone consist mainly of mudstones, commonly with between four and seven seatearth horizons and two to four thin coals. Grey fine-grained sandstones are prominent in Sidway Mill Borehole. The lowest of the coals is commonly over 0.3 m thick and can be widely correlated. The lower beds are commonly poorly exposed, but were formerly seen in cuttings on the M6 motorway and at Swallowcroft Wood (Figure 22), where four coals, individually up to 0.45 m thickness, were well exposed. One of the seams was also located in the M6 motorway cutting near Craddocks Moss. Above the lowest of these coals the mudstones commonly pass diachronously into red beds of the Keele formation, the passage being particularly marked in the south-east (Figure 22).

Over most of the district a 20 to 30 m-thick sandstone known as the Springpool Sandstone lies in the basal part of the Keele formation. However, near Sidway Mill Borehole this sandstone is developed in a greenish grey facies, so lies in the Newcastle formation, with the diachronous colour change to the purplish grey Keele formation located 12 m above the sandstone. Grey mudstones overlying the sandstone include a seatearth capped by mudstones with *Euestheria*, *Spirobis*, *Anthraconaia pruvosti* and ostracods.

KEELE FORMATION

The Keele formation consists mostly of red sandstone, with locally thick intercalations of mainly red mudstone near the base and top. The formation overlies, and locally passes laterally into, the Newcastle formation. The formation was originally named the Keele and Penkhull Beds by De Rance (1898), a name superseded by the terms Keele Series (Gibson, 1899; 1901) and Keele Group (Gibson and Wedd, 1905). Although no type section was defined, a representative section derived from railway cuttings and other exposures in the Keele area (Figure 24) was described by Gibson and Wedd (1905, p.41). Subsequently, Gibson (1925) and Whitehead et al. (1927) recognised that the rocks forming the upper part of this section were correlatives of the Enville Beds of south Staffordshire and Shropshire. Recent work, based largely on borehole geophysical logs, supports this conclusion (Figures 24, 27), and the upper part of the old Keele Group has now been separated off as the Radwood Formation. It differs from the underlying Keele formation in having a lower proportion of sandstone relative to mudstone, a different principal sandstone composition, and distinctive dominant colours and pedogenic textures (Table 2).

Definition of a formal type section for the Keele formation is made difficult by the discontinuous nature of exposures. The type area, near Keele Hall [819 448], only provides representative exposures of the sandstone-rich units. Cores from the Keele University Campus boreholes 9

Figure 23 Correlation of sections proved in shallow cored boreholes in the Keele area, showing the pattern of penetrative weathering in the Hanchurch Sandstone of the Newcastle formation, locally forming the transgressive base of the Keele formation.

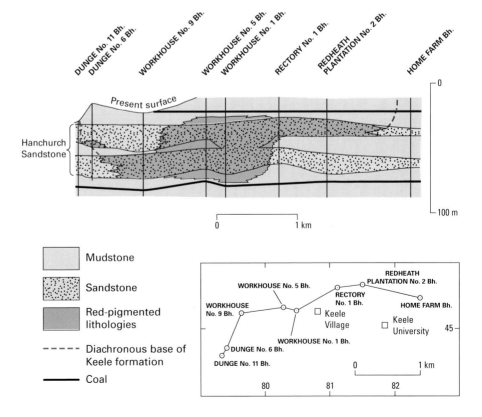

Figure 24 Comparison between the type outcrop section for the Keele formation in the Keele area by Gibson (1925) and geophysical logs of the type borehole section through the formation at Hey Sprink; grain-size indications for the outcrop section are schematic, and differentiate between lithologies recorded by Gibson as 'flaggy' and 'massive'. For key see Figure 21.

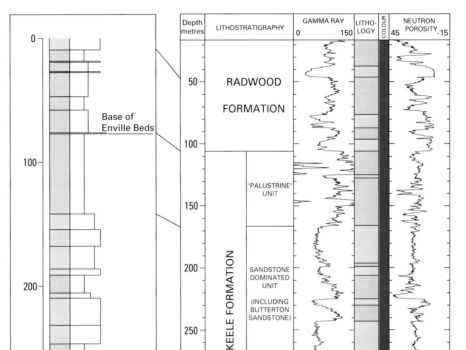

and 14 (Figure 26), drilled nearby, form good reference sections, as does the wireline log section from Hey Sprink Borehole (Figure 24).

The thick sandstones in the lower part of the formation locally form prominent scarps, as between Keele and Butterton, though elsewhere they form subdued topography. Exposures of the formation are limited to a few small excavations, mainly disused stone pits. Continuous cored sections are limited to a few deep boreholes, notably that at Sidway Mill. Cores were taken in the basal part of the formation in boreholes at Barker's Wood, Hobbergate, Meaford Hall No. 1, Newstead, and on Keele University campus. Cores from higher in the formation were taken in Beech Cliff Borehole. Most stratigraphic information relating to the formation has, however, been derived from borehole geophysical logs.

Gibson (1901) placed the base of the Keele Series at a fossiliferous grey mudstone horizon, with a thin coal, some 100 m above the base of the Newcastle formation, and 50 m above the base of the red beds. As this horizon is not mappable, the base of the Keele formation was subsequently placed at the base of the dominantly reddened sequence above the Newcastle formation. Defined in this way, the base of the Keele formation is strongly diachronous. It lies below the Hanchurch Sandstone in the south-east of the area, and above the Springpool Sandstone in Sidway Mill Borehole. Near Keele the base of the formation occurs towards the base of the Hanchurch Sandstone; correlations in closely spaced boreholes (Figure 23) show that the colour change at the base of the formation is related to a thick development of the Hanchurch Sandstone in channel facies. Laterally, where this facies passes into thin overbank sandstones, the sequence is grey. This pattern of reddening is interpreted as the result of penetrative weathering postdating burial and lithification, as described by Trotter (1954), Mykura (1960), and discussed by Besly et al. (1993). It is not clear how much, or if any, of the reddening is related to intra-Carboniferous weathering. The occurrence of red sandstones and mudstones in the Lower and Middle Coal Measures elsewhere in the district (Earp and Calver, 1961) suggests that Permo-Triassic reddening is more likely. For purposes of correlation, a useful datum, which commonly occurs close to the base of the formation in the present district, is provided by the base of the Springpool Sandstone (Figures 24, 25). The Keele formation has a characteristic geophysical log signature, comprising a sandstone-rich sequence in which sandstone bodies have distinctively high gamma radiation values (relative to the cleaner sandstones in the Radwood Formation) and cylindrical log shapes similar to those found in the underlying Newcastle formation, overlain by a mudstone-rich sequence (the 'palustrine' unit; see below) characterised by the presence of up to three high gamma radiation peaks, and containing only a few thin sandstone bodies (Figure 24).

Table 2 The lithological criteria allowing differentiation of the Keele and Radwood formations.

	KEELE FORMATION	RADWOOD FORMATION
Main colour	Dark brownish red	Brick red and purplish red
Reduction features	Rare	Common 'fish-eyes' and green-grey reduction traces
Detrital composition of sandstones	Dominantly lithic arenite containing micaceous metasedimentary debris; minor calcareous sublitharenite	Solely sublitharenite; abundant limestone grains in lower parts
Palaeosol types	Post-depositionally oxidised gley (seatearth); caliche in immature profiles	Mature caliche profiles dominant
Lacustrine facies	Rare; limestones only	Abundant, especially in lower part; limestones and laminated fossiliferous mudstones

No upper boundary of the formation has previously been recognised. It is here defined at the base of the Radwood Formation, and placed at the base of the first prominent sandstone bed overlying the mudstone-rich 'palustrine' unit (see below).

No stratigraphically significant fossils have been obtained from the Keele formation in north Staffordshire. Rare and poorly preserved plant fossils from correlative formations in the Stafford and Wolverhampton districts suggest a late Westphalian D age (*Lobatopteris vestita* zone of Wagner, 1984).

The formation is dominated by red beds, with most mudstones dark purplish or brownish red. Grey mudstones are locally common near the base of the formation, but become rarer upwards. The principal clay minerals in the mudstones are illite and kaolinite, as in the underlying Newcastle formation. Mudstones throughout the formation contain oxidised root traces, palaeosols showing strong colour mottling in shades of pink, red, ochre and purple, and poorly to moderately developed caliche nodules (stages I–II of the classification of Machette, 1985). Thin freshwater limestones are developed throughout mudstone sequences, as are thin coals near the base of the formation. Sandstones are mainly of a distinctive purplish brown or purplish red hue, and commonly contain intraformational red mudstone clasts and caliche-derived calcareous pellets. They fall into two distinct compositions that can easily be differentiated in borehole geophysical logs (Figures 24, 25). The dominant sandstones, occurring in bodies 20 to 40 m thick, are strongly micaceous lithic arenites containing a high proportion of grains of low-grade metasedimentary derivation, similar to those in the Newcastle formation. Textures of haematite in these indicate formation from a precursor siderite cement. These sandstones have unusually high gamma ray signatures, and, because of their large content of clay minerals, show anomalously high neutron porosities. The subordinate sandstones are more mature sublitharenites, characterised by polycrystalline quartz of metamorphic derivation and minor K-feldspar. They form laterally discontinuous sandstones, less than 15 m thick, in the middle of the formation. They have a characteristically low gamma ray response.

The lithic arenites that dominate the Keele formation are very similar in geochemistry to the sublitharenites that dominate the Radwood Formation (apart from Ca, which is higher in the sublitharenites and which resides mainly in cement), with an overlap between the ranges of values for each element (Haslam, 1993). However, of the immobile elements that may be provenance indicators, the ranges for Ti, V, Cr, Co, Ni and Nb (elements that tend to be concentrated in basic rocks) are somewhat higher in the lithic arenites, whereas the ranges of Y, Zr, La, Ce and Th (elements associated with more-acid rocks) are very much the same for both sandstone types. This suggestion, that there is a greater proportion of basic material in the lithic arenites, is borne out by the petrographic composition as seen in thin sections.

The Keele formation is interpreted as a sequence of fluvial deposits (Besly, 1988), deposited under conditions of alternately good and poor drainage, demonstrated by the occurrence of caliche and gley (seatearth) palaeosols in the overbank deposits. Oxidised siderite relics, which occur throughout the formation as hematised nodules in seatearths and as cements in sandstones, demonstrate that reddening occurred soon after deposition, as a result of groundwater fluctuation (Besly et al., 1993). The occurrence of caliche indicates the onset of increasingly arid climatic conditions. The limestones were probably deposited in bodies of short-lived standing water, probably small overbank lakes.

Isopachs for the formation can be constructed, using borehole penetrations, only in the southern part of the district. Because of the difficulty of recognising the colour change at the base of the formation, the base for this purpose has been taken at the base of the Springpool Sandstone (see above). As the top of the formation can only be recognised in a few boreholes, the largest gamma ray marker within the 'palustrine' unit (see below) has been taken as the upper datum. These constraints mean that the isopach map is generalised (Figure 7k); however, it shows a marked change from that for the underlying Newcastle formation. Although a consistent southward thinning is still present, pronounced zones of differential subsidence are highlighted by a marked thickening in the Tittensor area, and a marked thinning in the area of the

Hanchurch Hills. These zones have a north-east to south-west trend, and the zone of thinning corresponds to the position of the axis of the Clayton Anticline, which underwent a phase of uplift during deposition of the formation. In the west of the area the data are ambiguous. The formation is calculated to be 154 m thick in Bittern's Wood Borehole, only 1.5 km to the north-west of the 197 m thick sequence penetrated in Radwood Borehole. In the absence of other modern boreholes in the area, it is impossible to estimate the extent to which this thinning may result from faulting. However, the pattern of thinning in the area of the Western Anticline in other Westphalian intervals (Figure 7) suggests that the Keele formation may similarly thin towards this structure.

Several informal units may be recognised in the formation. These are most obvious on borehole geophysical logs (Figures 24, 25) but can, at least locally, also be inferred from outcrop data.

Beds below the Springpool Sandstone

In places, strata which typically occur within the underlying Newcastle formation are reddened, and thus were included in the Keele formation during the survey. For this reason, in the south-east of the district and in the Keele area, the Hanchurch Sandstone and some mudstones below it are included in the Keele formation (Figure 23). Reddened Hanchurch Sandstone is recorded in the shaft section of Holditch Colliery, and is exposed in Quarry Bank Quarry, Keele [8072 4610] and beside a minor road at Chesterton [8324 4906].

Springpool Sandstone

In many areas the Springpool Sandstone forms the basal 30 m of the Keele formation. It comprises red, fine- to coarse-grained, locally pebbly sandstones (Figure 26) containing abundant mudstone intraclasts. The sandstone can be readily correlated in borehole logs (Figure 25), though it thins and passes laterally into grey beds of the Newcastle formation in Sidway Mill Borehole. It is well exposed in a quarry near the M6 motorway [8210 4390], where several metres of cross-bedded, intraclastic sandstone is visible; the sandstone is further exposed in the adjacent motorway cutting [8220 4368]. The Springpool Sandstone may also be seen in a quarry south-east of Northwood Farm [8595 4195]. A detailed sedimentological log of the sandstone in the 1974 Keele University Site Investigation Borehole No. 14 is shown in Figure 26.

Coal-bearing unit

The Springpool Sandstone is overlain by a mudstone-dominated, locally coal-bearing unit. It is best developed around Keele University (Figure 26), where it contains two groups of thin coals, locally partly altered to limestone (cf. Mykura, 1960), together with red seatearths and laminated mudstones which show textures that suggest initial deposition as grey facies (cf. Trotter, 1954; Besly et al., 1993). Unreddened mudstones associated with the coals locally contain *Spirorbis*, ostracods and mussels. The coals can be recognised in cuttings and on density and neutron logs in some boreholes, and correlate with a carbonaceous mudstone seen in core in Allotment No. 1 Borehole in the Stafford district. In Sidway Mill Borehole the coals are represented by oxidised relics at 634.29 m depth and a carbonaceous mudstone at 651.24 m. The coal-bearing unit forms a mudstone-rich intercalation in the otherwise sand-dominated

Keele formation, and, as such, can be readily correlated in gamma ray logs, ranging in thickness from 25 to 35 m. Because it is mudstone dominated there are few natural exposures; however, it was intermittently exposed in excavations on Keele University campus (Exley, 1970; Boardman et al., 1972). A detailed log of the sequence in Site Investigation Borehole No.9 at the university (Figure 26) shows that the sequence here contains two distinct groups of thin coals.

Sandstone-dominated unit

The coal-bearing unit is overlain by a sequence 90 to 115 m thick, dominated by purplish red sandstones. Mudstone intercalations up to 15 m thick are present, and in places contain both seatearth and caliche palaeosols, as above a channel sandstone in the upper part of the sequence in Beech Cliff Borehole. Some sandstones form topographic features, but in general topography does not reflect the sand-dominated nature of the sequence; the sandstone locally mapped as **Butterton Sandstone** tends to represent a more resistant sandstone within the unit. Exposures south of Madeley [787 437, 774 425] suggest that such features may be preferentially formed by the sandstones of sublitharenite composition. Available exposures are small and do not illustrate the full variety of the sequence. Exposures of micaceous lithic sandstones are located on the university campus at Keele, at Clockhouse Drive [8181 4488], and in an adjacent disused quarry [8178 4487]. The disused quarry at the Higherlands, Newcastle-under-Lyme [8455 4555] provides a good section of upward-fining cross-bedded sandstone, with intraformational conglomerate lags developed near the base. Other exposures of sandstone are in Church Wood Quarry, Butterton [833 419], in a quarry south-east of Newstead Sewage Works [8930 4015], and in a road bank south-west of Madeley [7735 4300]. Sandstones of mature sublitharenite composition are exposed in disused quarries 300 m north-north-east of Netherset Hay Farm [787 437], and adjacent to the road between Madeley and Baldwins Gate [774 425]. The best exposure of the fine-grained overbank facies is in an excavation behind factory units in Turner Crecent, off Loomer Road, Chesterton [833 485]. Here poorly developed caliche palaeosols are present in thin mudstones interbedded with sandstones that form a prominent scarp feature. Evidence from mapping shows that a tonstein, the Trentham tonstein (Appendix 2), occurs towards the top of the sandstone-dominated sequence.

'Palustrine' unit

The sandstone-dominated unit is overlain by 50 to 80 m of mudstone, informally known as the 'palustrine' unit, after a similar facies described by Freytet (1984). This is best observed in borehole geophysical logs, being characterised by the presence of up to four prominent high gamma radiation peaks. The mudstones contain a characteristic development of intensely colour-mottled palaeosols, exhibiting red, pink, purple, grey and ochre pigments, and have textures reminiscent of the palaeosols in the Etruria Formation. No available borehole section gives both core and wireline data for this unit. In Sidway Mill Borehole, possibly the best section, the unit contains a significant bed of freshwater limestone at 474.35 m and an uncommonly thick mudstone unit between 500 and 504 m. The latter is probably the source of one of the high gamma ray anomalies. The 'palustrine' unit, being formed mainly of mudstone, occupies low drift-covered ground which generally lacks exposure. However, a silage pit at Netherset Hey Farm [7840 4355] exposed 9.40 m of mainly red-brown mudstone displaying a range of purple, ochre, bright red and grey pedogenic mottling textures, with minor development of caliche.

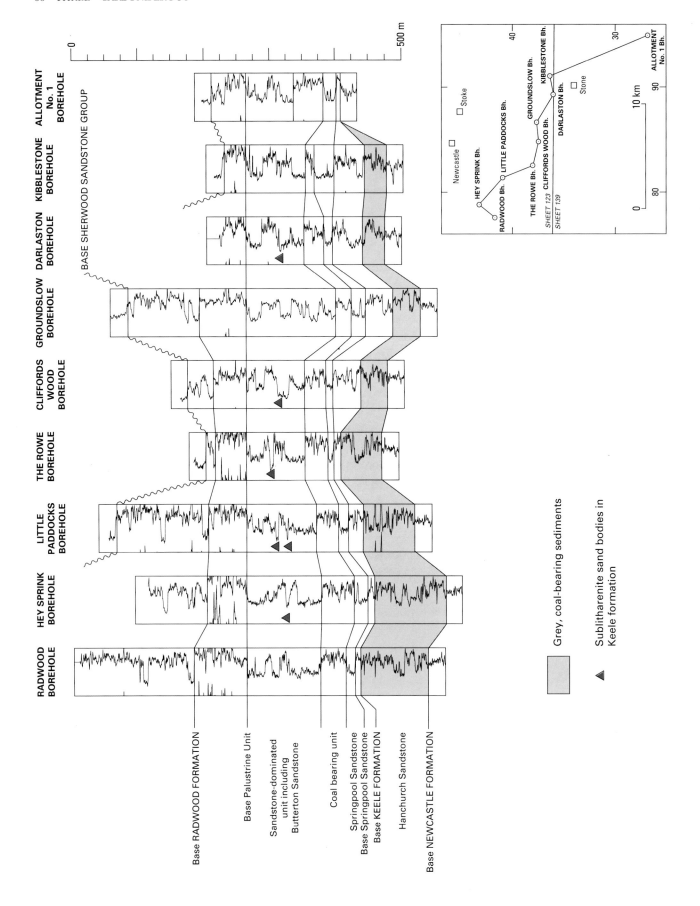

Figure 25 *(opposite)*   Correlation panel for the Newcastle, Keele and Radwood formations, from cutting records and gamma-ray logs in unfaulted borehole penetrations. Note the lateral passage from the grey, coal-bearing, Newcastle formation into the red Keele formation in the south-east and the pronounced thickness variation in the Keele formation.

? STEPHANIAN ROCKS

RADWOOD FORMATION

The Radwood Formation consists dominantly of red mudstone and siltstone with minor sandstones, and overlies the sandstone-rich red beds of the Keele formation. It is most extensively developed in the south and west of the district where it forms areas of undulating topography. Thin, resistant sandstones create poorly defined topographic features, as seen west of Dab Green [794 423] and south of Radwood Hall Farm [774 411]. However, thicker sandstones higher in the formation form well-defined topographic features, as at Willoughbridge Lodge [740 387] and east of Mucklestone. Small exposures occur in disused stone pits, in railway cuttings between Radwood and Stony Low, and in stream sections.

The strata now assigned to the Radwood Formation were previously included in the Keele 'Series','Group' and 'Formation' (Gibson, 1925; Wilson et al., 1992). However, Gibson already recognised that these beds differed in character from the lower strata at Keele, and proposed a correlation with the Enville Beds of south Staffordshire and Shropshire. The distinction is supported by recent borehole evidence, so the two sets of strata, now called the Keele and Radwood formations, have been mapped separately. However, as exposure is generally poor, recognition of the boundary between them is difficult. Normally, a substantial thickness of the boundary sequence requires study before the base of the Radwood Formation can be accurately positioned within it. The recent work of Besly and Cleal (in press) suggests that the Radwood formation is a correlative of the Salop Formation of the South Staffordshire Coalfield, and it is likely that the unit will eventually be referred to by that name.

The criteria by which the Keele and Radwood formations are differentiated are listed in Table 2. The base of the Radwood Formation is defined at the base of the first mappable sandstone overlying the 'palustrine' unit at the top of the Keele formation (see above). This sandstone forms a topographic feature at Netherset Hey Farm [7850 4340] and to the south of Madeley Old Manor [7722 4225]. It is exposed in the stream section immediately west of the latter and in the adjoining railway cutting. On gamma ray logs and neutron logs (Figure 24) the sandstone is distinctive. The type section is in the stream section running from Madeley Old Manor [7713 4224] southwards along the eastern edge of Radwood Copse [7738 4180], in which scattered exposures are present for at least 300 m south of the railway line. A section in the nearby Radwood Borehole (Figure 25) contains the thickest sequence of the formation for which wireline logs are available, though a thicker and

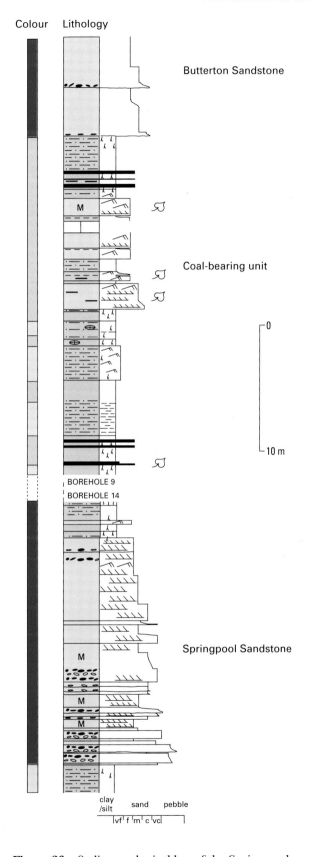

Figure 26   Sedimentological log of the Springpool Sandstone and coal-bearing unit of the Keele formation penetrated in the Keele University 1974 site investigation boreholes (numbers 9 and 14); for key see Figure 21.

more complete sequence through the formation was recorded from the continuously cored Sidway Mill Borehole (Figure 27). The formation is overlain unconformably by the Sherwood Sandstone Group. The entire thickness of the formation preserved in the district is probably greater than 700 m.

No biostratigraphically diagnostic taxa have been obtained from the formation. However, rare and poorly preserved plant fragments derived from correlative formations around Wolverhampton and in the Oxfordshire Coalfield (Cleal and Besly, 1994) suggest a latest Westphalian D to Cantabrian (early Stephanian) age (*Lobatopteris vestita* or *Odontopteris cantabrica* zones of Wagner, 1984).

The mudstones and siltstones of the Radwood Formation, like those of the Newcastle and Keele formations, contain the clay minerals illite, kaolinite and, locally, chlorite. They are mainly brick-red, chocolate brown and purplish brown. In Sidway Mill Borehole brick red mudstones are developed throughout the formation, with chocolate and purplish brown mudstones more common below 200 m depth (Figure 27). Green reduction spots with dark cores are common. In the lower part of the formation many mudstones show extensive pedogenic features (Figure 27), including colour mottling and common caliche nodules; however, gleyed palaeosols of seatearth type appear to be absent. Palaeosols are less common in the higher part of the formation.

In Sidway Mill Borehole fossiliferous mudstones occur sporadically throughout the formation, but are more abundant towards the base (Figures 27, 28). The fauna includes indeterminate thin shelled mussels, the pulmonate gastropod *Anthracopupa britannica*, the annelid *Spirorbis, Leia, Estheria* and ostracods. This fauna is identical to that described from the Windrush Formation of Oxfordshire by Poole (1969). Thin freshwater limestones are developed in mudstone-rich facies throughout the formation.

The sandstones are sublitharenites, characterised by polycrystalline quartz of metamorphic derivation and minor K-feldspar, and resemble the thin sandstones in the middle of the Keele formation. They are mainly pale brownish red; where locally reduced they are pale greenish grey in streaks and along laminae. Intraformational mudstone pellets and caliche pebbles are common; in places, distinctive thin grain-supported conglomerates consisting entirely of reworked caliche nodules ('pellet rock') occur. A few small extraformational pebbles of quartzite and yellow chert are present in sandstones in the upper and middle parts of the formation in Sidway Mill Borehole (Figure 27). In the lower part of the formation the sandstones generally contain large amounts of detrital Carboniferous Limestone, partly as rounded grains containing characteristic fossils and partly as grains formed by fragments of brachiopod shell. Local high Ca values in chemical analyses (Haslam, 1993) are attributed to the presence of these calcareous grains. The sandstones show characteristically clean responses in both gamma ray and neutron logs, and are therefore easily identified in borehole geophysical logs (Figures 24 and 25).

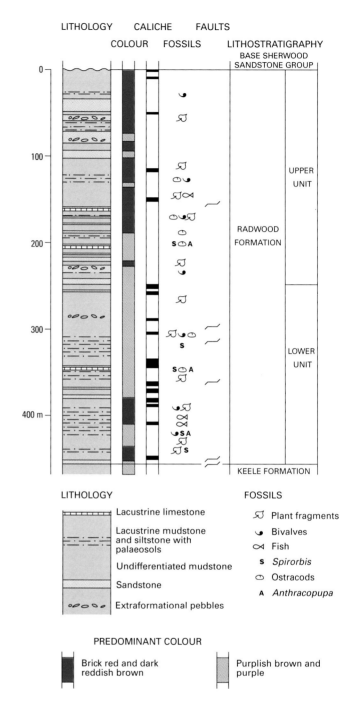

Figure 27 Summary log of the main stratigraphical features of the Radwood Formation in Sidway Mill Borehole. Note that the base of the formation is faulted in this borehole.

The Radwood Formation, like the Enville Formation to the south (Besly, 1988), is interpreted as a sequence of fluvial deposits, containing minor conglomeratic fans and lacustrine sequences. The abundant development of caliche palaeosols shows that deposition occurred under semi-arid climatic conditions, and may also reflect the availability of calcium carbonate from the denudation of Dinantian limestones in the sediment source area. The

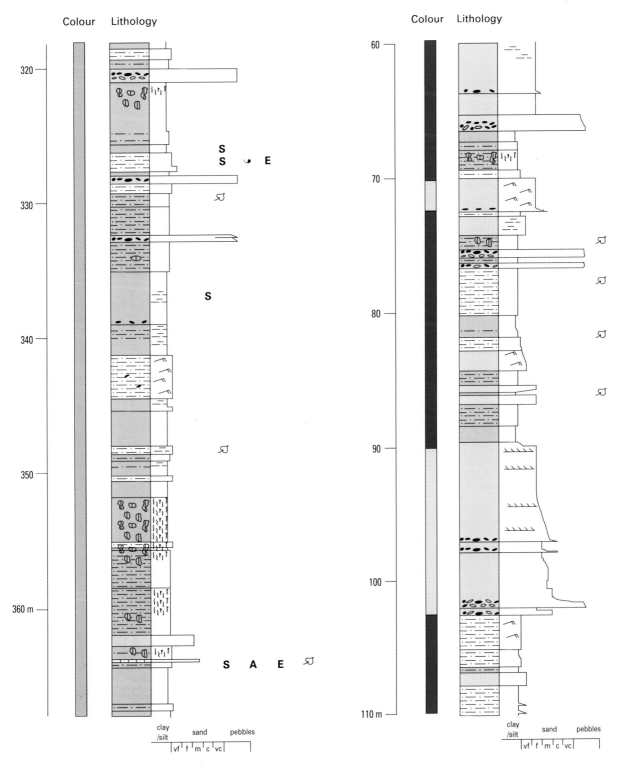

Figure 28 Comparison of sequences typical of the lower and upper units of the Radwood Formation in Sidway Mill Borehole. For key see Figure 21.

abundance of caliche decreases upwards, possibly indicating the decreasing availability of detrital carbonate.

The formation, as proved in Sidway Mill Borehole, contains many fluvial sequences, represented by sharp-based, upward-fining sandstone bodies between 8 and 16 m thick, associated with pedogenically modified overbank deposits in which faunas are generally lacking. Lacustrine rocks are best developed in the lower part of the formation. They are characterised by 10 to 30 m thick mudstone and siltstone sequences capped by palaeosols. The lower parts of these commonly contain freshwater faunas, are cross-laminated or horizontally laminated, and apparently lack root disturbance. Each sequence represents the infill of a freshwater lake, followed by emergence and pedogenic modification of the fill, without any channel influence. The development of thin limestones in some lacustrine sequences suggests the occasional cut-off of clastic supply.

The formation in Sidway Mill Borehole (Figure 27) can be broadly split into upper and lower parts. At present these cannot be recognised outside the district.

The **lower part**, between 266.20 and 473.58 m in the borehole, consists mainly of purple, chocolate-brown and brick-red mudstone, with subordinate beds of sandstone (Figure 28a). The mudstones are locally silty or sandy, show rooted and slickensided textures, and in many cases contain caliche nodules. A fauna of *Spirorbis*, *Anthracopupa britannica*, fish debris, ostracods, *Anthraconaia prolifera*, and *A. saravana* is locally present. Thin limestones are sporadically developed. The best exposure of the lower part of the formation is in the railway cutting to the east of Stony Low [793 438], where hard buff-brown calcareous sandstones are exposed around the southern portal of the tunnel. Formerly, a more complete sequence was exposed, listed in detail by Gibson (1925). Comparison of Gibson's section with the wireline logs of the nearby Hey Sprink Borehole (Figure 24) suggests that the Stony Low exposure lies a few metres above the gamma ray marker bands of the 'palustrine' unit at the top of the Keele formation. Sandstones at a similar horizon are also exposed east of Netherset Hey Farm [786 432]. Red mudstones and mainly fine-grained sandstones at the base of the formation are exposed in the stream sections between Madeley Manor [771 422] and Radwood Copse [7725 4190], and also in the railway cutting to the north-east of the copse.

The **upper part** of the formation, between 18.36 and 266.20 m depth in Sidway Mill Borehole, contains a higher proportion of rocks attributable to a fluvial origin and a smaller part attributable to a lacustrine origin. This is illustrated by a log extract of the upper part of the formation in Sidway Mill Borehole (Figure 28b). Representative exposures are confined to stream sections and small quarries near Willoughbridge. The most extensive section occurs [between 7327 3775 and 7355 3750] in Cowleasow Wood, Mucklestone. Intermittent exposures of poorly bedded red mudstone enclose beds of reddish brown, fine- to medium-grained sandstone, one of which is 5 m thick and shows upward-fining and abundant haematised intraclasts. A sandstone-rich part of the sequence forms a prominent topographic feature at Willoughbridge Lodge. A small quarry south-east of the lodge [7444 3849] exposes 6 m of purple, fine- to medium-grained, cross-bedded sandstone containing red and yellow mudstone and caliche intraclasts. The argillaceous rocks of the formation are poorly exposed in a claypit 300 m south-south-west of Willoughbridge Wells [744 392]. The mudstones are brick red and reddish brown, and contain greenish grey reduction spots and streaks. The bedding is disturbed, possibly by recent root action.

FOUR

Permian

PERMO-TRIASSIC SETTING

Deposits of Permian age are confined to the deeper parts of the Cheshire Basin, so do not crop out in the district, but the Permian earth movements have had a profound effect on the geology of the region. Following the early phases of compression in the late Carboniferous, the Variscan orogeny probably culminated, on regional evidence (Besly, 1988), in early Permian times. The compression, directed broadly to the north-east, was caused by closure of the Rheno-Hercynian basin to the south. It generated folds on the margins of Carboniferous basins and reactivated many Caledonian faults as compressional structures (Chapter 8). Several of these faults were again reactivated, during the Permian and Triassic, to form extensional sedimentary basins (Wills, 1956; Audley-Charles, 1970b). These basins, developed by intra-plate continental rifting in response to east–west tensional stresses in the region between Scandinavia and Canada, trapped much of the material eroded from the landscapes created by Variscan uplift.

The Stoke-on-Trent district includes parts of three such basins, separated by roughly north–south structural highs; the Cheshire Basin in the west, the Stafford Basin in the south, and the Needwood Basin in the south-east (Figure 29). As the Permo-Triassic basin-fills have densities significantly lower than the older rocks, they are associated with gravity lows.

The **Cheshire Basin** is associated with one such major gravity anomaly. Where the south-eastern margin of the basin crosses the district there is a zone of steep gradients (Pontesford Lineament, Figure 3) with a range of values from -8 mGal to +17 mGal (Figure 2). The precise limits of the basin changed through Permian and early Triassic times and its margins at any particular time are uncertain. For the purposes of defining stratigraphical nomenclature (Figure 29) the eastern margin is defined by the Hodnet Fault as far north as Mucklestone, beyond which it corresponds with the Carboniferous inliers on a line between Mucklestone and Sidway Mill [7572 3986].

The **Stafford Basin** is marked by a north–south gravity low (Figures 2 and 3) the minimum of which indicates that the thickest Triassic rocks occur towards the west of the basin. For the purposes of stratigraphical nomenclature, the margins of the basin are defined by the Sidway Fault in the west and the Hopton Fault in the east (Figure 29).

The north-north-west trend of the faults (Chapter 8) that bound the **Needwood Basin** is reflected in the orientation of the gravity low associated with the basin. The marked gravity gradient along the south-west margin is associated with the Sandon Fault, which is taken as the

margin of the basin for the purposes of stratigraphical nomenclature (Figure 29).

The three basins are separated by horsts. That between the Cheshire and Stafford basins is known as the Market Drayton Causeway (Wills, 1956). This name is adopted here to avoid confusion with the Market Drayton Horst (Chapter 3), which occupied a similar geographical position in the Carboniferous, but also included most of the area of the Permo-Triassic Stafford Basin. The horst is represented by a gravity high in the south-west of the district (Figure 2). The margins of the north-north-west-trending horst between the Needwood and Stafford basins appear to be defined by the Hopton and Sandon faults. However, a pronounced gravity high, of the same orientation, suggests that the form of the horst at depth is controlled by other faults, farther west, which approximate to the western limit of the outcrop of the Keele formation at the southern limit of the Stoke-on-Trent district.

In the future, the Triassic sequences on the horsts between the basins may warrant separate nomenclatures from those deposited in the adjoining basins. For present purposes, however, the sequence on the horst between the Cheshire and Stafford basins is given the same nomenclature as the equivalent sequence in the Cheshire Basin and that on the horst between the Stafford and Needwood basins is given the same nomenclature as the equivalent sequence in the Stafford Basin.

During the Permo-Triassic, Britain migrated northwards from near the equator to north of the tropics. The sediments of this age, deposited in an arid to semi-arid climate, are mostly red because of the alteration of detrital ferromagnesian silicates and iron-bearing clay minerals to iron oxide (hematite) (Walker, 1976). They overlie an eroded surface of Carboniferous rocks that were folded and faulted during the Variscan orogeny (Chapter 8). These rocks are commonly red-stained below the unconformity (Chapter 3). The Permo-Triassic rocks have been faulted by reactivated Variscan structures in the Mesozoic and Palaeogene and have been gently folded during the latter (Chapter 8).

PERMIAN ROCKS

Rocks deposited during the Permian (about 300 to 251 Ma: Forster and Warrington, 1985; Claoué-Long et al., 1991; Hess and Lippolt, 1986), only occur at depth in the central part of the Cheshire Basin (Figures 29 and 30), in the north-west part of the Stoke-on-Trent district. They are also present in the Stafford Basin, but only to the south of the district. No boreholes intercept these formations in the Stoke-on-Trent district, so much of this

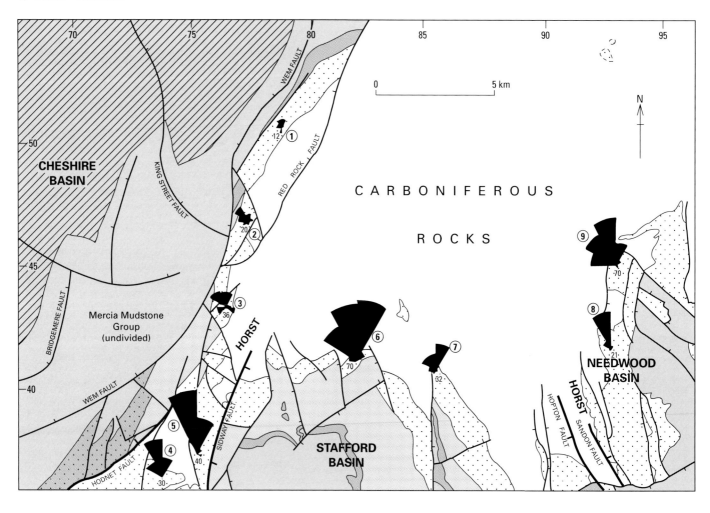

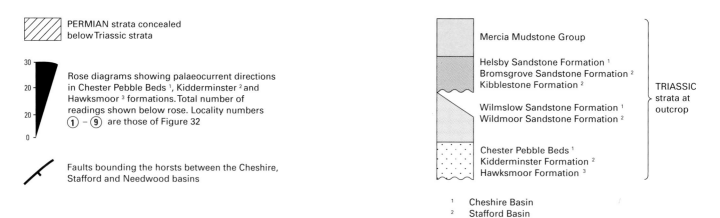

Figure 29 Map showing the position of the Permo-Triassic Cheshire, Stafford and Needwood basins, named basin-margin faults and outcrop of Sherwood Sandstone Group.

Palaeocurrent data from main exposures of the Chester Pebble Beds (Cheshire Basin), Kidderminster Formation (Stafford Basin) and Hawksmoor Formation (Needwood Basin), detailed sections of which are shown in Figure 32: 1 Kent Hill Quarry, 2 Heighley Castle, 3 Bar Hill, 4 Tadgedale Quarries, 5 Willoughbridge Quarries, 6 Acton Quarries, 7 M6 Motorway at Trentham, 8 Lightwood Quarries, 9 Hulme and Willfield Quarries. Palaeocurrents of localities 6, 7 and 9 from Steel and Thompson (1983).

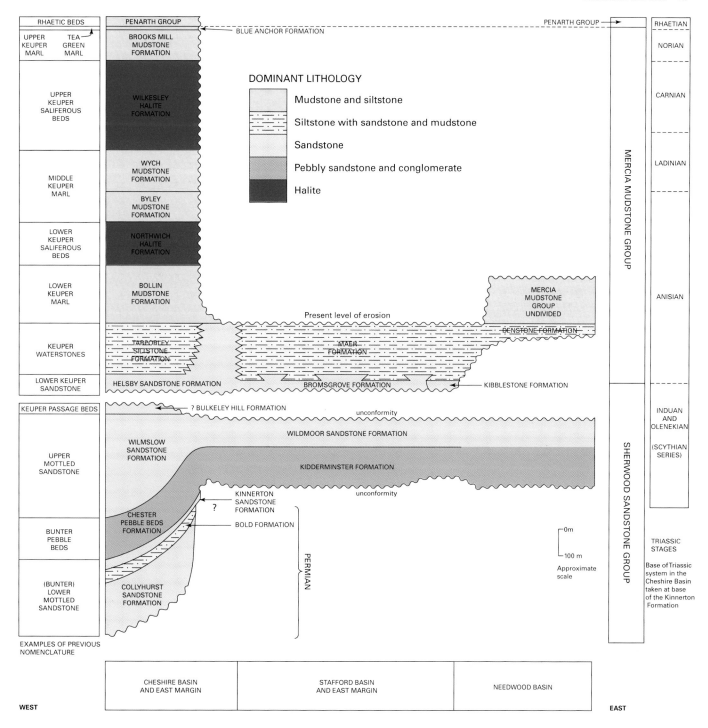

Figure 30 The lithostratigraphical and chronostratigraphical status of formations in the three Permo-Triassic basins. Chronostratigraphy after Wilson (1993) and Benton et al., 1994

account is based on the boreholes at Knutsford and Prees (Figure 31). A network of seismic reflection lines, which link the Knutsford and Prees boreholes, allows interpretation of the Permian rocks in the Stoke-on-Trent district. Thick, concealed, Permian successions are preserved west of the Bridgemere and Wem faults in the north-western part of the district (Figure 29). East of this fault system Permian rocks are thin or absent.

In the northern part of the Cheshire Basin the lowest three Permo-Triassic formations, in upward order, are the Collyhurst Sandstone, Manchester Marl and a sandstone referred to here (Chapter 5) as the Kinnerton Sandstone. Palaeontological evidence of Late Permian age is available for the Manchester Marl but the age of the other formations is uncertain (Pattison, 1970; Pattison et al., 1973; Smith et al., 1974; Warrington et al., 1980; Wild, 1987).

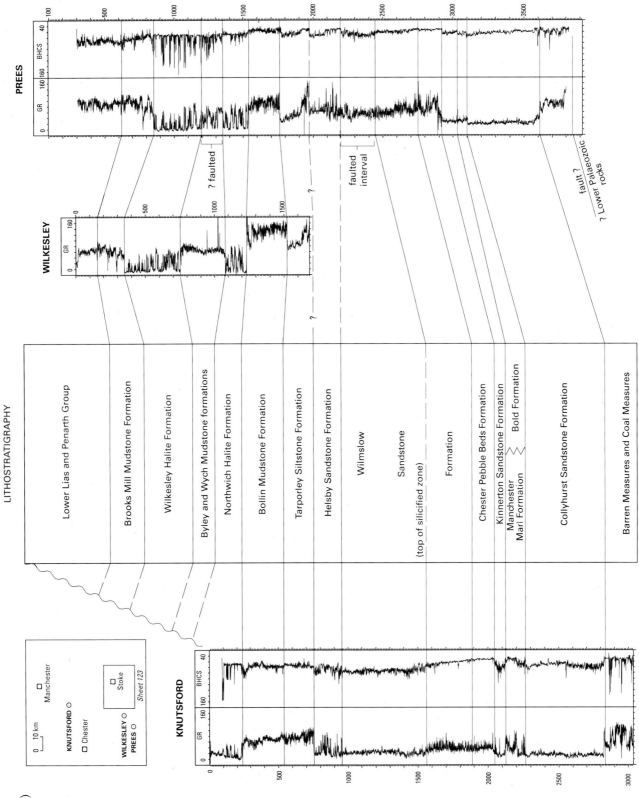

Figure 31
Gamma-ray (GR) and sonic (BHCS) logs of major boreholes through Permo-Triassic strata in the Cheshire Basin. Depths in metres.

The Collyhurst Sandstone is commonly regarded as Early Permian. The position of the Permo-Triassic boundary, above the Manchester Marl, is uncertain. It is here placed arbitrarily at the base of the Kinnerton Formation, which is described in Chapter 5. The Collyhurst Sandstone and the Bold Formation (lateral equivalent of the Manchester Marl in the southern part of the basin) are reviewed here. The Collyhurst Sandstone consists of aeolian and alluvial-fan sediments deposited in a continental interior. The sandstones of the Bold Formation (see below) probably represent the deposits of a coastal plain associated with the 'Bakevellia Sea' in which the Manchester Marl was deposited farther north.

Collyhurst Sandstone Formation

This formation is best known around Manchester, in the northern part of the basin (Tonks et al., 1931; Poole and Whiteman, 1955). There it was deposited unconformably on Coal Measures and Barren Measures, as in the Knutsford and Prees boreholes (Figure 31). The top of the formation is defined by the base of the overlying Bold Formation (or, to the north, the Manchester Marl). Where these formations cannot be recognised, as on the western margin of the Cheshire Basin, the Collyhurst Formation passes laterally into the Kinnerton Sandstone Formation (Chapter 5). The Collyhurst Sandstone is dominated by red-brown, cross-stratified, pebble-free sandstones with a high proportion of rounded frosted grains. The formation is mostly aeolian in origin (Thompson, 1985). However, the basal 70 m of the formation in the Knutsford Borehole, which is notably pebbly and greyer in colour than the rest of the formation, probably comprises interbedded aeolian and fluvial, possibly alluvial fan, deposits.

On seismic sections the formation forms a transparent zone between the strong reflections from the top of the Carboniferous and the Bold Formation (Figure 45). It is the most transparent of the Permo-Triassic formations in the Cheshire Basin but does produce laterally impersistent reflections of variable amplitude. The formation is about 510 m thick in the Prees Borehole and about 556 m thick in the Knutsford Borehole. It probably thickens to the south-east and east, reaching 600 to 700 m in the Stoke-on-Trent district.

Bold Formation

The Bold Formation is a sandstone-dominated lateral equivalent of the Manchester Marl. Where the latter is well developed, in the north-east part of the Cheshire Basin (Tonks et al., 1931), it comprises red, horizontally bedded mudstones and siltstones with subordinate sandstones and sporadic shelly limestones of Zechstein EZ1 to ?EZ2 age (Pattison, 1970). However, to the west the character of the formation changes, becoming increasingly siltstone- and sandstone-rich and, in boreholes close to the village of Bold, near St Helens, the sequence is dominated by fine-grained sandstones in tabular beds less than 2 m thick. This sequence can be readily distinguished from the underlying and overlying Collyhurst and Kinnerton Sandstone formations by virtue of the many mudstones and siltstones within it. This sandstone-rich equivalent of the Manchester Marl has been named the Bold Formation (Birmingham University, Geological Sciences and Civil Engineering Departments, 1981; Thompson, 1985; 1989).

Borehole and seismic evidence suggests that an argillaceous, sandstone-rich unit between the Collyhurst and Kinnerton sandstones extends southwards into the Cheshire Basin. For instance, in the Knutsford Borehole (Figure 31), it comprises mainly sandstone and contains few mudstone-dominated intervals over 10 m thick. Thompson (1989) proposed that the Bold Formation should be recognised at least into the Chester district. The equivalent unit in the Prees Borehole (Figure 31) appears similar to, though perhaps more sand-rich than, the Bold Formation farther north, and this name is therefore used there and in the present district also.

The formation has a distinctive seismic character, showing high-amplitude reflections with good continuity, and contrasts well with the underlying, seismically transparent, Collyhurst Sandstone. It also has a distinctive gamma response which reflects its argillaceous character. These features enable it to be traced in the subsurface at least as far south as Whitchurch (Gale et al., 1984), the Prees Borehole, and into the Stoke-on-Trent district (Evans et al., 1993). The Bold Formation is about 70 m thick in the Prees Borehole, 148 m thick in the Knutsford Borehole, and probably between 100 and 150 m thick in the Stoke-on-Trent district.

FIVE

Triassic

Rocks of Triassic age (251 to 205 Ma : Forster and Warrington, 1985; Claoué-Long et al., 1991) crop out over most of the western and southern parts of the district in the Cheshire, Stafford and Needwood basins (Chapter 4). The Triassic successions in these basins have separate stratigraphical nomenclatures, although the same broad divisions are recognised in each. The strata are classified, following Warrington et al. (1980) into three major units, the Sherwood Sandstone, Mercia Mudstone and Penarth groups. The Sherwood Sandstone Group and Mercia Mudstone Group replace the 'Bunter' and 'Keuper' units, which carried a spurious implication of chronostratigraphical equivalence with the rocks of the same name in Germany.

SHERWOOD SANDSTONE GROUP

The Sherwood Sandstone Group is dominated by sandstones, in part conglomeratic, and is composed of mature, variably sorted sediment. The sandstones are generally red-brown or ochreous, and commonly planar or trough cross-stratified. Most are composed of over 75 per cent quartz and are subarkoses, though sublitharenites and quartz arenites also occur (Appendix 2). The conglomerates are clast- or matrix-supported, are commonly cross-bedded, and many have a mauve hue when observed from a distance, due to the colour of the pebbles. The composition of the pebbles varies little over the district, or through the group, and comprises quartz (about 25 per cent), quartzite (about 55 to 60 per cent) and rarer siliceous lithologies (about 15 to 20 per cent) including sandstones, cherts, rhyolites, agates and rhyolitic tuffs (Thompson, 1970b). Most of the pebbles are well rounded and oblate, or equant; some have concave impressions caused by pressure solution. The abundance of polycrystalline quartz in the sand fraction, and the overall pebble composition, suggest low-rank metamorphic source terrains, probably sited in the Midlands, south-west England, and the English Channel (Greenwood, 1918; Wills, 1947; 1956; Campbell-Smith, 1963; Fitch et al., 1966; Ali, 1982). Much of the group was deposited in low-sinuosity braided rivers which flowed in a north to north-westerly direction (Figure 29), towards the Irish Sea Basins. Parts of the group, particularly the Kinnerton and Kibblestone formations, and some parts of the Wilmslow, Wildmoor, Bromsgrove and Helsby formations, have an aeolian origin.

The group consists of two parts separated by an unconformity. The lower unit containing the Chester Pebble Beds, Kidderminster Formation and Hawksmoor Formation coarsens upwards to a conglomerate-dominated sequence and then fines off again. Steel and Thompson (1983) suggested that the coarsening upward trend represents increasing tectonism and climatic change in the source areas, whilst the subsequent fining represents more stable tectonic conditions, and decreasing marginal relief in these areas.

The upper unit, comprising the Helsby, Bromsgrove and Kibblestone formations, was deposited after part of the underlying unit had been eroded during syn-extensional regional uplift (Evans et al., 1993), (Chapter 8). The resulting unconformity, in places angular, but generally only disconformable, probably equates with the Hardegsen Disconformity in Germany (Audley-Charles, 1970a; Warrington, 1970a; Wills, 1970a; Zeigler, 1975; Evans et al.,1993). These younger sandstones were derived from sources similar to the rocks below the unconformity (Fitch et al., 1966), the main difference being the presence of more acid igneous detritus (Ali, 1982).

No biostratigraphical information has been obtained from the Sherwood Sandstone Group within the district. A sample from the Hawksmoor Formation in Fulford Quarry [948 386] was examined for palynomorphs but proved barren. However, the upper age limit of the group is constrained by palynological evidence of an Anisian (early Mid-Triassic) age for the succeeding Maer and Denstone formations of the Mercia Mudstone Group.

Indirect evidence of the age of the upper part of the group comes from surrounding areas, but there is no evidence for the age of the older formations in the group.

Miospores indicating an Anisian (early Mid Triassic) age occur in the middle and upper parts of the Bromsgrove Sandstone Formation (Finstall and Sugarbrook members) in its type area in north Worcestershire (Benton et al., 1994; Barclay et al., in press), and in the lower part of the Helsby Sandstone Formation in north Cheshire (Benton et al., 1994). In north Worcestershire and in Warwickshire the Bromsgrove Sandstone has also yielded vertebrate (amphibian and reptilian) fossils. Comparable remains occur in successions in parts of the Cheshire and Needwood basins adjacent to the district. These occurrences, in the Grinshill Sandstone and overlying Tarporley Siltstone formations in the Wem district (Benton et al., 1994) and the Hollington Formation in the Ashbourne district (Chisholm et al., 1988; Benton et al., 1994), support the assignment of an Anisian age to the representatives of the Helsby Sandstone and Bromsgrove Sandstone formations within the Stoke-on-Trent district.

In the following account, the sequences in the three basins will be treated separately.

Cheshire Basin

The nomenclature of the Sherwood Sandstone Group in the Cheshire Basin is based on that of Warrington et al. (1980). As used here, the names Kinnerton Sandstone,

Chester Pebble Beds, Wilmslow Sandstone and Helsby Sandstone formations are equivalent to the 'Lower Mottled Sandstone', 'Bunter Pebble Beds', 'Upper Mottled Sandstone' and 'Lower Keuper Sandstone' respectively of Hull (1860; 1869). The Bulkeley Hill Formation is now tentatively recognised in the district. There is a marked contrast in the thicknesses of some formations between the central parts of the Cheshire Basin and its eastern flanks. The character of individual formations also changes; for instance, all or parts of the Kinnerton, Wilmslow and ?Bulkeley Hill formations are absent or are only poorly recognisable near the eastern margin of the basin, east of the Bridgemere and Wem faults (Figure 29).

KINNERTON SANDSTONE FORMATION

The old name 'Lower Mottled Sandstone' was used in two senses; in the area where Manchester Marl is identifiable it was applied to the sandstones between the Manchester Marl and the Chester Pebble Beds, and where the Manchester Marl could not be identified it covered all the Permo-Triassic sandstones below the Chester Pebble Beds. Warrington et al. (1980) introduced the name Kinnerton Sandstone Formation to replace Lower Mottled Sandstone in its second meaning, but recent usage has extended the name to cover the first meaning also. The term is used here in this widened sense, as a full synonym of Lower Mottled Sandstone. The formation is dominated by red-brown, non-pebbly, cross-stratified, fine- to medium-grained sandstones. It has an extensive, though mostly subdrift, crop between Audley and Madeley. However, it is thickly developed only in the central parts of the Cheshire Basin.

On fault blocks towards the eastern margin of the basin, the formation unconformably overlies faulted and folded rocks of the Keele Formation, as below Heighley Castle, north-west of Madeley. It is possible that the base of the formation is marked by alluvial-fan conglomerates, like those of the Rushton Spencer Breccia of the adjoining Macclesfield district, which are composed largely of Carboniferous clasts (Thompson, 1985). Conglomerates such as these are not ubiquitous however, and at Hodnet Pumping Station [621 281], in the adjoining Wem district, Wild (1987) noted only rare extraformational clasts in the basal 5.2 m of the formation.

Towards the centre of the Cheshire Basin, the Kinnerton Sandstone Formation conformably overlies the Bold Formation (Chapter 4). West of the present district, where the latter cannot locally be identified because its argillaceous component cannot be recognised, the formation is indistinguishable from the underlying Collyhurst Sandstone and the combined unit is termed the Kinnerton Formation; in such areas the lower part of the formation is probably of Permian age. The top of the Kinnerton Sandstone is placed at the channellised base of the overlying Chester Pebble Beds Formation; in the central part of the basin the boundary is concordant, but towards the eastern margin of the basin it may be locally discordant, or marked by an angular unconformity.

The formation comprises mainly red-brown, texturally mature sublitharenites composed largely of well-rounded, well-sorted sands with little mica (Wild, 1987). Most sands are fine to medium grained; coarse sand is commonest towards the bases of beds. The sandstones commonly form cross-stratified asymptotic sets with reactivation surfaces. The formation is interpreted to have accumulated by wind-ripple, sandflow and grainfall processes on dune complexes and on dry interdune areas in an aeolian environment (Wild, 1987).

The formation varies considerably in thickness. In the central parts of the Cheshire Basin it ranges from 70 to 210 m, and at the eastern margin of the basin it is mostly absent through non-deposition or erosion prior to deposition of the Chester Pebble Beds.

The formation is only well exposed below Heighley Castle [772 468] where about 15 m of fine- to medium-grained sandstone forms a crag behind farm buildings. The sandstones are cross-stratified, with fine, inverse graded, 'pin-stripe' laminae suggesting an aeolian origin (Clemmensen and Abrahamsen, 1983; Fryberger and Schenk, 1988). Bedforms migrated towards the south-west. Minor mud-flake lags, representing deflation surfaces, are visible.

CHESTER PEBBLE BEDS FORMATION

This formation is dominated by ochreous to foxy red-brown, cross-stratified, pebbly, medium- to coarse-grained sandstones, with conglomerates that commonly vary both in degree of sorting and in thickness. Erosion surfaces and reactivation surfaces are common. The formation was formerly termed 'Bunter Pebble Beds' (Hull, 1860; 1869). In the Stoke-on-Trent district it forms eastward-facing scarps along the margin of the Cheshire Basin, including prominent hills such as Bar Hill, Knowl Bank and the hills north and south of Heighley Castle and Aston.

The formation conformably overlies the Kinnerton Sandstone in the central parts of the Cheshire Basin, but the boundary may become unconformable towards the eastern margin of the basin. In the marginal areas, the Chester Pebble Beds commonly rest unconformably on faulted and folded rocks of the Keele and Radwood formations. This relationship is suggested near Audley [7882 5059], where 0.9 m of reddish brown, pebbly, fine-grained sandstone is separated from 0.6 m of probable Keele formation by a 0.9 m gap. The top of the formation is defined at the conformable base of the succeeding Wilmslow Sandstone Formation or where that formation is absent, as in the Audley area, at the unconformable base of the Helsby Sandstone Formation.

Several boreholes have been drilled in the Chester Pebble Beds at the eastern margin of the Cheshire Basin. However, the records from these usually only distinguish between conglomerates, pebbly sandstones, sandstones and mudstones. Whilst the recognition of conglomerates and pebbly sandstones serves to identify the formation, such basic descriptions are of little use in interpreting depositional environments: This has been done chiefly by investigation of exposed sections. Following widespread study of exposures in the district, and an earlier classification of the rocks in the district into lithofacies (Thompson, 1970a; 1970b; Buist and Thompson, 1982), Steel and

Thompson (1983) divided the rocks of the 'Bunter Pebble Beds' in the Stafford and Needwood basins (the Kidderminster and Hawksmoor formations, respectively) into five lithofacies, each related to a fluvial depositional environment. Although they did not extend this classification to the equivalent Chester Pebble Beds in the Cheshire Basin, the same subdivision can be readily applied to it and is adopted here. The five lithofacies are:

(A)   horizontally stratified conglomerates, ranging from simple disorganised clast- or matrix-supported conglomerates to complex, rhythmically graded types, in which imbrication is common. These represent longitudinal or diagonal river bars.

(B)   planar or trough cross-stratified conglomerates with bimodal grain-size distributions, commonly showing rhythmically developed textures and structures on the foresets, which represent the downstream avalanche surfaces of such bars or the migration of transverse gravel bars.

(C)   planar or trough cross-stratified, pebbly, medium- to coarse-grained sandstones, in which intraformational pebbles, cobbles and blocks lie above erosion surfaces, and reactivation surfaces are common. These represent the migration of dunes, sandwaves, megaripples and transverse bars in a sandy river.

(D)   cross-bedded, argillaceous, medium- and fine-grained cross-bedded sandstones containing soft-sediment deformation structures. Intraformational pebbles are common but only rarely form lenses and stringers. These sandstones represent sandwave deposits and sheet-sands.

(E)   interbedded, horizontally stratified, parallel-bedded, argillaceous, fine-grained sandstones, siltstones and mudstones, which represent finer sediments deposited during periods of lower river discharge or in abandoned parts of the river system.

The distribution of the lithofacies (A–E) in the principal exposed sections of the district is illustrated in Figure 32.

Several features indicate deposition by braided rivers (Steel and Thompson, 1983). These include the predominance of sheet-like conglomerates and pebbly sandstones, the rarity of argillaceous rocks, the unimodal transport direction indicated by palaeocurrents (Figure 29) and the lack of evidence for point-bar deposition. The gravels are interpreted to have been deposited in deeper water than in most modern braided streams; braided streams with relatively confined channels and considerable bar or channel relief are better depositional analogues (Steel and Thompson, 1983).

Aeolian rocks are rare in the Chester Pebble Beds. However, 'pin-stripe' laminated sandstones, typical of aeolian rocks, are exposed at Merelake [8100 5388] where they are interbedded with fine- and medium-grained, slightly pebbly sandstones of fluvial origin. Wedd (1899) noted that in the nearby railway cutting these are cemented by barytes and celestine. This association passes up into more pebbly sandstones (shown as the base of the 'Bunter Pebble Beds' by Hull, 1869) and

then into mudstones. The section at Merelake thus also includes finer-grained lithologies than are common elsewhere in the Pebble Beds, and may represent a lower energy, somewhat marginal, part of the fluvial system where aeolian rocks could be preserved.

The Chester Pebble Beds may be split informally into lower, middle and upper divisions, though this tripartite subdivision is less well marked than in the Stafford Basin (Figure 33). The lower and upper divisions are dominated by pebbly sandstones with few conglomerates, whereas the middle division contains several conglomerates. The only borehole in which the middle division can be clearly differentiated is that at Pipe Gate, where it is 18.3 m thick (Figure 33).

Over most of the district the formation is between 125 and 220 m thick, though in parts of the district on the eastern flank of the Cheshire Basin it is less than 90 m thick.

The formation is described from south to north along the outcrop, where exposures in quarries lie near the eastern margin of the basin. In the south-eastern part of the district, on the horst between the Hodnet Fault and the Sidway Fault (Figure 29) (the Market Drayton Causeway of Wills, 1956), the formation forms rolling terrain, with many small quarries of little note. Conglomerates and pebbly sandstones have been worked recently south of Mucklestone, in Tadgedale Quarries [733 366] (Figures 29 and 32), though these are now being backfilled. The best exposure in the district is at Willoughbridge quarries [748 390 to 747 372], from which large quantities of sand and gravel are currently being worked, principally from conglomerates in the middle part of the formation. The quarries demonstrate very well the lateral impersistence of beds; several of the conglomerates pass laterally into pebbly sandstones, and pebbly sandstones into pebble-free sandstones. A generalised composite log of the quarries is shown in Figure 32.

The only notable exposure of the formation where it is approaching its typical thickness is at Bar Hill [762 435]. The outcrop here is marked by an escarpment cut by numerous faults. A patchily exposed section was logged in a cutting of the A525 road and adjacent quarries (Figure 32). The Chester Pebble Beds form a marked escarpment at Bryn Wood [7739 4599] with exposures in 6 m of brown, medium-grained sandstone and 3.6 m of soft, reddish brown sandstone with a conglomerate bed [7737 4564], both close to the base of the formation. Higher parts of the formation are exposed in Beck Wood [7679 4614] where 7.5 m of sandstone with some scattered pebbles are exposed. Close to Lower Thornhill, 12 m of brown, cross-bedded, medium-grained sandstone exposed in a quarry [7641 4629] has well-marked jointing parallel to the nearby Wem Fault. The escarpment of the Chester Pebble Beds is well marked at Heighley where it is crowned by the remains of a castle. Exposures in the moat [7723 4684] show reddish brown, medium-grained sandstone with scattered pebbles and rare beds of conglomerate. More extensive exposures, mainly of sandstone with scattered pebbles, occur in quarry and track sections nearby [7676 4685 to 7726 4701] (Figure 32).

North of Heighley, because no detailed lithological record was made of the Plum Tree Borehole (Figure 33), the Chester Pebble Beds are only well known from the Audley Waterworks Borehole (Figure 33). This proved pebbly sandstone, with conglomerate beds up to 5.2 m thick, which forms an escarpment capped by the storage reservoir east of the pumping station. An old quarry and underground workings (Middleton, 1986a) for gravel are situated close to the crest of Kent Hill [7891 5093]

(Figure 32). Nearby [7890 5104], 2.4 m of cross-stratified, medium-grained sandstone with scattered pebbles, overlying 1.5 m of conglomerate, are separated from coarse-grained sandstone with scattered pebbles by a fault.

North of Audley, the Mill End Borehole (Figure 33) recovered only chippings of very fine- to medium-grained sandstone with no indication of pebble content. Near Alsager, the higher beds of the formation form a marked ridge on the golf course and, may be seen at Merelake in a road cutting [8100 5388] (Hull, 1869, fig. 20) and a railway cutting [8122 5411]. These consist of reddish brown sandstones with pebbly horizons and include a red mudstone, probably some 3 m thick, which has been worked in a line of old clay pits. The lower part of this section was included by Hull (1869) in the 'Lower Mottled Sandstone'; however, the section is here included in the Chester Pebble Beds because Gibson (1925) recognised conglomerates lower in the formation.

WILMSLOW SANDSTONE FORMATION

This formation typically comprises orange-red to dark brick red, planar cross-stratified, non-pebbly, fine- to medium-grained sandstones. It is equivalent to the 'Upper Mottled Sandstone' of Hull (1860, 1869). The formation usually forms lower ground than the contiguous Chester Pebble Beds and Helsby Sandstone formations. It has a discontinuous outcrop between Betley and Mucklestone, where it is covered by thick drift and is very poorly exposed. Most information comes from boreholes.

The Wilmslow Sandstone conformably overlies the Chester Pebble Beds. The current definition of the base of the formation is poor (Warrington et al., 1980), and the criterion taken here is where dominantly pebbly sandstones (of the Chester Pebble Beds) are overlain by at least 20 m of sandstones in which pebbles are rare.

In the central part of the Cheshire Basin the Wilmslow Sandstone is conformably overlain by the Bulkeley Hill Formation (Poole and Whiteman, 1966), but this dies out towards the basin margin, and the Wilmslow Sandstone there is thought to be overlain unconformably by the Helsby Sandstone. Whilst it is also possible that the Wilmslow Sandstone passes laterally into parts of the Chester Pebble Beds (which would explain the relationship of the formations without the need for an unconformity) this context seems less plausible, considering the widespread recognition of an unconformity at the base of the Helsby Sandstone and its equivalents elsewhere in western and central England (Geiger and Hopping, 1968; Warrington, 1970a; Warrington et al., 1980; Evans et al., 1993), which has been equated with the Hardegsen Disconformity in Germany (Audley-Charles, 1970a; Warrington, 1970a; Wills, 1970a). The unconformity appears to cut out the Wilmslow Sandstone completely at the north end of the outcrop, as in a road cutting at Merelake [810 538], where Helsby Sandstone directly overlies Chester Pebble Beds.

The formation is dominated by pebble-free sandstones; pebbly sandstones, conglomerates and mudstones are very rare. Two different facies are present. The first comprises reddish brown, poorly cemented, poorly sorted, fine- to medium- grained sandstones, alternating with paler, coarser, calcareous, better-sorted sandstones.

All contain angular grains, are locally interbedded with siltstones, and are interpreted to be of fluvial origin. They are more common in the lower part of the formation. The second facies comprises numerous friable, frosted, well-rounded, well-sorted, cross-stratified sandstones, invariably with 'pin-stripe' lamination due to ripple-drift migration (Fryberger and Schenk, 1988). These are interpreted to be of aeolian origin, and are more common in the upper part of the formation (Thompson, 1970b). Fluvial sandstones in the upper part of the formation include the deposits of the low-sinuosity 'Midland River' of Thompson (1970a), which periodically flooded, leaving features such as wave-rippled surfaces in flat interdune areas.

In the central part of the Cheshire Basin the formation can be divided into two parts based on the sonic log signature (Figure 31). The lower part equates with a zone with uniformly low porosity values. The cause of this low porosity is somewhat enigmatic; Colter and Barr (1975) considered it to be a zone of greater silicification (hence it has become known as the 'silicified zone') though it may coincide with the part of the formation that is dominated by rocks of fluvial origin (D B Thompson, personal communication). The 'silicified zone' remains remarkably constant in thickness across the Cheshire Basin from the Prees Borehole (505 m) to the Knutsford Borehole (459 m) (Figure 31) and into the Irish Sea Basin, where it is mostly between 300 and 500 m thick (Colter and Barr, 1975). It is unclear whether this zone persists to the eastern flank of the Cheshire Basin, though 10 to 20 m of harder sandstones occur at the base of the formation in boreholes at Bearstones Waterworks [7243 3899].

Of the Triassic formations in the district, the Wilmslow Sandstone is the most variable in thickness. In the central part of the Cheshire Basin the formation is about 900 m thick, but on the eastern flanks of the basin it is only up to 150 m thick and in some areas, such as west of Audley, is absent. The thin sequence on the eastern margin of the basin may have resulted in part from a lower rate of sedimentation. However, it is probable that some of the formation was eroded away prior to deposition of the Helsby Sandstone (see above). The great contrast in thickness between the centre and flanks of the Cheshire Basin may reflect syndepositional rifting (Figure 45).

? BULKELEY HILL FORMATION

This formation is preserved only in the central part of the Cheshire Basin. Its presence in the Stoke-on-Trent district is inferred from seismic evidence, and it is tentatively correlated with the trough cross-bedded, micaceous and argillaceous sandstones known at Bulkeley Hill in the northern part of the adjoining Nantwich district (Poole and Whiteman, 1966; Evans et al., 1993).

Seismically, the formation is represented by laterally persistent, moderate- to high-amplitude reflections. These are interpreted to represent interbedded fluvial sandstones and mudstones and possibly aeolian sandstones.

It appeared to Poole and Whiteman (1966) that the formation conformably overlies the Wilmslow Sandstone and is unconformably overlain by the Helsby Sandstone. The formation is most fully developed in the hanging-wall block of the fault zone marking the eastern margin of the Cheshire Basin, where it is interpreted from seismic lines to be up to 200 m thick.

HELSBY SANDSTONE FORMATION

The Helsby Sandstone Formation is dominated by pebbly sandstones. These are interbedded with subordinate silt-stones and mudstones which appear to pass basinwards into the Tarporley Siltstone Formation. The Helsby Sandstone Formation is largely equivalent to the 'Lower Keuper Sandstone' of Hull (1860; 1869).

The formation outcrops near the eastern margin of the Cheshire Basin, where it dips steeply to the west. The sandstones form conspicuous ridges, separated by topographical lows that represent siltstones and mudstones. In the central part of the Cheshire Basin the formation is recognised only on seismic evidence. The formation is about 250 m thick across the district.

The base of the formation is separated from underlying formations by an unconformity equivalent to the Hardegsen Disconformity in Germany. It is best seen on seismic sections in the central part of the Cheshire Basin (Evans et al., 1993). The top of the formation is most commonly defined by the base of the Tarporley Siltstone Formation. However, where the latter has passed laterally into the Helsby Sandstone Formation (near the basin margin) the base of the succeeding Bollin Mudstone Formation defines the top of the Helsby Sandstone Formation.

In the Macclesfield district, Greenwood (1918) and Fitch et al. (1966) showed that the Helsby sandstones are broadly similar in composition to the Sherwood sandstones below the unconformable base of the Helsby Sandstone, suggesting a similar provenance. However, they noted that the rocks above the unconformity contain less microcline, chert and orthoquartzite than the rocks below, but more oligoclase, and more rock fragments of schistose, rhyolitic, chloritic and sericitic nature.

The formation consists of two major facies as associations, one of fluvial origin, the other aeolian. The fluvial facies comprises reddish brown to white, medium- to coarse-grained sandstones, which are moderately well sorted, micaceous, and cross-stratified. Intra- or extra-formational pebbles may be common, and locally form conglomerates. These sandstones are commonly associated with siltstones or mudstones, and contain desiccation surfaces exhibiting polygonal cracks, as in medium-grained sandstones exposed near Napley Heath [7202 3844]. The sandstones are interpreted as river channel or sheet-flood sands and the siltstones and mudstones as overbank or within-channel abandonment deposits. The fluvial facies is dominant in the lower part of the formation near the western and northern margins of the Cheshire Basin, where it is known as as the Delamere Member (Thompson, 1970a), and also in the basin centre. In these areas the formation is commonly dominated by pebbly sandstones and conglomerates interpreted as sinuous river deposits (Thompson, 1970a;b).

The aeolian lithofacies comprises well-sorted, non-micaceous, pebble-free, fine- to medium-grained sandstones which commonly contain well-rounded, 'millet seed' grains. They are commonly cross-stratified, with asymptotic sets of cross-laminae, reactivation surfaces and a distinctive 'pin stripe' lamination (Hunter, 1977). The pore spaces of these sandstones commonly contain bladed crystals of barite, as in an exposure opposite Norton Forge Farm [7042 3789]. The 'pin-stripe' laminae are interpreted to represent wind-ripple migration. Aeolian sandstones probably occur sporadically throughout the formation, though in the centre and north of the Cheshire Basin they are most common in the upper part, known there as the Frodsham Member (Thompson, 1969). This member comprises mainly well-sorted, non-pebbly, cross-stratified, medium-grained sandstones and is interpreted in the Frodsham area, Cheshire, to have formed from large dunes, migrating from east to west, which were modified into dome shapes by strong winds (Thompson, 1969; Brookfield, 1977). Subordinate flat-bedded strata are interpreted as interdune deposits that resulted from both flooding and drying-out, the latter process giving rise to aeolian sheet flow sands.

The following account describes the outcrop of the Helsby Sandstone from south-west to north-east. The best exposures are in the south-west of the district, west of Mucklestone, where the formation consists of up to six sandstones interbedded with mudstones containing thin sandstones. The thickness in this area is about 250 m and the dip is to the north-west. The sandstones are medium-grained, commonly brown in colour, and are exposed in scattered old quarries and a few natural exposures. The intervening interbedded mudstones and sandstones are poorly exposed. The lowest sandstone is yellowish brown, medium-grained and well-bedded where seen in small exposures near Napley Heath. The second sandstone, also well exposed near Napley Heath [7202 3844], comprises medium-grained sandstones, separated by mudflake conglomerates at three or more horizons. The third sandstone forms ridge-like features at Napley Heath, but appears to die out south-westwards. However, from the vicinity of Oakley Hall it can be traced southwards to the Hodnet Fault. Adjacent to and opposite Oakley Hall [7010 3704 to 7002 3702] the lower 3 to 5 m of the sandstone consist of cross-stratified, 'pin-stripe' laminated, aeolian sandstones. These are overlain by over 5 m of fine- to medium-grained sandstones, with mudflake conglomerates and erosional bases, of fluvial origin. The fourth sandstone is exposed south of Bearstone [7220 3925] and contains disseminated copper minerals (Garner, 1844; Smithson, 1947; Anon, 1947). The fifth sandstone is seen near Arbour Cottage [7123 3773, 7041 3787] and in Bearstone village [7236 3982]. It is a brown, fine- to medium-grained sandstone which, west of Norton Forge Farm [7042 3789], is aeolian in origin. The sixth sandstone is poorly exposed, and south of the Norton in Hales area has not been separated on the maps from the fifth sandstone. The strata between the sandstones are rarely seen. The best exposure, near Oakley Hall, is at a level where the lowest two sandstones are thought to have died out laterally. An old clay pit at Marlpit Wood [6984 3636], and several exposures in the valley of the River Tern, show brown, very fine-grained sandstones, up to 0.45 m thick, alternating with silty mudstones up to 1.5 m thick.

North of Madeley the topmost 24 m of sandstones in the Plum Tree Borehole are grey-brown rather than reddish brown in colour, suggesting that they represent the Helsby Sandstone, not the underlying Wilmslow Sandstone. In the Alsager area, the formation is largely under drift though both sandstones and mudstones are locally exposed. Pale, 'pin stripe' laminated, aeolian sandstones crop out north of Merelake [8069 5395] and a prominent feature at Linleywood [8191 5443] relates to 15 m of fine- and medium-grained sandstones with a barytes cement (Gibson, 1905). Borehole provings hereabouts include that at Town House Farm [8020 5487], containing red and grey sandstone, much of which is thought to be Helsby Sandstone, and three closely spaced boreholes [798 539] drilled for Alsager Rural District Council, which intercepted dominantly red and grey sandstone but contained different thicknesses of mudstone partings; the thickest set of beds rich in mudstone partings lies between 42.1 and 77.1 m depth in No. 3 borehole.

Stafford Basin

The nomenclature of the Sherwood Sandstone Group for the north of the Stafford Basin is, except for the newly recognised Kibblestone Formation, the same as that erected for the south of the basin by Warrington et al. (1980, area 12, p.20) (Figure 30). The set of names used by those authors for central Staffordshire (part of area 11, p.20) has become redundant in the Stafford Basin (Buist and Thompson, 1982; Steel and Thompson, 1983; Thompson, 1985), yet may relate to the Cannock area on the horst between the Needwood and Stafford basins, where the Wildmoor Sandstone Formation is absent. The inferred northern extension of this horst enters the Stoke-on-Trent district but, for simplicity, the nomenclature of the Stafford Basin is applied here also. The Sherwood Sandstone Group therefore comprises the Kidderminster, Wildmoor Sandstone and Bromsgrove Sandstone formations, plus a newly defined unit, the Kibblestone Formation, which occurs at the top of the group south-west of Moddershall. The first three units correspond approximately to the 'Bunter Pebble Beds', 'Upper Mottled Sandstone' and 'Keuper Sandstone' respectively of Hull (1860; 1869).

KIDDERMINSTER FORMATION

This, the basal formation in the Stafford Basin, is dominated by ochre-brown to red-brown, trough and planar cross-stratified, pebbly sandstones and conglomerates. It is equivalent to the unit formerly designated 'Bunter Pebble Beds' in its type area around Kidderminster, Worcestershire (Warrington et al., 1980). In the Stoke-on-Trent district, scarps of the formation form the northern rim of the Stafford Basin, including the Maer Hills, the hills north-west and north-east of Whitmore, the Hanchurch Hills, Kingswood Bank and Tittensor Chase. The unit also occurs in a small outlier near Keele and on the horst adjoining the Needwood Basin, south-west of Moddershall. It is typically poorly cemented and forms light, sandy, acidic soils.

In the Stoke-on-Trent district the Kidderminster Formation rests unconformably on Carboniferous strata. The intervening Bridgnorth Sandstone, present towards the south of the Stafford Basin, dies out northwards and does not enter the district. The existence of palaeovalleys at the base of the Kidderminster Formation (Gibson, 1905) is now in doubt; an example interpreted beneath Kingswood Bank [860 395] by Gibson has been disproved during remapping. The top of the formation is defined by the base of the overlying Wildmoor Sandstone Formation. South-west of Moddershall, where the latter is absent, the Kidderminster Formation is overlain, probably unconformably, by the Kibblestone Formation.

As with the Chester Pebble Beds in the adjoining Cheshire Basin, most borehole records of the Kidderminster Formation are very crude, and lithological information has been derived mainly from exposures. The detailed studies by Buist and Thompson (1982) and Steel and Thompson (1983) on lithofacies of the 'Bunter Pebble Beds' were based largely on exposures of the Kidderminster Formation, especially those at Acton and the M6 motorway at Trentham (Figure 32).

Several lithofacies associations, including upward-fining cycles (Figure 36), are recognised in the formation. Most are discussed in some detail by Steel and Thompson (1983). These change very rapidly vertically and laterally (Figure 33). However, the formation is informally split into three divisions, none of which may be traced in all areas of outcrop. The lower division mainly comprises upward-coarsening pebbly sandstones, the middle division is dominated by conglomerates, and the upper division is characterised by pebbly sandstones. The rocks of the middle division tend to form steep, distinctively rounded slopes at outcrop (Malkin, 1985).

The thickness of the formation appears to decrease westwards from the Groundslow Borehole near Tittensor, where it is about 150 m thick, to the Whitmore area where, in Whitmore No. 4 Borehole, it is about 100 m thick. Farther west, adjacent to the Madeley Fault, the formation in The Wellings Borehole is only about 70 m thick (Figure 33).

The formation is commonly poorly exposed, as it is typically covered by a thin blanket of head. Gullies in the Maer Hills [7725 3970] provide examples. The base of the formation is not exposed, though Gibson (1925) noted at Spring Valley, Trentham Park, a conglomerate of pale angular to rounded limestone clasts at the base of the formation. The lower division of the formation is usually 5 to 20 m thick, coarsens upwards, and is dominated by yellow-brown medium-grained sandstones with few pebbles. It is well exposed at Acton [8237 4163], Trentham M6 motorway cutting (Buist and Thompson, 1982) and Trentham Park [8714 3877], where finer-grained sandstones are red-brown. These grade upwards into the middle division of the formation which is dominated by conglomerates and yellow-brown to yellow-red, pebbly, medium- to coarse-grained sandstones. The conglomerates are composed mostly of pebbles, though boulders over 0.6 m in diameter occur. Rocks of the middle division of the formation were previously well exposed at the now backfilled Acton quarries [818 412] (Figure 32; Steel and Thompson, 1983). These quarries also formerly exposed finer-grained lithologies such as siltstones in which the small crustacean *Euestheria minuta* has been found (Wilson, 1962). The middle division is presently best exposed in the M6 motorway cutting at Trentham [8526 4073 to 8534 4019], and in the adjacent Kingswood Quarry

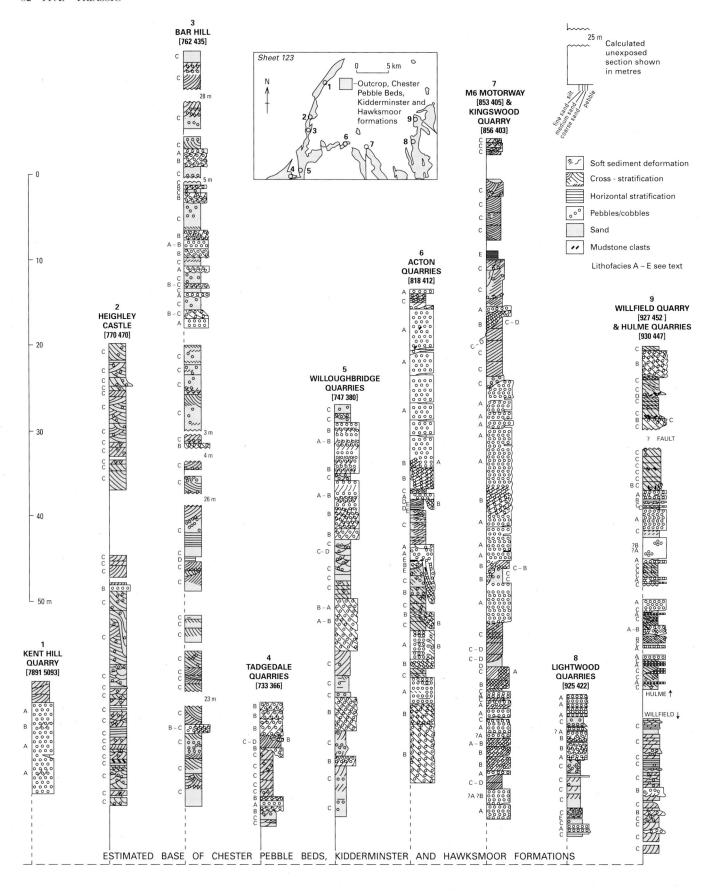

Figure 32 *(opposite)*   Detailed sections of the Chester Pebble Beds, Kidderminster and Hawksmoor formations, showing lithofacies as recognised by Steel and Thompson (1983). Sections 6, 7 and 9 from Steel and Thompson (1983) and Thompson (1985). Palaeocurrent data from these localities are shown in Figure 29.

[8563 4031] (Figure 32; Buist and Thompson, 1982; Steel and Thompson, 1983). Conglomerates are well exposed and accessible at the quarry. They include clasts of Dinantian limestone and Silurian sandstones from which numerous fossils have been recovered (Gibson, 1925). Great lateral variation, and erosional contacts between lithofacies, occur in the middle division, as in the quarry [7653 3982] 800 m east of Sidway Hall. Where conglomerates are absent, as in boreholes at The Wellings, some 2 km to the south, it is impossible to recognise the middle division. The upper division of the formation is the most variable in thickness, and the most difficult to differentiate. In places where conglomerates occur at the top of the formation, the highest division is absent, as in the Whitmore Rectory Borehole [8038 4093]. South of the Maer Hills, the division is at least 15 m thick in the vicinity of the quarry at Blackbrook [7697 3884], where mudstone intraclasts occur at the base of pebbly sandstones with large cross-beds. At Hatton Waterworks (Figure 33) the upper division is over 45 m thick. The middle and upper divisions together are usually about 100 m thick.

WILDMOOR SANDSTONE FORMATION

This formation comprises foxy coloured, bright orange-red to dark brick-red, argillaceous, fine- to medium-grained sandstones with few pebbles. Red-brown siltstones also occur. Sandstones are commonly mottled with bleached zones, usually less than 1 m in length, which accentuate bedding, laminae, joints and faults. Conglomerates and pebbly sandstone horizons are few; most clasts are intraformational. The formation is equivalent to the 'Upper Mottled Sandstone' of the Central Midlands Triassic succession, well documented by Wills (1970a; 1970b; 1976). As most of the sandstones are soft, the formation occurs in the lower ground between hills formed by the more resistant Kidderminster and Bromsgrove Sandstone formations. The formation has been extensively quarried for moulding sand in the district (Gibson, 1905).

Warrington et al. (1980) place the lower boundary of the Wildmoor Sandstone Formation above the highest pebbly sandstone of the Kidderminster Formation. This definition is difficult to apply in the Stoke-on-Trent district as the junction between the formations is transitional and because pebbles, and pebbly sandstones, occur in the Wildmoor Formation (albeit rarely) as well as in the Kidderminster Formation. The base of the Wildmoor Sandstone is here taken at the top of the sequence which is dominated by pebbly sandstones and conglomerates, and which is overlain by at least 20 m of sandstones in which pebbles are rare. The Wildmoor Sandstone is overlain by the Bromsgrove Sandstone Formation, which has an erosional, probably unconformable base.

Much of the Wildmoor Sandstone, particularly its upper part, is composed of horizontally interbedded, micaceous, argillaceous, fine-grained sandstones, siltstones and mudstones. The sorting of the sandstones is variable, though it varies little within any one bed. The exposures at Red Hill Quarry [7874 3962] (Figure 34), near Baldwin's Gate, are typical. Thompson (1985) interpreted the finer-grained beds here as fluvial overbank or abandoned channel deposits, and the coarser-grained beds as braided river deposits. Coarser-grained channel deposits occur sporadically, as in the upper part of the formation at Hill Chorlton [8067 3917], where trough cross-bedded, pebbly, medium- to coarse-grained sandstones and imbricated sandy conglomerates are well exposed. The pebbles in these are rounded, and compositionally similar to those of the Kidderminster Formation, from which they may have been reworked. The pebble composition of the formation in this district differs from that at Wildmoor pumping station, in the type area in Worcestershire, where Wills (1976) noted that angular, locally derived pebbles predominate.

Aeolian sandstones also occur commonly within the formation, especially near the top. At least two units of these are well exposed at Red Hill Quarry (Figure 34). Thompson (1985, figs 31 and 32) interpreted the higher unit as deposited by easterly winds. It comprises orange-red, mottled white, clean, well-sorted, medium-grained sandstones, with large-scale trough cross-bedding, 'pin-stripe' laminae, reactivation structures and rhythmically developed foresets. The lower unit in Red Hill Quarry contains zig-zag thrust-deformed foresets, and exhibits more diverse palaeocurrents that may indicate wind reversals during some periods.

The formation is estimated to be about 200 m thick in the area to the north and west of Maer, and west of Tittensor Chase. However, it thins to about 30 m near Knowl Wall [853 393], north of Beech, and is absent near Oulton Heath, in the easternmost part of the Stafford Basin. The areal variation in thickness of the formation may reflect lateral facies changes and/ or depositional processes, as well as erosion at the time of the Hardegsen Disconformity.

Most rocks of the formation are poorly exposed, as they are covered by a variable thickness of pebbly head derived from the Kidderminster Formation. In the case of boreholes, lithological information has been recorded in little detail. Most lithological data come from the few exposures.

The best extant exposure is the quarry at Red Hill [7874 3962], where sandstones of the formation are unconformably overlain by pebbly sandstones of the Bromsgrove Sandstone Formation (Figure 34). Deep orange-red, flaggy, flat- and cross-bedded, medium- to fine-grained sandstones with mottling developed along bedding, are also well exposed in the cutting south-west of Castle Hill [7617 3713], near Ashley. Some of these sandstones exhibit 'pin-stripe' laminae, indicative of aeolian deposition. A former exposure of note was the moulding sand pit [7988 3955], 300 m south-east of Chorlton Moss, where over 10 m of cross-bedded, micaceous, silty, fine-grained sandstones with mudstone clasts, and largely of fluvial origin, were exposed (recorded in part by Knowles, 1985b). Another pit, to the south-east of Hatton Waterworks

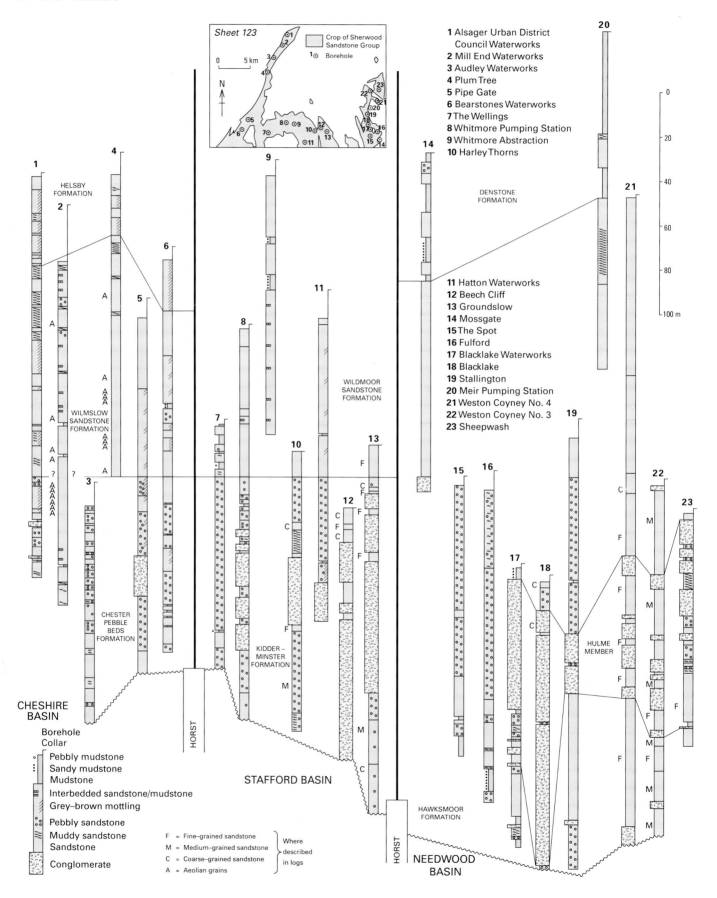

Figure 33 Correlation of the main Triassic borehole sections in the district.

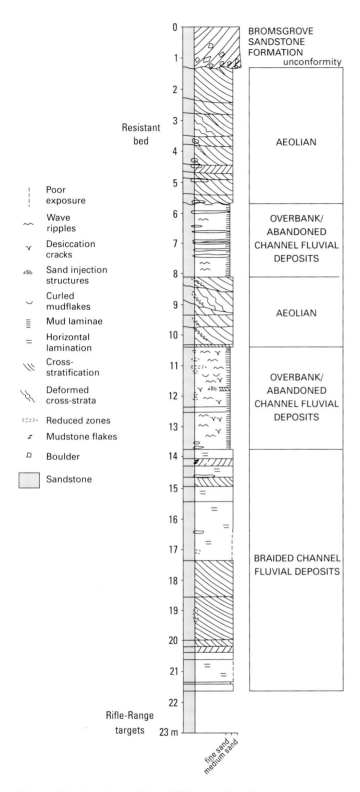

Figure 34 Section of the Wildmoor Sandstone Formation exposed at Red Hill [7874 3962].

[8315 3670], exposed over 29 m of pebble-free sandstone. This has been mostly infilled, though mottled, fine- to medium-grained sandstones, interlaminated with siltstones and mudstones, are still exposed around the top of the pit.

Pebbly sandstones and thin conglomerates are most common in the lower part of the formation, which is transitional with the underlying Kidderminster Formation. These are only distinguished from similar lithologies in the latter because they are underlain by at least 20 m of sandstones in which pebbles are rare. Such is the case in the borehole at Dog Lane [8268 3922], near Harley Thorns, where a conglomerate and pebbly sandstones are underlain by over 20 m of very argillaceous sandstone. Conglomerates and pebbly sandstones occur near the middle of the upper part of the formation and crop out near Ashley [7554 3630, 7558 3626, 7556 3624], at Castle Hill [7650 3750], on the hill east of Maer [798 387], on the slope north of Maerfield Gate [796 392], at Hill Chorlton [8067 3917], at Shelton under Harley [8165 3958] and east of The Rowe in the valley of Meece Brook [825 382]. At least 44 m of pebble-free sandstones underlie these locally, as has been proved by the water-bore at Hill Chorlton [7989 3947]. Apart from these pebbly developments in its upper part, the formation generally fines upwards, as is demonstrated in the Whitmore Pumping Station Borehole [8125 3980], where there is a gradation from medium- and locally coarse-grained sandstones at the base of the 120 m section to fine-grained sandstones and mudstones near the top.

BROMSGROVE SANDSTONE FORMATION

This formation is dominated by flat-bedded or cross-stratified, pale yellow to dark red-brown, calcareous, fine- to medium-grained sandstones and dark red-brown to green-grey siltstones. It is equivalent to the unit formerly known as the 'Lower Keuper Sandstone' (Hull, 1860; 1869; Whitehead et al., 1927) or 'Upper Sandstone' (Gibson, 1925). The sandstones tend to be well cemented and resistant, and form scarps near Ashley, Maer and Chapel Chorlton, and at Beech. Steep-sided hills are formed by outliers of the formation at War Hill [7839 3932], Red Hill [7870 3945], Berth Hill [7875 3905], and a hillock crowned by an unnamed tumulus [7815 3955].

Few boreholes have been drilled in the formation, and most lithological details are derived from the exposures. The base of the formation is well exposed at Red Hill [7874 3962] (Figure 35). The Bromsgrove Sandstone near Chapel Chorlton is poorly exposed, though variably sorted micaceous sandstones are visible in small quarries [8123 3860; 8151 3763]. The largest building stone quarries there having been infilled, the best exposure of the upper part of the formation in the district is at Beech [8543 3820] (Figure 35) where man-made caves (Middleton, 1986b) have been excavated.

Over most of the Stafford Basin, the Bromsgrove Sandstone overlies the Wildmoor Sandstone. The contact is presently only exposed at Red Hill (Figure 35), where it is erosional and probably unconformable (Hull, 1869). Pale pebbly sandstones and conglomerates containing some boulders of Wildmoor Sandstone lie in erosional channels that cut down at least 7.5 m into the top of the Wildmoor Sandstone. More widespread erosion of the Wildmoor Sandstone is indicated by the occurrence of well-rounded red sands, typical of this formation, in the Bromsgrove Sandstone south of Maer,

Figure 35   Representative sections of the Bromsgrove Sandstone Formation.

and below Beech (see below). Derivation from older rocks of the Sherwood Sandstone Group is reflected by the composition of pebble lithologies at Red Hill. The pebbles are similar to those of the underlying Kidderminster Formation, but vein quartz and quartzite are more common. Thompson (1985) suggested that this is due to recycling of the sediment. He also noted the common occurrence of ventifacts, which are unknown in the underlying Kidderminster and Wildmoor formations. These basal beds apart, conglomerates and pebbly sandstones containing extraformational pebbles are few.

Most sandstones of the Bromsgrove Sandstone are pale yellow-brown and micaceous, contain moderately well-rounded grains, and are flaggy or cross-stratified. The sandstones are compositionally similar to others in the Sherwood Sandstone Group, though in the Central Midlands, Ali (1982) noted that the formation contains more sanidine, orthoclase, biotite and apatite than underlying formations. This difference, particularly the abundance of biotite, and an increase in the abundance of igneous and plutonic rock fragments, he attributed to rapid uplift and erosion of acid igneous and high-grade metamorphic sources prior to deposition of the formation. Intraformational conglomerates composed principally of mudstone clasts, in a medium-grained sandstone matrix, are common. Mudstone clasts are also well exposed at Beech (Figure 35), and in pale red-brown mottled, cross-stratified sandstones near the base of the formation at Red Hill (Figure 35).

Locally, at a higher level in the basal beds of the formation near Maer [7887 3834], the sandstones are non-micaceous, inverse laminated, contain reactivation surfaces and are interpreted as aeolian in origin. Aeolian, large-scale trough cross-bedded sandstones at a similar horizon in the formation were once well exposed in the M6 motorway cutting south of Beech (D B Thompson, personal communication). They are now only modestly exposed, though still reveal their aeolian origins. Although most of these sandstones are similar in colour and texture to those of the Kibblestone Formation (see below), others consist of well-rounded red grains that may have been derived from the Wildmoor Sandstone. The siltstones in the formation tend to be parallel laminated, micaceous, and interlaminated with mudstones, as may be seen at Beech.

Most rocks were deposited in channels of sand-charged, low-sinuosity rivers. The basal structureless pebbly sandstone at Red Hill is interpreted by Thompson (1985) as a mass-flow deposit resting in a newly and deeply cut channel. Most sandstones represent sandwaves or dunes formed on the river bed, and the siltstones mostly represent overbank deposits. Palaeocurrents measured at Beech and Red Hill (Figure 35) (Thompson, 1985) show that the channel bedforms migrated in a west-north-westerly direction. The less-common aeolian sandstones in the formation probably consist of material reworked from fluvial sediments, as

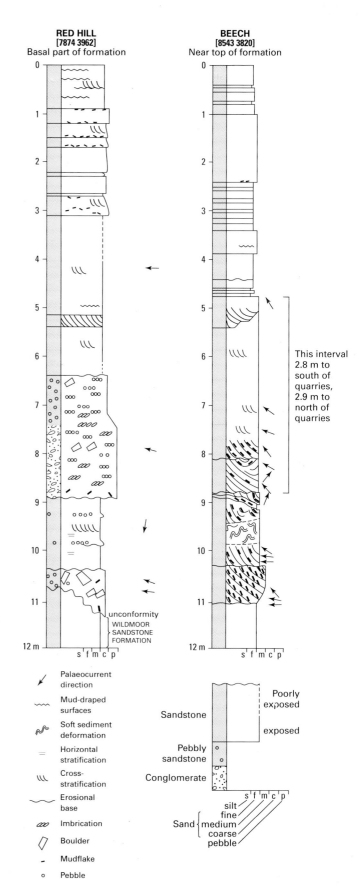

well as aeolian grains derived from the underlying Wildmoor Sandstone.

The succession of sandstones and siltstones in the formation varies considerably from west to east across the basin. In the west, near the Madeley Fault, the formation contains three main sandstones but eastwards, over a distance of only 2 km, the uppermost sandstone dies out and the two lower sandstones amalgamate. In the area between Maer and Chapel Chorlton the formation comprises a single sandstone. Farther east again, to the south of Beech, a sequence of sandstones separated by mudstones and siltstones is more like the formation in its type area (Old et al., 1991).

The formation thins considerably from west and east towards Maer. At The Wellings [7685 3805], adjacent to the Madeley Fault, it is about 41 m thick, near Bates Farm [7850 3800] it is about 32 m thick, and south of Maer [7923 3845] it is about 15 m thick. East of Maer, near Chapel Chorlton and at Beech, it is about 35 to 40 m thick.

KIBBLESTONE FORMATION

South-east of Kibblestone [914 362], near the eastern margin of the Stafford Basin, a sequence dominated by grey sandstones occupies the stratigraphical position of the Bromsgrove Sandstone (between the Kidderminster/Wildmoor formations and the Maer Formation). These beds are believed to be dominantly aeolian in origin; because they are lithologically distinctive, and their stratigraphical setting is uncertain, they are distinguished here as a separate formation. However, aeolian rocks do occur in the Bromsgrove Sandstone, and are common in the Wildmoor Sandstone to the west (see above), though there they are red-brown in colour and are not seen to pass upwards into the Maer Formation.

The formation gives rise to steep scarps and cliffs along the north–south valley, south-east of Kibblestone, which forms the type area. Notable exposures here include the sections opposite and behind the mill at Kibblestone scout camp [9157 3595] (Figure 36), in the present district, and in Vanity Lane [9158 3562], behind Ivy Mill [9158 3543] and beside the Stone Road [9140 3527 to 9159 3531], in the Stafford district to the south. Notable exposures outside the type area occur at Edge Hill [8949 3528] and in the railway cutting [893 351] north of Stone.

The base of the formation is exposed behind Ivy Mill [9158 3543] (Figure 36). The nature of the exposure does not show whether an unconformity or disconformity exists, though the very fine-grained sandstones immediately below the Kibblestone Formation have been disrupted, possibly by plant colonisation or pedogenesis. The top of the formation is exposed in the valley east of Hayes House [9217 3529] (Figure 36), just south of the district, where sandstones typical of the Kibblestone Formation are overlain by purple-brown flaggy sandstones assigned to the Maer Formation. The junction is probably transitional, as sandstones like those in the Kibblestone Formation occur within the basal few metres of the Maer Formation.

The formation is dominated by pale, ochre to yellow-grey, locally red-mottled, non-pebbly, well-sorted, fine- to

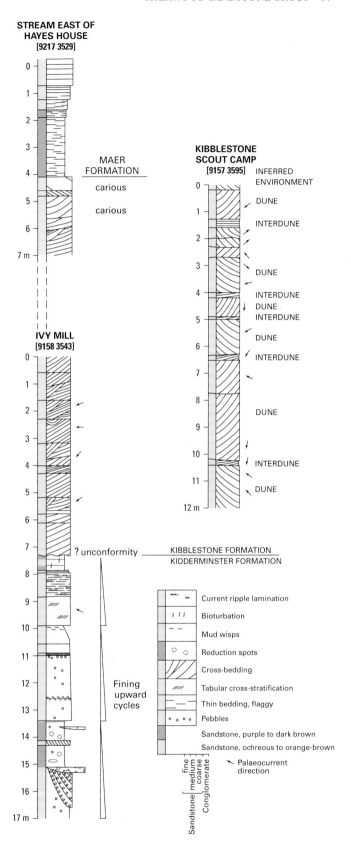

Figure 36 Logs of the Kibblestone Formation in the type area at Kibblestone.

medium-grained sandstones with a calcareous cement. The grains consist mainly of quartz, and are moderately well rounded, though 'millet seed' grains are sparse; mica is rare. The finer-grained sandstones occur in sets with low-angle cross-stratification. These commonly contain distinct very fine laminae of pale, fine-grained sandstone, and of darker, red, very fine-grained sandstone or siltstone. These thin beds are overlain typically by beds with large, steeply dipping, asymptotic cross-sets, which reach over 3.6 m in height at Edge Hill in the adjoining Stafford district. Cross-sets within both thin and thick beds dip mainly to the south-west (Figure 36). These sets are sometimes interbedded with small-scale trough and tabular cross-stratified sandstones, with undulating erosional bases, as exposed at Edge Hill.

Several features suggest that most of these sandstones are aeolian and were probably deposited as transverse/barchanoid ridge dunes. These features include the well-rounded, well-sorted, poorly micaceous nature of the sands which form most of the sequence, and the large asymptotic dune-forms indicating an easterly to north-easterly wind direction, consistent with deposition in the northern trade-wind belt. The 'pin-stripe', inverse graded, very fine laminae are a feature of aeolian ripple migration on slopes of up to $20°$ on modern and ancient aeolian dunes (Clemmensen and Abrahamsen, 1983; Fryberger and Schenk, 1988; Hunter, 1977). Some interbedded sandstones, such as the small-scale cross-stratified sandstones at Edge Hill, some of the interdune deposits shown in Figure 36, and the very fine-grained sandstones in the transition upwards into the Maer Formation were water-laid. Interdune surfaces locally supported vegetation, as shown by rootlets found in boulders derived from near the top of the formation near Stone in the Stafford district.

The formation may extend eastwards into the Needwood Basin, as similar sandstones occur at the top of the Sherwood Sandstone Group near Draycott in the Moors in the Ashbourne district [993 404]. It is possible that the formation developed as a dune-field on higher ground associated with the horst between the Stafford and Needwood basins.

The thickness of the formation is about 15 to 20 m in the type area, increasing southwards to about 30 m north of Stone in the adjacent Stafford district.

Needwood Basin

The stratigraphy of the Sherwood Sandstone Group of the Needwood Basin in this district is the same as that established for the Ashbourne district to the east (Charsley, 1982; Chisholm et al., 1988), except that the Huntley and Hollington formations are not recognised. The rocks in the Stoke-on-Trent district are all assigned to the Hawksmoor Formation.

HAWKSMOOR FORMATION

The Hawksmoor Formation consists mainly of red-brown to yellow-brown, commonly cross-bedded, very fine- to coarse-grained sandstone. Thin micaceous siltstones and mudstones occur sporadically and some conglomerates are present, notably in the Hulme Member (described separately below). The formation is equivalent to the 'Bunter Pebble Beds' and 'Lower Keuper Sandstone' of Gibson (1925). In the Stoke-on-Trent district it forms conspicuous scarps on the edge of the Needwood Basin, as between Lightwood and Moddershall, and there are small outliers in the Endon and Rownall areas. The sandstones give rise to well-drained soils.

The formation unconformably overlies folded and faulted Carboniferous strata, as is well illustrated in the Cellarhead area [958 475] where gently dipping rocks of the Hawksmoor Formation unconformably overlie steeply dipping Carboniferous rocks. Some parts of the Coal Measures, which are normally grey, have been stained red by percolation of oxidising waters prior to, or during, deposition of the Hawksmoor Formation. Derived clasts of Carboniferous formations under the unconformity, including Coal Measures, Etruria and Keele formations, are common at the base of the Hawksmoor Formation between Cellarhead and Moddershall (Gibson, 1925). Field evidence suggests that a palaeochannel with an east-north-easterly axis is eroded into the Coal Measures at the base of the formation south of Werrington [933 462 to 948 369].

Few borehole logs of this formation have recorded the lithofacies in any detail. However, good exposures occur in the Hulme and Lightwood areas, and these have been described in detail by Steel and Thompson (1983) and Thompson (1985). The lithofacies used by these authors are described in the account of the Chester Pebble Beds. Like the Chester Pebble Beds and the Kidderminster Formation, the Hawksmoor Formation is interpreted as the deposits of migrating channels, bars and dunes within a large braided river system.

A single lenticular conglomeratic unit, the Hulme Member (defined below) is distinguished within the Hawksmoor Formation in this district (Piper, 1982). The stratigraphical status of the sandstones overlying the member has been problematic in the past (Gibson, 1905) and earlier workers assigned the sandstones above the conglomerates to the 'Keuper Sandstone' (now the Hollington Formation of Charsley, 1982). However, the characteristic features of this unit are not developed in this district, and in the recent remapping the sequence between the Hulme Member and the Denstone Formation of the overlying Mercia Mudstone Group was included in the Hawksmoor Formation. These upper sandstones tend to be finer grained than those lower in the formation, and they are commonly better cemented.

The thickness of the Hawksmoor Formation is estimated to be about 330 m. No borehole has penetrated the whole formation, though over 310 m were penetrated in Weston Coyney No. 4 Borehole, and in Cresswell No. 1 Borehole, in the adjoining Ashbourne district, the Hawksmoor Formation (or undivided Sherwood Sandstone Group) is over 325 m thick.

The contact between the Hawksmoor Formation and Carboniferous rocks is commonly faulted or covered by drift, typically a pebbly sandy head. In the small quarry at Copshurst Farm, Lightwood [9249 4085], the boundary is obscured by quarry spoil; the sandstones above it consist of micaceous, variably pebbly and pebble-free, fine- to coarse-grained sandstones. Locally, sandstones in the lower part of the formation, as exposed to the south of Hulme [932 440], are mottled and contain few pebbles. However, much of the sequence below the conglomeratic Hulme Member comprises cross-stratified, pebbly sandstones such as are exposed at Lightwood quarries [9249 4216] (Figure 32). The upper part of the formation generally contains finer-grained sandstones, less pebbly than in the lower part of the formation, though these are rarely well exposed. Sections with few pebbles include a borehole at The Dams, Caverswall [9499 4309], in which over 140 m of sandstone, argillaceous sandstone and siltstone occur, and Weston Coyney No. 4 Borehole, which proved over 210 m of pebble-free sandstone.

HULME MEMBER

The Hulme Member (Piper, 1982) is dominated by conglomerates and contains many pebbly medium- to coarse-grained sandstones, though finer-grained lithologies, such as wave-rippled fine-grained sandstones, also occur (Steel and Thompson, 1983). The member lies in the middle part of the Hawksmoor Formation (Figure 33), and commonly creates steeper, more rounded, slopes than the other parts of the formation. Areas of outcrop are commonly covered with pebbly sandy soils.

The base of the member is taken at the base of the lowest conglomerate of the pebble-dominated part of the Hawksmoor Formation, and the top is placed at the top of the uppermost conglomerate.

The distribution and thickness of the member vary considerably because of the lateral impersistence of the constituent conglomerates. In the borehole at Blacklake (Figure 33) it is over 115 m thick, and in others between Blacklake and Spot Acre [9378 3892, 9349 3661] it is over 50 m thick. In equivalent borehole sections to the south and east of Fulford [9471 3825] and near The Spot [939 377] the member appears to be thin or absent.

The conglomerates and pebbly sandstones, like their equivalents in the Cheshire and Stafford basins, were probably deposited in high-energy river systems (Steel and Thompson, 1983). The greater conglomerate content of the Hawksmoor Formation, compared with the Kidderminster Formation and the Chester Pebble Beds, suggests that the axial channels of the braided system responsible for the transport of the conglomerates were most often located near the western margin of the Needwood Basin.

The member is best exposed in the disused quarries at Hulme [930 447], where sections have been documented by Steel and Thompson (1983) (Figure 32) and Thompson (1985). Numerous small exposures also occur in westward-facing slopes between Hulme and Moddershall.

MERCIA MUDSTONE GROUP

The Mercia Mudstone Group consists dominantly of reddish brown mudstones and siltstones, and in Cheshire includes two halite formations. The group was formerly known as the 'Keuper Marl' (Hull, 1869) but was renamed by Warrington et al. (1980). The outcrop of the group is generally low-lying, is restricted to the Cheshire, Stafford and Needwood basins, and covers about one third of the district (Figure 29).

The base of the group appears to be conformable with the top of the Sherwood Sandstone Group: the junction is transitional and diachronous in most places, lying within an upward-fining sequence at the top of the Helsby, Bromsgrove and Kibblestone formations. The base is generally drawn where siltstones and mudstones predominate over sandstones. However, above the lithologically distinctive Kibblestone Formation the lowest part of the group locally comprises purple to dark brown very fine-grained sandstones; these are typical of several in the Mercia Mudstone Group but differ markedly from sandstones in the Kibblestone Formation, being micaceous, flaggy, and considerably finer-grained. The base of the group here is placed at the bottom of the lowest metre-thick sandstone of this type. The top of the group is taken at the base of the Penarth Group. In the Cheshire Basin, the only part of the district where the full sequence is preserved, the group is up to 1500 m thick.

The basal formations of the Mercia Mudstone Group consist mostly of siltstone: the Tarporley Siltstone Formation in the Cheshire Basin, the Maer Formation in the Stafford Basin and the Denstone Formation in the Needwood Basin. Within the Cheshire Basin in the district the former is overlain sequentially by the Bollin, Northwich, Byley, Wych, Wilkesley, Brooks Mill and Blue Anchor formations. All are dominated by mudstone except the Northwich and Wilkesley formations, which consist of halite (Figures 30 and 37).

Four principal lithofacies (A–D) are recognised in the Mercia Mudstone Group.

(A) Interlaminated and interbedded siltstones, mudstones and sandstones typify the Tarporley, Maer and Denstone formations (Figure 30). The siltstones are micaceous and laminated, or interlaminated with mudstones or sandstones; the mudstones are commonly blocky. The sandstones are usually fine to medium grained, well sorted and flaggy, or cross-stratified. Most lithologies are red-brown, though green-grey mottles, often parallel to laminae or beds, are common. Most beds are tabular and laterally extensive, though channel sandstones typically have incised, erosional bases. The upper surfaces of sandstones and coarse siltstones are mostly wave rippled, but commonly current rippled; in section the sandstones are commonly ripple-laminated, and exhibit sporadic rootlets. Most lithologies are interpreted as deposited in fluvial, floodplain environments. The beds are apparently devoid of marine macrofauna or trace fossils (such as those noted at Daresbury, Cheshire, by Ireland et al., 1978, and Pollard, 1981), though some marine influence is indicated by the presence of pseudomorphs after halite. These may represent marine incursions, or possibly recrystallised windborne salt (Holland, 1912). The presence of pseudomorphs, desiccation cracks and rootlets shows that at times the sediments were subaerially exposed.

(B) Interlaminated reddish brown or greenish grey mudstones and dolomitic siltstones are chiefly found in the Bollin, Byley and Blue Anchor formations (Figures 30 and 37). Cut and fill structures, current ripple-laminations, cross-laminations and convolute laminae, in siltstone units usually 20 to 50 mm thick, suggest deposition in shallow water, probably in transient lakes (Klein, 1962). Periodic exposure is indicated by numerous desiccation cracks and halite pseudomorphs in siltstone. Although nodules of anhydrite or gypsum are rare, veins of gypsum are common, indicating an abundance of calcium sulphate in the lake. The generally arid conditions were probably inimical to life, judging by the general lack of macrofossils, despite the presence of mudstones in which they might have been preserved. However, mudstones in the Blue Anchor Formation do contain fish remains (Poole and Whiteman, 1966), and ichnofaunas have been documented from the Bollin Formation (Pollard, 1981). Palynomorphs occur in grey beds at several levels in the Bollin and Byley formations and include acritarchs (of marine origin), suggesting a periodic connection with the sea, probably across a broad coastal plain.

(C) Blocky mudstones are typically reddish brown and lack lamination. They commonly contain nodules of anhydrite or gypsum up to 0.3 m across, which are thought to have grown within the soft sediment. Nodules are likely to have grown in slightly elevated areas where gypsiferous solutions were drawn through the ground by evaporitic pumping induced by temperature variations (Shearman, 1970). The mudstones resemble loess and are probably largely of aeolian origin (Taylor et al., 1963). Arthurton (1980) suggested that relict structures in the Byley Mudstone Formation (Figures 30 and 37) resemble those found in some present-day semi-arid areas where a blistered terrain has formed by the displacive growth of halite or gypsum crystals within the sediments; when wet these uneven surfaces act as traps for windblown dust. This largely aeolian origin proposed for parts of the Byley Formation, where there is an alternation of blocky and laminated facies, is also believed to apply to thicker sequences of the blocky facies, such as those which comprise most of the Wych and Brooks Mill formations (Figures 30 and 37).

(D) Halite is virtually restricted to the Northwich and Wilkesley formations (Figures 30 and 37), where it is interbedded with mudstone. The beds of halite vary from nearly pure salt to 'haselgebirge', a mudstone rock with halite crystals in varying proportions. Arthurton (1973), in a study of the Northwich Halite Formation some 15 km north of the present district, concluded that the halite formed in shallow brine pools, both at the water surface and on the bed of the pool. In the Northwich Halite at Meadowbank Mine, Winsford, some 10 km north-west of the present district, Tucker (1981) and Tucker and Tucker (1981) described salt polygons, between 6 and 15 m across, which are bounded by sediment-filled fissures about 1.5 m deep. Tucker believed that the polygons were caused by diurnal temperature variations in emergent conditions, which caused halite crystals to grow along fissures and also caused buckling of adjacent sediments. The source of the halite is thought to be chiefly sea water, judging by the Br, Sr, K and Mg content of the analysed material (Thompson, 1989, quoting the unpublished work of Tucker). It is likely that there were frequent incursions of sea water across the plains in which the brine pools were situated.

Biostratigraphical information on this group within the Stoke-on-Trent district has been obtained from palynological studies there and in the adjacent Chester (Sheet 109), Macclesfield (110), Nantwich (122), Ashbourne (124) and Wem (138) districts.

The Mercia Mudstone Group present in the west and north-west of the district comprises the full Cheshire Basin succession (Figure 37). Within the district palynological studies have been carried out on 43 core samples from the Byley Mudstone Formation and on 11 core samples and five cuttings samples from the succeeding Wych Mudstone Formation proved in Crewe Heat Flow Borehole. Palynological information on those and the remaining formations, with the exception of the Wilkesley Halite Formation, has also been obtained from sections in adjacent districts.

On the basis of palynological evidence from adjacent districts (Warrington, 1970a; 1970b; Earp and Taylor, 1986; BGS ms. records) the Anisian sequence was recognised as extending upwards from the Helsby Sandstone Formation into the Mercia Mudstone Group above the Northwich Halite Formation, with the Anisian–Ladinian boundary occurring (Figure 30) around the level of the Byley Mudstone/Wych Mudstone contact (Wilson, 1993; Benton et al., 1994); in Crewe Heat Flow Borehole that formation boundary is at 157 m depth. Eighteen samples from the Byley Mudstone between 233.9 and 298.5 m in that borehole proved productive but those from higher in the borehole proved largely barren, with productive material being recovered only from the Byley Mudstone at 227.70 to 227.95, 208.55 to 209.30, 167.7 to 167.9, 163.65 to 167.80 and 158.32 to 158.42 m, and from the Wych Mudstone at 117.0 to 117.1, 105.8 to 106.0, 65.75 and 60.3 m. In Crewe Heat Flow Borehole *Tsugaepollenites oriens* is recorded within the Byley Mudstone below 209 m and *Retisulcites perforatus* occurs in samples from the upper part of that formation, above 164 m, and in the Wych Mudstone at 105.9 to 106.0 m; possible specimens of *Echinitosporites iliacoides* occur in the Wych Mudstone at 105.8 m. These occurrences indicate, by comparison with records from Triassic successions elsewhere in Europe (Visscher and Brugman, 1981; Brugman, 1986; Brugman et al., 1988), that the Anisian–Ladinian boundary occurs within the upper part of the Byley Mudstone Formation, between 164 and 209 m in Crewe Heat Flow Borehole, and thus at a slightly lower level than suggested previously (see above) on the basis of fewer samples from scattered sections in the neighbouring districts.

No palynological samples have been examined from the succeeding Wilkesley Halite Formation. However, the lower part of the overlying Brooks Mill Mudstone Formation in boreholes [5842 4134, 5816 4330] some

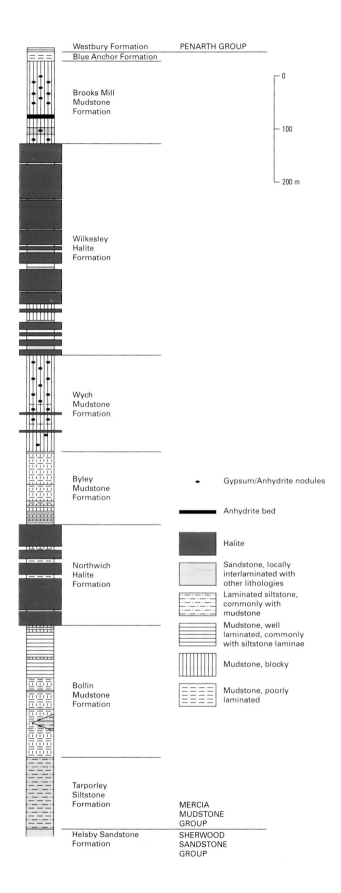

Figure 37 The Mercia Mudstone Group in the Cheshire Basin, based chiefly on the Wilkesley Borehole.

9 km farther west, in the Nantwich district, has yielded sparse miospore assemblages; these include *Ovalipollis pseudoalatus* and *Vallasporites ignacii* and are assessed as Carnian (early Late Triassic) in age. In the absence of biostratigraphical evidence from the higher part of the Wych Mudstone, and from the Wilkesley Halite the position of the Ladinian–Carnian boundary cannot be closely determined. The age of the Wilkesley Halite may be late Ladinian to Carnian, or entirely Carnian, on the basis of indirect palynological evidence from the contiguous mudstone formations.

A borehole [6721 3917] close to the eastern margin of the Nantwich district proved the Blue Anchor Formation, at the top of the Mercia Mudstone Group; palynology samples from that formation yielded a few miospores, possibly including the pollen *Gliscopollis*, and suggestive of a Late Triassic age.

In the south-central part of the district the Mercia Mudstone Group comprises the lower part of the succession preserved in the Stafford Basin. Samples from the basal part of the Maer Formation at Maer [789 382] were examined for palynomorphs and proved to be productive but others from higher beds in the formation at Maer [7814 3782] and Ashley [765 369] proved barren. The assemblages from Maer comprise miospore and organic-walled microplankton associations comparable with those recorded from the Denstone Formation in the Ashbourne district (Warrington, *in* Charsley, 1982) and include the following:

Miospores:

Accinctisporites radiatus, Alisporites cf. *circulicorpus, A. grau-vogeli, Angustisulcites gorpii, A. grandis, A. klausii, Apicu-latasporites plicatus, Aratrisporites rotundus, A. saturni, Calamospora* sp., *Colpectopollis ellipsoideus, Cycadopites coxii, C. subgranulosus, C. trusheimii, Cyclogranisporites* sp., *Cyclotriletes microgranifer, C. oligogranifer, Illinites chitonoides, I.* cf. *kosankei, Lunatisporites* sp., *Microreticulatisporites* sp., *Perotrilites minor, Protodiploxypinus fastidiosus, P. sittleri, Punctatisporites triassicus, Rugulatisporites mesozoicus?, Scabratisporites scabratus, Spinotriletes echinoides, Stellapollen-ites thiergartii, Striatoabieites balmei, Sulcatisporites kraeuseli, Tsugaepollenites oriens, Triadispora crassa, T. falcata, T. plicata?, Verrucosisporites applanatus, V. jenensis, V. pseudo-morulae, V. thuringiacus, Voltziaceaesporites heteromorpha.*

Organic-walled microplankton:
?Tasmanites sp., *Veryhachium reductum.*

The presence of *Angustisulcites gorpii, Perotrilites minor* and *Stellapollenites thiergartii* is indicative of an Anisian (early Mid-Triassic) age for these associations. The organic-walled microplankton, represented by acritarchs (*Very-hachium reductum*) and tasmanitid algae, reflect a marine influence on the depositional environment. Younger beds, including a halite-bearing unit, are preserved farther south in the Stafford Basin but biostratigraphical information is not yet available from those beds; the halite-bearing sequence is, however, regarded as a correlative of that present in the Needwood Basin, and of the Wilkesley Halite Formation in the Cheshire Basin (Warrington et al., 1980).

Westbury Formation   PENARTH GROUP
Blue Anchor Formation

Brooks Mill Mudstone Formation

Wilkesley Halite Formation

Wych Mudstone Formation

Byley Mudstone Formation

Northwich Halite Formation

Bollin Mudstone Formation

Tarporley Siltstone Formation   MERCIA MUDSTONE GROUP

Helsby Sandstone Formation   SHERWOOD SANDSTONE GROUP

0
100
200 m

- Gypsum/Anhydrite nodules

Anhydrite bed

Halite

Sandstone, locally interlaminated with other lithologies

Laminated siltstone, commonly with mudstone

Mudstone, well laminated, commonly with siltstone laminae

Mudstone, blocky

Mudstone, poorly laminated

In the south-east of the district the Mercia Mudstone Group comprises the lower part of the succession preserved in the Needwood Basin. Samples from the basal Denstone Formation exposed in a railway cutting near Caverswall [9565 4200 to 9578 4200] and from higher levels in that formation at Caverswall [9422 4156] and Meir airfield [939 409] were examined for palynomorphs but proved barren. Farther east, in the adjacent Ashbourne (124) district, assemblages of miospores (spores and pollen of land plants) have been recovered from this formation (Warrington, *in* Charsley, 1982; Chisholm et al., 1988). These assemblages include *Angustisulcites gorpii* and *Stellapollenites thiergartii*, an association which, by comparison with those known from independently dated Triassic successions elsewhere in Europe (Visscher and Brugman, 1981; Brugman, 1986) is indicative of an Anisian (early Mid-Triassic) age. Miospore assemblages from succeeding unnamed beds in the Mercia Mudstone Group of the Ashbourne district (Warrington, *in* Charsley, 1982) include *S. thiergartii* and *Perotrilites minor*, an association which also indicates an Anisian age. In these higher assemblages the miospores are associated with acritarchs and tasmanitid algae, indicating a marine influence on the depositional environment. The palynological results from the Ashbourne district indicate that the Denstone Formation and the succeeding (undivided) Mercia Mudstone Group beds in the south-east part of the Stoke-on-Trent district are of Anisian (early Mid-Triassic) age. Younger beds are preserved farther east in the Needwood Basin (Stevenson and Mitchell, 1955), where miospores have been recovered in Bagot's Park Borehole [SK 089 269] at a depth of 188.37 m, 27.53 m below the top of a halite-bearing unit (Warrington, 1970a; *in* Charsley, 1982). Re-examination of this assemblage has shown the presence of *Camerosporites secatus*, *Duplicisporites verrucosus*, *Echinitosporites iliacoides*, *Ovalipollis pseudoalatus* and *Retisulcites perforatus*, an association which, by comparison with those known from independently dated Triassic successions elsewhere in Europe (Mostler and Scheuring; Visscher and Brugman, 1981; van der Eem, 1983) is indicative of a Ladinian (late Mid-Triassic) to early Carnian (early Late Triassic) age. The Mercia Mudstone Group in the Needwood Basin is succeeded by the Penarth Group (Stevenson and Mitchell, 1955) which has yielded macrofaunas and a microflora indicative of a Rhaetian (late Late Triassic) age.

Cheshire Basin

In the Cheshire Basin, where the Mercia Mudstone Group is almost totally concealed by glacial deposits, our knowledge of the group is largely gained from boreholes and the distribution of salt solution subsidence features. The group comprises eight formations (Figure 37), all but the lowest of which are believed to have been deposited over the entire basin.

TARPORLEY SILTSTONE FORMATION

The Tarporley Siltstone Formation (Warrington et al., 1980) was formerly known as the 'Keuper Waterstones'

(Hull, 1869). It typically consists of interlaminated siltstones and mudstones, with beds of sandstone about 1 m thick (lithofacies A). The outcrop of the formation in the district is cut out by the Wem Fault; the best exposures occur in the River Dane, east of Congleton [900 656 to 892 657], some 10 km north of the district boundary (Wilson, 1993).

The base of the formation is defined at the base of the sequence where siltstones and mudstones predominate over sandstones; the latter belong to the Helsby Sandstone Formation. This junction is complex, being both transitional and diachronous, and locally there is a lateral passage between the two formations. For instance, near Norton in Hales, siltstones, mudstones and thin sandstones (like those of the Tarporley Siltstone) alternate with thicker sandstones of Helsby Sandstone type; here the finer-grained lithologies are included within the latter formation. This sequence passes laterally into one where sandstones are subordinate, and this is assigned to the Tarporley Siltstone Formation. Lateral passage from a sequence dominated by lithofacies A into one dominated by thick sandstones, such as this, is not restricted to the base of the formation; for instance, west of the present district, near the Wilkesley Borehole, the upper part of the formation passes into a sandstone of aeolian origin, the Malpas Sandstone (Stephens, 1961; Poole and Whiteman, 1966), which is succeeded by similar facies to those that preceded it. The top of the Tarporley Siltstone Formation is defined by the base of the Bollin Mudstone Formation. Locally the Tarporley Siltstone is absent and the Bollin Mudstone forms the basal formation of the Mercia Mudstone Group (see below). The thickness of the Tarporley Siltstone in the district is typically 100 to 150 m.

BOLLIN MUDSTONE FORMATION

The Bollin Mudstone Formation (Wilson, 1993) comprises dominantly reddish brown and greenish grey mudstones (lithofacies B), which were formerly known as the 'Lower Keuper Marl' (Hull, 1869; Pugh, 1960). The outcrop of the Bollin Mudstone in the present district is restricted by the Wem Fault.

The base of the Bollin Mudstone is defined at the base of the sequence in which mudstones predominate over siltstones and sandstones, usually of the Tarporley Siltstone Formation; this junction is diachronous, and the latter is locally absent. For instance, a borehole at Town House Farm, 4 km west of Kidsgrove, showed that the basal 31.8 m of the Mercia Mudstone Group consists almost totally of mudstones. Although these are overlain by 16.4 m of 'shaly rock' (probable Tarporley Siltstone facies), the latter are so attenuated that they have been mapped as part of the Bollin Mudstone. Thus in this vicinity, the Bollin Mudstone Formation, which is 79 m thick, forms the lowest unit of the Mercia Mudstone Group. This usage is similar to that in the Stockport area, where the 'Upper Keuper Sandstone', of Tarporley Siltstone facies, was included in the 'Lower Keuper Marl' (Bollin Mudstone) (Taylor et al., 1963). The top of the formation is defined by the conformable base of the Northwich Halite Formation. The greatest thickness of

Bollin Mudstone in the district is in the north-west, where it probably approaches 300 m (Wilson, 1993, fig. 5).

The best local record of the formation occurs in the Wilkesley Borehole, some 4 km west of the district, where it is 257 m thick (Figure 37). The grey-green and red laminated beds, which dominate the upper part of the formation, show desiccation cracks, current ripple lamination and pseudomorphs after halite. The lower part of the formation is dominated by reddish brown mudstones with greenish grey layers; anhydrite nodules and desiccation cracks occur at a few levels.

NORTHWICH HALITE FORMATION

The Northwich Halite Formation (Warrington et al., 1980) comprises halite interbedded with subordinate mudstone (lithofacies D). Formerly known as the 'Lower Keuper Saliferous Beds' (Pugh, 1960), it can be traced laterally for more than 30 km in adjacent districts (Evans et al., 1968, fig. 12). The Northwich Halite is believed to have been faulted out at surface by the Wem Fault, except near the northern fringe of the district near Alsager and probably in a small triangular outcrop [747 470] south of Betley. The form of these outcrops is imprecisely known, due to a dearth of boreholes and a lack of solution subsidence features. Solution by groundwater has affected the Northwich Halite, however, and it does not occur at or near surface, or immediately below drift. Instead, it is covered by a breccia of collapsed mudstones, similar to that developed above the Wilkesley Halite (see below).

The base of the Northwich Halite is defined at the base of the lowest bed of halite greater than 2 m thick (Wilson, 1993), and the top by the base of the overlying Byley Mudstone Formation. The formation is 283 m thick in the Byley Borehole, 15 km north of Crewe, but thins rapidly eastwards; it is likely that, near Alsager, the formation is about 110 m thick.

The only known proving of the formation in the district was in Alsager Waterworks Borehole [7938 5517]:

		Depth m
DRIFT	to	19.50
NORTHWICH HALITE		
Mudstone with 0.3 m bed of halite at base (probable collapsed beds)	to	68.60
Mudstone, red	to	94.50
Mudstone and halite	to	97.54
Mudstone	to	107.00
Halite	to	109.40
Mudstone, red	to	110.40

About 1 km farther north, at the former Lawton salt works [805 572] in the adjoining Macclesfield district, there was another incomplete penetration of the Northwich Halite, totalling 50 m with a high proportion of mudstone in the upper strata.

BYLEY MUDSTONE FORMATION

The Byley Mudstone Formation (Wilson, 1993) is dominated by a characteristic alternation of blocky, reddish brown mudstones (lithofacies C) and laminated, greenish grey or reddish brown mudstones (lithofacies B). It equates with the lower part of the 'Middle Keuper Marl' (Pugh, 1960). Although it has an extensive outcrop it is, because of thick drift cover, virtually unexposed.

The base of the formation is commonly sharply defined by the bottom of the sequence where mudstones predominate over halite. The top of the formation is defined by the top of the highest unit of laminated greenish grey mudstones below the Wych Mudstone Formation (Wilson, 1993).

In the Crewe Heat Flow Borehole, though 143 m of Byley Mudstone were penetrated, the borehole terminated some 20 m above the estimated position of the base, giving a total formational thickness of some 163 m (Wilson, 1993, fig. 8). The sequence resembles that described by Arthurton (1980) from boreholes in the Middlewich area where there is a comparable close alternation of laminated and blocky beds, with a tendency for blocky mudstones to become dominant in the uppermost strata. The laminated strata (lithofacies B) contain cut-and-fill features, load casts and many desiccation cracks picked out by gypsum veins. The red-brown blocky mudstones (lithofacies C) have, in places, a wispy, discontinuous internal fabric and contain greenish grey blotches, particularly near green-grey mudstone beds.

Other records of the Byley Mudstone in the district are very tentative. They include the Lower Thornhill Borehole, 1 km north-west of Madeley, which is said to have penetrated 152.4 m of Mercia Mudstone without any halite, the Onneley Borehole which proved, below drift, 19.8 m of mudstone with thin sandstones, and a small exposure of weathered mudstone, below drift, in nearby Beech Wood [7524 4427].

WYCH MUDSTONE FORMATION

The Wych Mudstone Formation (Wilson, 1993) comprises dominantly blocky, reddish brown mudstones with sporadic greenish grey blotches (lithofacies C). These contain nodules of anhydrite and gypsum up to 0.2 m in diameter, and numerous gypsum veins. The formation equates with the upper part of the 'Middle Keuper Marl' (Pugh, 1960). The outcrop is extensive, though almost totally mantled in drift. The boundaries of the formation are defined by the top of the Byley Mudstone Formation (see above) and the base of the Wilkesley Halite Formation (see below). The full thickness of the formation is likely to be up to about 196 m, as proved in the Wilkesley Borehole, 4.4 km west of the present district (Figure 37).

In the present district, the Crewe Heat Flow Borehole penetrated the upper 151 m of the formation, of which 106 m were cored (Wilson, 1993, fig. 12). Interlaminated mudstones and siltstones occur at several levels; these are chiefly greenish grey in colour, and are associated with desiccation cracks. Pale brown halites at two horizons (Figure 37) are concordant with bedding, and a faint banding is present in the lower bed. Interestingly, there are gaps in the core recovery in Wilkesley Borehole, at levels closely corresponding to the halites in the Crewe Heat Flow Borehole. The halites could be laterally extensive, because there are parallels as far away as Blackpool, 90 km distant, where up to four very thin beds of halite or halite with anhydrite occur at comparable levels in the equivalent Breckells Mudstones Formation (Wilson and Evans, 1990, fig. 13).

Other provings of the formation in the district include 62 m of mudstone with gypsum nodules in Crewe Gates No.1 Borehole, and 35.05 m of mudstone below drift in a non-cored borehole at Checkley New Farm. Near Betley, a stream section [7565 4796] in Wrench's Wood reveals reddish brown clay with grey siltstone fragments, likely to be high in the Wych Mudstone. Good surface exposure occurs in Wych Brook [4887 4469 to 4931 4530], south of Malpas, 19 km west of the present district.

WILKESLEY HALITE FORMATION

The Wilkesley Halite Formation (Warrington et al., 1980) consists of interbedded halite and mudstone (lithofacies D), formerly known as the 'Upper Keuper Saliferous Beds' (Pugh, 1960). The formation is likely to have a widespread outcrop in the area between Weston, Wybunbury and Buerton, to judge by the distribution of probable salt solution subsidence hollows (Figure 38) and the evidence of reflection seismic surveys. Other surface indications of the Wilkesley Halite include provings of brine within 1 km of the western boundary of the district [6647 4673; 6678 4821], and former brine springs near Balterley Heath [749 501] (D B Thompson, personal communication). The halite is not preserved near the surface, nor directly below drift cover, because it has been removed through solution. The halite is likely to be overlain by a collapse breccia of mudstones, described under salt subsidence below.

The base of the formation is placed at the base of the lowest thick halite bed (Wilson, 1993), and the top above the highest halite bed below the Brooks Mill Mudstone Formation. The full thickness of the Wilkesley Halite has only been proved in the Wilkesley Borehole, in the adjoining Nantwich district, where it is 404.5 m thick. The character of the formation is best illustrated in this borehole (Figure 37), where halites up to 26 m thick are interspersed with numerous partings of dominantly reddish brown mudstone, up to 12 m thick, which contain crystals and veins of halite at many levels (Wilson, 1993, fig. 15). In contrast to the Northwich Halite, the Wilkesley Halite also contains sandstones, some of which are over 1.0 m thick.

SALT SUBSIDENCE

The subsidence mechanisms related to solution of the halites in the Mercia Mudstone Group have been described by Evans et al. (1968) and Wilson (1993). These authors showed that halite does not outcrop in the same way as non-saliferous formations because in near-surface areas it has always been dissolved away. Instead, the intact halite is overlain by a collapse breccia composed of mudstones that originally overlay or were interbedded with the halite; the base of the breccia is known as 'wet rock head' and is usually marked by the presence of brines. The inclusion in the breccias of material from the overlying formation is illustrated in the case of the Wilkesley Halite by a number of boreholes in the district.

Borehole AU16, 3 km south of Buerton (Figure 38), in the south-west of the district proved:

		Depth m
DRIFT	to	26.20
COLLAPSE BEDS		
Mudstone, brecciated, reddish brown	to	46.02
Mudstone, silty in places, brown, with satin spars and small gypsum nodules	to	97.93
Mudstone, green	to	98.50
Mudstone, brown	to	99.19
Sandstone, mottled green and brown	to	99.59
Mudstone, brown and green, with selenite crystals and gypsum nodules	to	103.72
Mudstone, brecciated, brown with green patches, gypsum nodules throughout	to	110.24
Mudstone, brown with blackish green mottling, with selenite crystals and spars	to	114.38
Mudstone, brown, soft and 'washed'	to	116.66
?WILKESLEY HALITE		
Mudstone, brown, with honeycomb of remnant halite	to	116.73

The presence of breccias and selenite in the mudstone-dominated sequence indicates that it does not consist of in-situ strata, but of collapse beds caused by halite solution. The presence of gypsum nodules and a sandstone suggests derivation from the Brooks Mill Mudstone. The 'washed' appearance of the debris at the base of the formation is suggestive of a brine run above the halite.

Nearby, a borehole near Holly Farm penetrated 71.93 m of very hard, dark red mudstone below drift. Whilst no breccias were recorded in this, local seismic survey and subsidence evidence suggest that these mudstones also probably represent collapsed beds over dissolved Wilkesley Halite.

A borehole at Pewit Lane, 2 km west of Bridgemere, proved:

		Depth m
DRIFT	to	6.40
PROBABLE COLLAPSE BEDS		
Mudstone, red	to	11.88
Sandstone	to	12.80
Mudstone, red	to	14.32

It seems likely that the sandstone is one of those recorded in the Brooks Mill Mudstone, between 22 and 34 m above the Wilkesley Halite in the Wilkesley Borehole.

Exposures in the area of probable salt solution collapse are restricted to the banks of Birchall Brook, near Buerton, where three small exposures were noted. The best of these [6882 4476] was in 1.8 m of reddish brown and greenish grey mudstone, which was faulted, probably due to collapse by salt solution.

Halite subsidence usually has some surface effect in the form of linear subsidence hollows and crater-like depressions of circular or oval shape (Figure 38). Some

Figure 38   Distribution of the main subsidence features resulting from solution of the Wilkesley Halite.

crater subsidences are problematic because they resemble hollows left by the melting of ice masses contained within glacial deposits.

One of the most convincing crater subsidence hollows is about 15 m deep and of ovoid shape, measuring some 600 by 300 m [690 445]. The nearby valley of Birchall Brook runs directly past the end of the depression, which is only slightly above valley level and filled with at least 1.2 m of peat. It would appear that the subsidence occurred after Birchall Brook had begun to cut its valley; a second and smaller depression is seen on the north side of the stream.

Linear subsidence hollows are best developed in the Buerton area (Figure 38) where they reach 8 m in depth. Two areas of linear subsidence appear to terminate against the Bridgemere Fault.

Several hollows appear to have had a recent history of subsidence:

Part of a closed hollow 260 m in length at Wheel Green [7140 4480] is flooded, and newly submerged bushes were noted in 1964.

A hollow, close to the large group of subsidence hollows near Doddington Park, contains a peaty flat in which fencing posts have become flooded in places [7015 4613].

After a sand pit, south-east of Wynbunbury [708 494] and near a 1.5 km-long linear subsidence, became disused and flooded, it was pumped out, reputedly causing a subsidence 2.7 m deep in the pit floor.

St Chad's Church, Wybunbury was located on top of a sandy bluff overlooking a major probable subsidence hollow. The building has long suffered instability and had to be partly demolished a few years ago, leaving only the tower, held together by metal rods and mortar between the sundered stonework.

Active subsidence in fields and a sports ground was reported on Demesne Farm, Doddington. In 1966, this 20-m wide depression [7043 4773] was said to have been actively subsiding over a period of 50 years.

A cricket ground near Betley [7362 5139] has suffered subsidence in the last decade.

Small scarps, possibly related to salt solution, have been reported from the valley of Basford Brook [731 513] near Weston (D B Thompson, personal communication).

BROOKS MILL MUDSTONE FORMATION

The Brooks Mill Mudstone Formation (Wilson, 1993) is dominated by blocky, red to chocolate-brown mudstones with a few greenish horizons and blotches (lithofacies C), but also includes anhydrite beds and nodules and thin sandstones (Figure 37). It was previously known as the 'Upper Keuper Marl' (Pugh, 1960). The outcrops lie near the western edge of the district, near Rooms Farm [679 391] and Highfields Farm [675 410], and are almost wholly covered in drift.

The base of the formation is placed at the top of the highest halite bed in the Wilkesley Halite Formation (Wilson, 1993), and the top at the red-brown to greenish grey colour change in mudstones at the base of the overlying Blue Anchor Formation. The thickness of the Brooks Mill Mudstone in the Wilkesley Borehole, the only complete penetration of the formation, is 161.30 m.

The lower part of the formation is best known from the Wilkesley Borehole and consists of 52 m of blocky, reddish brown mudstones with a few anhydrite nodules. Four sandstones up to 1.05 m thick, with local desiccation cracks, occur. The uppermost of these lies at a similar level in the sequence to the Hollygate Sandstone of the Nottingham succession and the Coolmaghera Sandstone of Northern Ireland (Wilson, 1993). These beds are overlain by a distinctive sequence, 6.4 m thick, containing anhydrites up to 0.9 m thick and mudstone interbeds up to 0.3 m thick.

The remainder of the formation, above these anhydrites, is 103.6 m thick in the Wilkesley Borehole and comprises reddish brown, blocky mudstones with nodules and a few thin beds of anhydrite. The anhydrite nodules in boreholes AU15 and AU17 (Wilson, 1993, fig. 16), drilled less than 500 m from the western boundary of the district, are up to 0.3 m in diameter and are commonly interconnected by gypsum veins. The uppermost 18 m of the formation contains greenish grey interbeds, and, in the Wilkesley Borehole, a sandstone and a limestone (Wilson, 1993, fig. 16).

The formation is poorly exposed in Howbeck Brook [6869 4934], near Walgherton, where 0.3 m of reddish brown, poorly laminated mudstone with reduction spots may be seen. Better exposures [6296 4370; 6315 4384] occur beside Barnett Brook near Brooks Mill, 4 km west of the present district.

BLUE ANCHOR FORMATION

The Blue Anchor Formation (Warrington et al., 1980) consists of green, greenish grey and grey calcareous mudstones (lithofacies B), and was formerly known as the 'Tea Green Marl' (Etheridge, 1865) (Figures 30 and 37). The outcrop of the formation in the district is entirely drift-covered and restricted to two small areas near the western boundary of the district [676 397; 674 419].

The base of the formation is placed at the conformable junction between the underlying red-brown mudstones of the Brooks Mill Formation and the greenish grey mudstones of this formation (Wilson, 1993), and the top at the conformable base of the Penarth Group.

The formation has been penetrated in boreholes AU15, AU17 (Wilson, 1993, fig. 16), and in the Plattlane and Wilkesley boreholes further to the west. In the two last boreholes, it is 16.7 and 15.1 m thick respectively and consists of green calcareous mudstones and siltstones, with sandy horizons containing millet-seed sand. Desiccation cracks, plant debris, worm tubes, fish remains and the small crustacean *Euestheria minuta* are preserved in these. Sporadic exposures of the formation occur between 100 to 200 m south of Brooks Mill [633 436], in the Nantwich district (Poole and Whiteman, 1966).

Stafford Basin

In the Stafford Basin, only the lowest formation of the Mercia Mudstone Group, the Maer Formation, is present in the Stoke-on-Trent district. The overlying sequence, in the adjoining Stafford district, includes the Stafford

Halite Formation which is equivalent to the Wilkesley Halite of the Cheshire Basin. The outcrop of the Maer Formation is mantled by thin, patchy drift.

MAER FORMATION

The Maer Formation is here newly named after the village of Maer [792 382] at the northern limit of its outcrop. It consists of mainly red-brown siltstones and mudstones, interbedded with sandstones (lithofacies A). The formation is lithologically similar to many units formerly referred to as the 'Waterstones' of the Midlands (Hull, 1869; Gibson, 1925). These are diachronous (Warrington, 1970a), so chronostratigaphical equivalence with similar units, such as the Tarporley and Denstone formations of the Cheshire and Needwood basins respectively, may be only partial.

The base of the formation is taken where the dominantly arenaceous sequence of the Sherwood Sandstone Group passes into the mainly argillaceous sequence of the Mercia Mudstone Group, and is exposed in the type section (see below). It is placed at the base of the lowest 1 m thick siltstone or mudstone of the dominantly argillaceous sequence. Where it overlies the Kibblestone Formation the base of the formation is locally defined by sandstone character (see below). The top of the formation does not occur in this district, though is marked by an increased proportion of mudstones, with fewer siltstones and no sandstones, as at 234.7 m depth in Coton Field No. 1 Borehole near Stafford, 12 km south of the district. The thickness of the formation in this district is probably some 50 m or more, but no borehole proves this. Farther south, the formation is about 60 m thick in Ranton No. 1 Borehole, 7 km west of Stafford.

Exposures in the area between Ashley and Chapel Chorlton are common, though they tend to be small. A few exposures occur near Beech. The basal stratotype occurs in a cutting north of Maer [7909 3803], through which the footpath to Ashley passes.

	Thickness m
MAER FORMATION	
Sandstone, fine-grained, yellow-brown with sporadic mudflakes	1.2
Gap	1.2
Mudstone, silty, dull purple-brown, micaceous, with siltstone laminae	3.6
Siltstone, dull purple-red, gradational base	1.2
Siltstone, sandy, brown	1.8
Siltstone, black-brown, poorly exposed	4.2
Siltstone, dark brown, purple and yellow	1.2
Siltstone, pale yellow, with dark carbonaceous spots (?rootlets)	0.9
Siltstone, grey and violet	0.1
BROMSGROVE SANDSTONE FORMATION	
Sandstone, fine-grained, friable	0.1
Sandstone, fine- to medium-grained, pale yellow	1.1
Sandstone, fine- to medium-grained, micaceous, greenish grey, friable, with laminae of grey-green siltstone and partings of brown mudstone	1.8
Sandstone, fine-grained, micaceous, pale yellow, cross-stratified	6.1

The base of the formation, where it overlies the Kibblestone Formation, is exposed in a stream section [921 353] east of Hayes House, in the adjoining Stafford district (Figure 36). Here, red-brown and purple-brown very fine-grained sandstones, typical of the Maer Formation, overlie pale medium-grained sandstones of the Kibblestone Formation. The junction between the two formations is probably transitional as pale, medium-grained sandstones, more typical of the Kibblestone Formation, occur within the basal 3 m of the Maer Formation.

Needwood Basin

The rocks of the Mercia Mudstone Group in the Needwood Basin in the Stoke-on-Trent district are subdivided into the Denstone Formation, below, and undivided measures, above. These strata are partly drift-covered.

DENSTONE FORMATION

This formation (Charsley, 1982) comprises micaceous siltstones, with mudstones and sandstones (lithofacies A), and is equivalent to the 'Waterstones' of Hull (1869). It forms the lower ground near Blythe Bridge, to the east of the Hawksmoor Formation crop, and the only exposures comprise degraded pits dug for sandstone. The soils overlying the formation tend to be poorly drained and micaceous.

The base of the formation is not exposed though it was penetrated in boreholes (Figure 33) at Meir pumping station, and at Mossgate, near Fulford. The formation is not fully preserved in the district, but 75 m of it were recorded in the Meir pumping station borehole.

MERCIA MUDSTONE GROUP (UNDIVIDED)

The undifferentiated, higher part of the Mercia Mudstone Group comprises mainly mudstone of lithofacies B, and was previously assigned to the 'Lower Keuper Marl' (Hull, 1869). Parts of the sequence occur in three fault-bounded areas: in the graben east of Caverswall, in a downfaulted tract east of Fulford, and in a strip south of the Swynnerton Fault. The base of this sequence is not exposed, and its preserved thickness is unknown.

PENARTH GROUP

The Penarth Group (Warrington et al., 1980), formerly the 'Rhaetic', is dominated by dark grey mudstones of marine origin and marks a marine transgression. In the Stoke-on-Trent district the group occurs only in the Cheshire Basin, where it forms a very small outcrop, concealed by drift, at the extreme western edge of the district [676 398].

The base of the group is placed where the dark grey mudstones of the Westbury Formation rest on the green-grey mudstones of the Blue Anchor Formation, whilst the top is defined by the base of the Lias Group.

In the Nantwich district the group comprises two formations, the Westbury Formation and the overlying Lilstock Formation (Warrington et al., 1980). The former consists of dark grey fissile mudstones with the bivalves *Rhaetavicula contorta* and *Protocardia rhaetica*, and fish remains. The Lilstock Formation comprises greyish green calcareous mudstones with *Euestheria minuta* and fish fragments (Poole and Whiteman, 1966).

The thickness of the group in the Wilkesley and Plattlane boreholes, in the adjoining Nantwich district, is about 13.5 m (Westbury Formation: c.7.8 m, Lilstock Formation: c. 5.7 m). Probably the thickness of the group preserved in this district is less.

No biostratigraphical information has been obtained from the small outcrops of the Penarth Group in the district. Micro- and macrofaunas have been recorded from sections farther west, in the Plattlane and Wilkesley boreholes (Anderson, 1964; Ivimey-Cook, *in* Poole and Whiteman, 1966). The palynology of those sections has not been studied, but samples from the group in a borehole [5816 4330] some 9 km west of the Stoke-on-Trent district yielded palynomorph assemblages which include the following:

Miospores:

Acanthotriletes varius, Chasmatosporites magnolioides, Cingulizonates rhaeticus, Classopollis torosus, Convolutispora microrugulata, Geopollis zwolinskai, Gliscopollis meyeriana, Lunatisporites rhaeticus, Microreticulatisporites fuscus, Ovalipollis pseudoalatus, Quadraeculina anellaeformis, Rhaetipollis germanicus, Ricciisporites tuberculatus, Vesicaspora fuscus.

Organic-walled microplankton:

Cymatiosphaera polypartita, Dapcodinium priscum, Micrhystridium lymense var. *gliscum, Rhaetogonyaulax rhaetica.*

The presence of taxa such as *Rhaetavicula contorta* in the macrofaunas, and of the miospores *Quadraeculina anellaeformis, Rhaetipollis germanicus* and *Ricciisporites tuberculatus* and the dinoflagellate cyst *Rhaetogonyaulax rhaetica* in the palynomorph associations, indicates that the Penarth Group is of Rhaetian (late Late Triassic) age.

SIX

Palaeogene

Intrusive dykes of olivine dolerite and related rock types occur in the southern part of the district between Keele University and Swynnerton. Their composition and age suggest that these are the most south-easterly intrusions of the British Tertiary Igneous Province (Musset et al., 1988).

BUTTERTON-SWYNNERTON DYKES

The Butterton–Swynnerton dykes (Garner, 1844; Sowerbutts, 1988) extend from Keele University in the north, through Swynnerton, to Norton Bridge in the Stafford district to the south (Geological Survey of Great Britain, 1857; Cherry, 1877; Kirton and Donato, 1985). On a regional scale this swarm of dykes is remarkably straight and occurs within a corridor over 15 km long but only 750 m wide (Figure 39). In places the swarm includes up to ten dykes, which are broadly parallel to each other. The dykes were first identified in the early 1840s, probably by James Kirkby and Charles Darwin independently, in quarries and natural exposures in the Butterton–Hanchurch area (Kirkby, 1894). From exposures such as that at the southern end of Church Wood [8332 4197] (figured in Gibson, 1905, 1925; Gibson and Wedd 1905; Cope, 1966), single or multiple intrusions were traced by topographic feature and surface debris (Kirkby, 1894; Gibson, 1905), post-holing (Cope, 1966), trenching (Kirkby, 1894), occurrence in boreholes (Cope, 1966), and provings in motorway cuttings (Exley, 1970). However, it is the use of geophysical methods which has allowed the dykes to be mapped in some detail over their length. In one of the first such surveys in the UK, McLintock and Phemister (1928) describe the results of a gravity survey using an Eötvös torsion balance over the exposed dykes. More important in mapping have been magnetic surveys (Hallimond, 1929; Sowerbutts, 1987, 1988). Sowerbutts' detailed mapping of the dykes (reproduced on BGS 1:10 000 scale geological maps) is based on a vertical gradient magnetic survey using a microcomputer-based data gathering system (Sowerbutts, 1987).

Where the dykes have not been recorded at surface along the linear corridor, they probably occur at depth. This is illustrated in the grounds of Keele University, beyond the northernmost mapped limit of the dykes (Figure 39), where a 3.4 m thickness of dyke rock occurs at 77.1 m depth in Springpool No. 10 Borehole. The dykes vary considerably in width, from about 1 m to over 20 m; for instance in the Yarnfield area, in the adjoining Stafford district, one dyke is over 27 m wide (Kirkby, 1894). The dykes are near-vertical and they mostly hade to the west at up to 15° (Cope, 1966). There is no evidence to suggest that the dykes were intruded along faults, but they locally appear to be offset across faults.

There are examples north of Butterton (Cope, 1966) and at the Swynnerton Fault in the adjoining Stafford district (Whitehead et al., 1927). However, the throw of such faults commonly appears to be small where dykes show displacement (Hampton, in Kirkby, 1894; Sowerbutts, 1988). The tectonic setting of the dykes is reviewed by Thompson and Winchester (1995).

Compositionally the dykes are alkaline basaltic types, though they vary between areas, and between parts of the same intrusion (Allport, 1874; Teall, 1888; Hampton, in Kirkby, 1894; Flett in Gibson, 1905; Scott, 1920, 1925; Cope, 1966; Thompson and Winchester, 1995). They range from ultrabasic rocks, including limburgites and augitites (Scott, 1925) through nepheline basanites (Flett, in Gibson, 1905) to olivine dolerites free of feldspathoids. The rocks are fine grained and contain phenocrysts of olivine (commonly over 2 mm in diameter) and less conspicuous euhedral augite, in a matrix containing augite, plagioclase and magnetite with lesser quantities of spinel, nepheline, analcite, and ilmenite (Flett, in Gibson, 1905; Scott, 1925; Cope, 1966). The state of the olivine is very variable and commonly has been replaced by serpentine, initially around the margins of crystals (Flett in Gibson, 1905). Serpentine is mainly associated with rims of magnetite crystals that pick out the shape of former olivine phenocrysts (Flett in Gibson, 1905; Cope, 1966), which are commonly rounded because of early alteration. The plagioclase comprises labradorite (Hampton, in Kirkby, 1894; Flett, in Gibson, 1905) which in places forms two generations of distinctly different size (Cope, 1966). The larger pyroxene crystals commonly have more basic rims (Scott, 1925). Augite crystals, sometimes occurring in stellate groups, partially overgrow plagioclase crystals, giving a subophitic texture (Flett, in Gibson, 1905). Apart from serpentine (see above), other alteration minerals of the dolerites are biotite (after magnetite) and leucoxene (after ilmenite) (Scott, 1925). The chilled margins of intrusions contain smaller crystals and olivine phenocrysts that include crystals of magnetite and spinel (Flett, in Gibson, 1905). The geochemistry of the dykes, though not detailed here, is described by Thompson and Winchester (1995).

The dykes contain marked cooling joints, which lie perpendicular and parallel to contacts with country rocks; examples were noted by Cope (1966) at Keele University sewage works [8242 4412]. These joints have commonly been the focus for spheroidal weathering, as at Butterton Church Wood quarry (see above) (Gibson, 1905). In many places, parts of dykes, particularly towards the centre, and even whole dykes, are more completely decomposed (Kirkby, 1894; Gibson, 1905, 1925; Scott, 1925) and commonly contain abundant calcite in the form of veins or vug fills (Gibson, 1905).

Country rocks adjoining the dykes are contact-metamorphosed though the extent of metamorphism on either side of the intrusions varies considerably (Gibson, 1905). In the Lymes Road and Springpool No. 10 boreholes (Figure 39) coals of the Coal Measures and Newcastle formation adjacent to the dykes are charred (Gibson, 1905) or in the case of the Great Row seam, completely burnt out, and mudstones of the Coal Measures and Etruria Formation are baked where they were in contact with the dykes (Cope, 1966). At Keele University sewage works (see above) the normally weathered yellowish grey mudstones of the Newcastle formation are brown and mottled in proximity to a dyke (Cope, 1966). Colour changes also occur in metamorphosed sandstones of the Keele formation. The normally red sandstones, where metamorphosed (up to 8 m from the intrusion), are olive-green. They are bleached white within 3 m of the intrusion in Butterton Church Wood quarry where Flett (*in* Gibson, 1905), recognised hornfels textures. The metamorphism of sandstones of the Kidderminster Formation extends only a few centimetres from the margin of the intrusion (Kirkby, 1894), and its effect is limited to discolouration and hardening (Gibson, 1905). Locally, fragments of country rock occur as xenoliths within a dyke (Gibson, 1925), as noted by Kirkby (1894) at Church Wood quarry.

Until recently, determinations of the radiometric age of the dykes were in broad agreement. J A Miller (*in* Cope, 1966) gave an age of 52.4 ± 1.4 million years, Fitch et al. (1969) gave an age of 52 ± 1.4 and Evans (1969) gave an age of 52 ± 2 million years for dolerites, suggesting a broadly Eocene age. This age appeared to be confirmed by Lewis et al. (1992) who, using apatite fission track analysis, determined an age of 52.8 ± 1.1 million years for the dolerites. However, these authors also carried out zircon fission track analysis on samples of country rock from within 2 m of the dyke. These gave a substantially older (Palaeocene) age at 61 ± 1.7 million years. Because of the annealing kinetics of zircons, compared with apatites, Lewis et al. believe that the age from the zircons more closely represents the age of intrusion. It is noted that although the dykes are reported (Dagley, 1969) to have the reversed direction of remanent magnetisation, found commonly for intrusions of this age in the UK, the forms of the magnetic anomalies indicate an overall normal magnetisation.

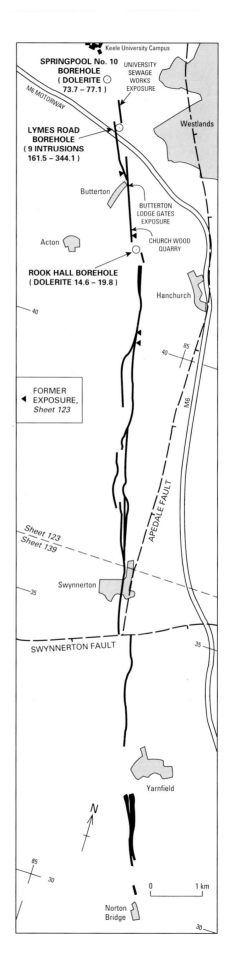

Figure 39 Map showing the location of the Butterton–Swynnerton dykes in the Stoke-on-Trent and Stafford districts.

SEVEN

Quaternary

Superficial (drift) deposits of Quaternary (Pleistocene and Recent) age which blanket much of the district were deposited in glacial, periglacial and temperate climates. The majority of glacial features and deposits relate to the retreat of ice sheets at the end of the late-Devensian glaciation. Discussion of these is followed by an account of the periglacial and postglacial deposits.

For the purposes of description of the drift deposits the district can be divided into two distinct physiographic areas, the Cheshire Plain and the Pennine margins (Figure 40). The Cheshire Plain is mostly underlain by Triassic rocks of the Cheshire Basin, and the Pennine margins are mostly underlain by Carboniferous rocks, plus Triassic rocks of the Stafford and Needwood basins. The Cheshire Plain is an area of subdued topography compared with the Pennine margins, which are typified by higher ground, generally more than 120 m above OD. The divide between the two areas coincides broadly with that between the drainage catchments for the Irish Sea and North Sea. The Trent River system, draining to the North Sea, has been influenced by bedrock structure, but locally it is discordant to this (Hind, 1906; Barke, 1920, 1929; Yates, 1956, 1957).

The greatest thickness of drift occurs on the Cheshire Plain (Figure 40), where there is virtually no bedrock exposure, and data on rockhead levels are derived almost entirely from boreholes. These are widely distributed, except between Blakenhall and Shavington (Figure 41).

On the Pennine margins, east of the plain, the drift on Carboniferous bedrock tends to be patchy and generally under 8 m thick. The main exceptions are in drift-filled valleys that trench through the interfluve between the Irish and North Sea basins. In these the valley fill locally attains 20 m (Wilson et al., 1992). The Triassic rocks east of the Cheshire Plain are generally free of drift, except for scattered patches and in drift-filled valleys at Maer and Whitmore.

GLACIAL DEPOSITS

The deposits of the Pleistocene glaciations are complex, having been deposited in a wide range of closely interacting glacial and periglacial environments, and they commonly grade compositionally and texturally into one another. Most deposits were formed in association with melting masses of ice, either directly, without water transport and sorting, or from water flowing within or away from the ice. Other deposits have an aeolian origin.

Most glacial deposits fall into one or other of two broad categories, till and glaciofluvial (meltwater) deposits. More detailed classifications of these are given by Goldthwait and Matsch (1988). **Till**, interpreted to have been deposited directly from melting ice, is typically a poorly sorted, unstratified mixture of rock fragments, up to boulder size, in a matrix of sand to clay grade material. Its composition varies considerably from area to area and commonly is closely related to bedrock formations over which the ice mass moved. **Glaciofluvial deposits** are dominated by sediments that were deposited by meltwater flowing on, in, or under the ice or on its margins. They mainly consist of sand and gravel, but include clays and silts. Clast content tends to mirror that of associated tills.

On the published 1:50 000 geological map, only a broad threefold subdivision of the glacial deposits is shown: till is distinguished from glaciofluvial deposits, and the latter are divided into a general category and a category showing sheet-like topography. The difference is likely to depend on location relative to the ice margin: deposits formed in and around the melting ice tend to show undulating topography, whereas deposits formed away from the ice tend to retain a sheet-like surface form.

Age of the glacial deposits

There has been a long debate about the number of glaciations the area has seen (review by Worsley, 1985; Knowles, 1985a). The view taken here is that most of the glacial sediments preserved in the district document the retreat of a single ice sheet which, on regional evidence, appears to be of late-Devensian age. During this glaciation, an ice sheet in the Irish Sea basin was fed by glaciers sourced on surrounding high ground (Eyles and McCabe, 1989). It advanced southwards across the Cheshire and Staffordshire lowlands, reaching its maximum extent, near Wolverhampton (Morgan, 1973), probably at about 22 000 years BP (Bowen and Sykes, 1988). Evidence from near Stafford (Morgan et al., 1977) indicates that mid-Staffordshire was ice-free from about 13 500 years BP, so that the timing of late-Devensian ice-retreat is fairly well constrained.

Deposits relating to a much earlier glaciation are also present in the region. The age of this pre-Devensian ('?Anglian') glaciation is unclear at present, though it is likely to have occurred before 250 000 years BP. The ice sheets that covered the Stoke-on-Trent district at that time probably were derived either from the Irish Sea basin or from Wales (Gemmell and George, 1972).

Pre-late-Devensian deposits

In the Stoke-on-Trent district there is evidence for at least one pre-late-Devensian glaciation. Boreholes sunk in connection with the A5020 link to the M6 motorway at

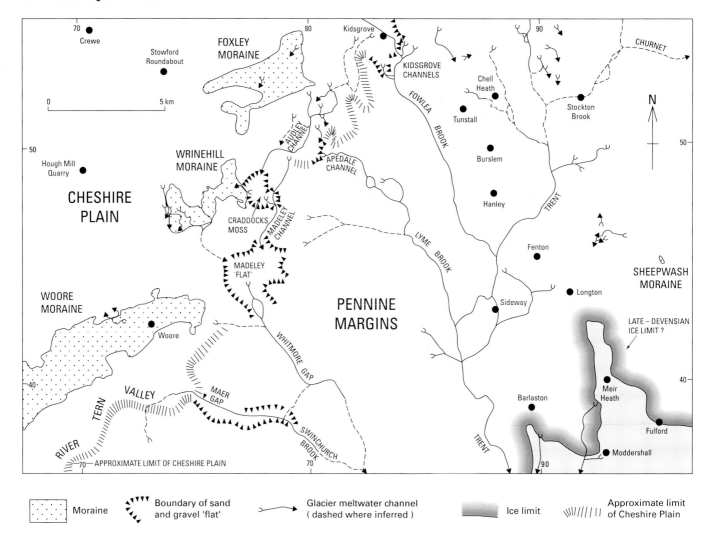

Figure 40 Map showing principal features associated with retreat of the late-Devensian ice sheet.

Stowford Roundabout [735 533] (Figure 40) proved organic silts, sands and peat underlying till of probable late-Devensian age. Three distinct tills (1 to 3, Figure 42) underlie the organic deposits, suggesting that the products of one or more pre-late-Devensian glaciations could be present here. Till 1 thins out eastwards over a ridge in the bedrock. Tills 2 and 3 die out in a westerly direction and could have been eroded before the overlying organic silts and local thin peat were laid down. The pre-late-Devensian sequence is notably rich in laminated clays compared with the late-Devensian sequence. The organic deposits may be equivalent in age to organic horizons recorded at Chelford (Worsley, 1980; Worsley et al., 1983), and at Burland near Nantwich (Bonny et al., 1986). The ages of these are still uncertain, except that they are pre- late-Devensian. Other organic horizons of possible interglacial or interstadial age, such as those at Stockton Brook or at The Bogs, Blackbrook (Knowles, 1985a), are unlikely to be pre-Devensian in age.

The thickness of pre-late-Devensian glacial deposits in the district is not clear. Drift of this age is likely to occur under the Cheshire Plain, but to the east such deposits, if present, are likely to be patchy. The drift on the Cheshire Plain is known to thicken substantially into buried valleys in the rockhead surface, though it is not known how much of the valley fill is pre-late-Devensian. A steep-sided valley with 102 m of drift was proved at Ettiley Heath just north of the Stoke-on-Trent district (Evans et al., 1968), and was tentatively extended southwards by McQuillin (1964) to the Nantwich district on the basis of electrical and gamma-ray measurements. He interpreted variations in single point resistivity values in terms of clay or sand beds within the drift, and concluded that shot hole 53, near Audlem, was entirely in drift to 10 m below OD.

Recent evidence from the Stoke-on-Trent district supports the existence of this buried valley (Figure 41). East of Crewe, and south of Ettiley Heath, borehole evidence suggests a buried valley at about 52 m below OD, contrasting with levels of 33 and 27 m above OD on the two flanks. The valley is associated with a well-defined gravity low at this point but the anomaly appears to diminish southwards. Gravity evidence for its continuation south-westwards may be provided by features near

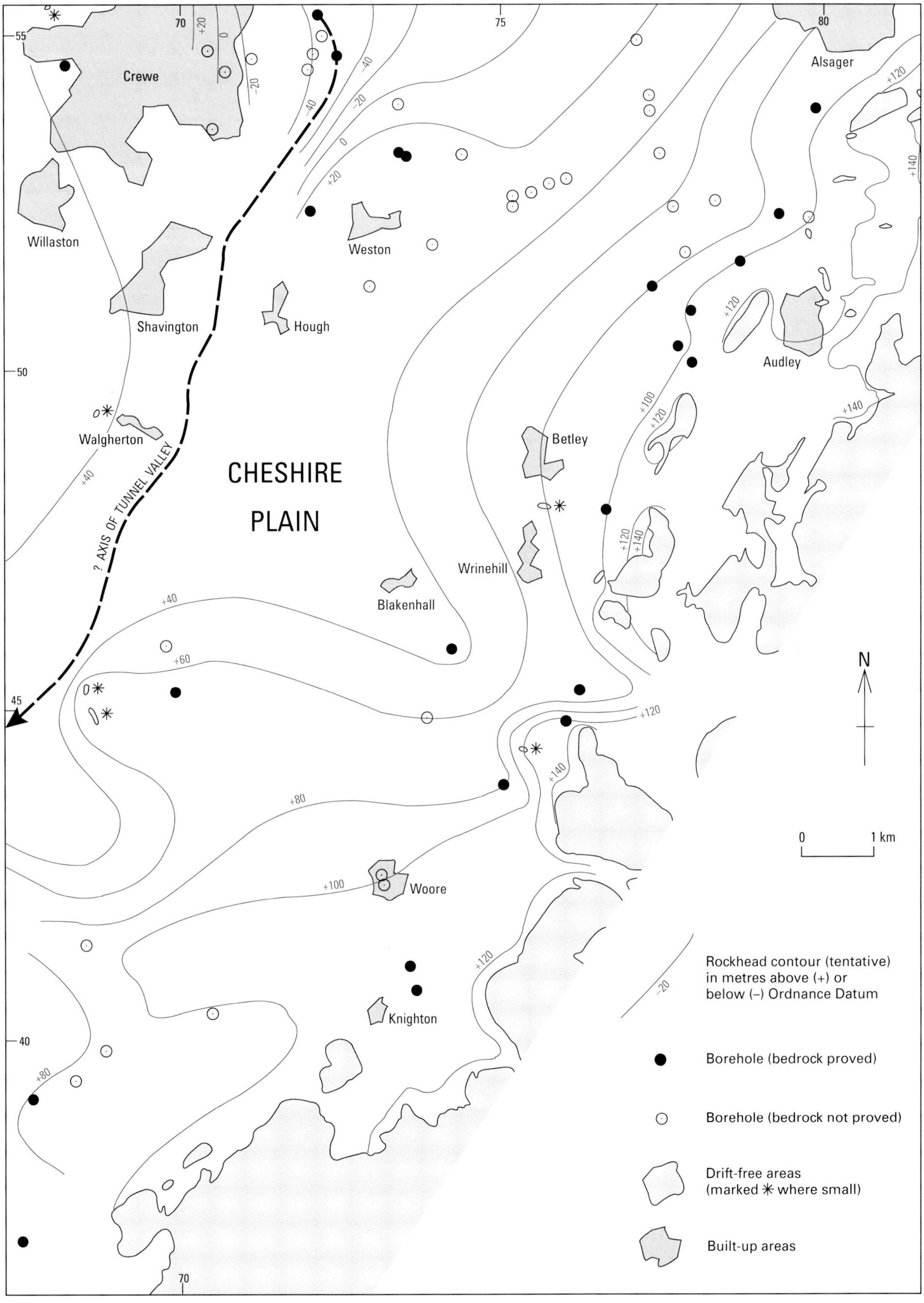

Figure 41   Rockhead map for the Cheshire Plain, showing the course of the probable tunnel valley south of Crewe.

Walgherton, where the south-south-east flank of a gravity high (Cornwell and Dabek, 1994) could reflect a sharply defined channel margin, with a calculated slope of about 40°. (The alternative model of Cornwell and Dabek for this feature implies the presence of non-stratiform bodies of different density in the Mercia Mudstone Group; this seems implausible as seismic evidence suggests that no such bodies exist locally.) In an attempt to isolate the anomaly due to the buried valley, Cornwell and Dabek (1994) made a residual anomaly map by removing a fifth-order polynomial field from the combined detailed and regional gravity data sets. Based partly on poorly defined gravity minima in this map, a valley is tentatively identified as extending south-west of Walgherton towards Audlem (Figure 41). The absence of

a consistent gravity low implies that the valley is variable in depth along its length, and that it was formed as a sub-glacial tunnel valley. It is likely that the valley was cut in pre-late-Devensian times, though it may also have a partial late-Devensian fill.

Deposits of likely, though unproven, pre-late-Devensian age are rare at surface in the Stoke-on-Trent district. However, in the adjoining Ashbourne district they are commonly preserved on interfluves and as remnant patches on valley sides, and are obviously dissected or degraded compared with the late-Devensian glacial deposits (Chisholm et al., 1988). The boundary between the two ages of deposit was tentatively drawn in the western part of the Ashbourne district and entered the Stoke-on-Trent district in the vicinity of Fulford. Based on

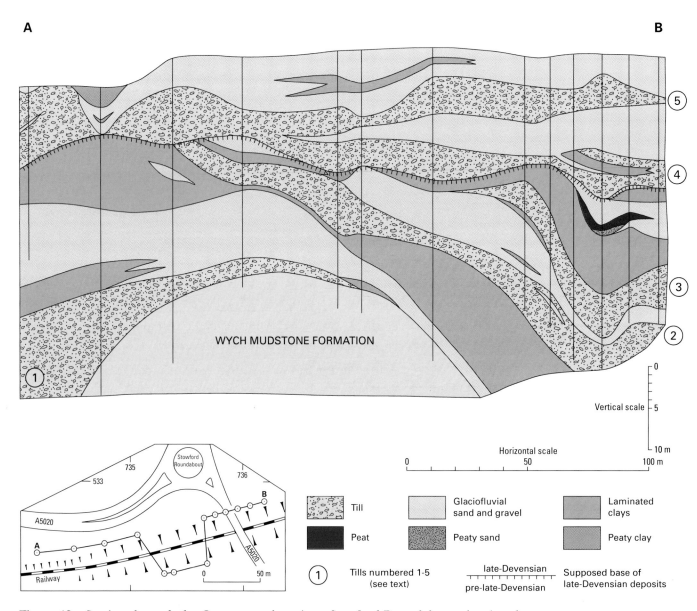

Figure 42 Section through the Quaternary deposits at Stowford Roundabout, showing the probable late-Devensian and older glacial deposits. Boreholes shown by vertical lines: inset shows location.

the criteria applied in the Asbourne district, the late-Devensian glacial limit in the present district probably extended west of Fulford and around the spur of high ground upon which Meir Heath lies (Figure 40). This high ground is covered by patches of dissected sandy till containing igneous pebbles. West of Moddershall, as far as the Trent valley, where it appears to extend into the Stafford district, the limit of the late-Devensian glaciation appears to have been controlled by the northern extent of the Kidderminster Formation outcrop. Three oversteepened valleys with extensive landslips, suggestive of incision by glacial meltwaters, extend southwards from this line. These lie west of Moddershall village [924 367], along the route of the Longton to Stone Road [918 364], and through Downs Banks [900 370].

Late-Devensian deposits

The deposits of the late-Devensian glaciation are considered here under the following headings; in each case the dominant depositional environment is shown in brackets: till sheet complexes (beneath active ice), morainic complexes (ice sheet meltout), glaciofluvial complexes (outwash plain), glaciolacustrine complexes (ice-marginal lake) and glacial valley fill complexes (glacial meltwater channel).

TILL SHEET COMPLEXES

These comprise the major expanses of till-dominated drift that blanket the district, and consist largely of lodgement tills which were deposited beneath the ice sheet. They commonly also contain deposits of glaciofluvial sand, gravel and silt indicative of deposition by water, especially towards the inferred margins of the ice sheet, where ablation tills prevail. The till sheet complexes are distinguished from morainic complexes by their topographically more subdued form. However, they commonly grade into them or into glaciofluvial complexes.

Till sheet complexes occur over large parts of the Cheshire Plain, where they form extensive spreads on low ground adjacent to the morainic complexes. They are especially extensive and unbroken on the gently shelving areas flanking the Woore Moraine. The constituent tills are commonly of greyish brown silty clay with numerous pebble- and cobble-sized erratics of greywacke and igneous rocks, and some sandstone and limestone. Studies of the lithology of erratics (Wedd, in Gibson, 1925; Yates and Moseley, 1958) show that many originated from the Lake District or Scotland. Small boulders cleared from the fields and left in piles are common, but larger erratics appear to be rare. At surface the sheets are characterised by numerous kettleholes and small sand mounds.

The complexity of the sheets is illustrated by the deposits proved in the boreholes sunk close to Stowford Roundabout, which reveal at least two till sheets above the probable pre-late-Devensian organic deposits described above. These sheets (4 and 5, Figure 42) join westwards and are probably a lodgement till of the late-Devensian glaciation. The uppermost till appears to be the feather edge of the extensive till sheet which directly underlies

Crewe and which contains beds of laminated clay in boreholes on the east side of the town. In some boreholes up to 6 m of laminated very silty clay with laminae of silt and fine-grained sand underlie pebbly till. The till sheets towards the margin of the Cheshire Plain appear to be less complex, the till commonly overlying bedrock directly, with no intervening sands and gravels (Yates and Moseley, 1958).

The till sheets south of the Woore Moraine (see below) tend to be thinner, and considerably more eroded, than those to the north, and show evidence of considerable solifluction; depositional forms are rare. Boulton and Worsley (1965) showed that the depth of carbonate leaching in the southern till sheets is substantially greater than in the northern sheets, indicating that the former are considerably older.

On the Pennine margins east of the Cheshire Plain lodgement tills are dominant, but where the glacial deposits are thick, ablation tills and glaciofluvial deposits prevail. It is likely that glacial deposits once covered the whole area, but that these were eroded back to their present distribution by subsequent mass wasting under periglacial conditions, and removal of material by stream flow. The till sheet ranges up to 13 m in thickness, but on average is about 3 m thick, so that many old quarries in bedrock formations are encircled by this thin till. In general the till sheets east of the Cheshire Plain appear to be thinner than those on the plain (Wedd, in Gibson, 1925). The till is commonly thickest close to buried valleys, as at Sideway [880 437] and in a strip from Carmountside Cemetery to Bucknall Park. Like the till sheets on the Cheshire Plain, those to the east contain lenses of glaciofluvial sand and gravel, such as those near Bucknall [9000 4777] and east of Hanley [8937 4650].

The till sheets on the Pennine margins commonly reflect the character of the bedrock over which the ice travelled. For instance, pale grey to ochre tills occur over most of the coalfield area and Namurian outcrop (as in Tunstall), red tills are present over much of the Barren Measures outcrop (as in Dresden), sandy tills occur over most of the Sherwood Sandstone Group outcrop (as at Tittensor), and red-brown tills overlie most of the Mercia Mudstone Group outcrop (as to the south of Woore). The boundaries between these till varieties is complex in most cases. The enclosed clasts are chiefly of debris of local derivation but also include far-travelled erratics of greywacke and igneous rocks. Local material chiefly comprises Carboniferous sandstones, with some mudstone pebbles, and debris from the conglomerates in the Sherwood Sandstone Group. In the area south of Hanchurch several large erratics occur at surface, and as noted by Wedd (in Gibson, 1925) these are dominated by grey granitic intrusive rocks, probably of Scottish origin. Other igneous erratics, such as the large porphyry boulder in Dresden Park [4296 4189] may have been derived from the Lake District (Wedd, in Gibson, 1925).

Several elongate ridges trending north-north-west have been recognised in the till sheet east of Barlaston, where they broadly parallel the structure of the bedrock formations. These elongate drumlinoids were probably caused by ice entrainment near the margin of the late-Devensian ice sheet, which terminated just to the south (see Figure 40).

MORAINIC COMPLEXES

The three main morainic complexes consist of broadly linear, topographically upstanding glacial sediments located along the edge of the Cheshire Plain (Figure 40). One minor example occurs farther east. They are believed to have been deposited at the margins of the ice sheet.

The **Woore Moraine** forms distinctive uneven terrain, elevated locally by as much as 50 m above the surrounding plain, along a ridge trending east-north-east through the village of Woore. It was first described by Lewis (1894) and subsequently by Wedd (*in* Gibson, 1925), Jowett and Charlesworth (1929), Poole and Whiteman (1961), Boulton and Worsley (1965) and in most detail by Yates and Moseley (1967) and Gemmel and George (1972). Boulton and Worsley linked the Woore Moraine with the moraines in the west of the Cheshire Plain, referring to the whole group as the Bar Hill-Whitchurch-Wrexham moraine. They attributed this to the late-Devensian glaciation at about 20 000 years BP. There are numerous mounds and ridges elongated on the same trend as the crest line of the moraine, some with slopes as steep as 17°, and a number of peaty kettleholes occur. Valleys parallel to the grain of the northern slopes, and some running down-slope, may have carried meltwater. The southern slopes of the moraine are more gradual than the northern.

The topography in the Woore Moraine is fresh (Yates and Moseley, 1967), and it is most likely that it represents a late-Devensian readvance of ice (Shotton, 1966; Boulton and Worsley, 1965), or possibly a period of still-stand during the shrinkage of the ice sheet from its maximum extent. Most of the morainic surface is composed of till, which commonly overlies pods or beds of underlying sand and gravel, particularly in the west around Holly Farm [689 413]. Some of the sand and gravel forms prominent hills, but many mounds of similar aspect are in till, or are made up partly of sand and partly of till.

McQuillin (1964) interpreted the resistivity logs of a line of seismic shot holes across the Woore Moraine. He suggested a complex interbedding of till and glacial sand and gravel on the northern flank of the moraine. His work provided some unexpected results; for instance in shot hole 59, on the crest of the moraine, the drift is interpreted to be only 34 m thick. It is not certain if high rockhead levels such as this (122 m above OD) are general beneath the crest of the moraine; other boreholes in Woore indicate unbottomed drift at 115 m above OD.

One of the best exposures in the Woore Moraine is in a sand pit [7209 4225] at Woore Hall, in which the face has been well preserved. It is on the flank of a triangular sand area within which lies a peaty depression. A 5 m exposure of fine-grained sand contains sporadic thin layers with pebbles, largely of greywacke, quartz and igneous rocks, in a sand matrix. Nearly horizontal bedding is picked out by sharply defined, slightly cemented layers about 3 cm thick, which stand out slightly from the softer sands. A few cemented surfaces are at a steeper angle and could be shear planes. The excavation is 30 m long, and at the northern end 1.5 m of sandy till with pebble-sized erratics overlie the sand with a sharp, steeply sloping contact which abruptly cuts across the layering of the sand. Notwithstanding this evidence for a readvance of ice, the generally fresh nature of the morainic landforms suggests that the entire moraine is likely to be a Devensian feature. The till on the lip of this pit is at the edge of one of several prominent clayey hills on the morainic ridge.

Numerous patches of sand, with some gravel, underlie the till on the Woore Moraine, but there are also several prominent sand hills, up to about 12 m in height, which appear to rest upon the till. On one group of hills [at 684 410] and on a hill nearby [6955 4030], there occur scattered small pebbles, some of them facetted and possibly wind-worn. A 600 m wide patch [695 413] of fine-grained silver sand with a few pebbles rests on till and forms a valley in the morainic belt.

Several anomalous features suggest that the moraine has influenced the course of local rivers, such as the Lea (Barke, 1920; Yates and Moseley, 1958) and the upper Tern (Wedd, in Gibson, 1925).

The **Wrinehill Moraine** was first recognised by Yates and Moseley (1958) as marking the position of the ice front at the time that an adjacent lake, at Craddocks Moss (see below), was formed. Unlike the Woore and Foxley moraines the Wrinehill Moraine lacks a distinct crest-line, but does include much moundy terrain with some enclosed hollows. Moundy terrain, much of it in sand, is well seen around Plumtree Park [765 480]. Farther north towards Balterley Hall [764 498] most of the moundy surface is in till. Towards the west, around Wrinehill itself, the moundy terrain is developed in a sand sheet at least 12 m thick. West of the London–Glasgow railway this takes the form of a sandy plateau dissected by glacial drainage channels. An excavation, one of several in the sand at Wrinehill [7541 4695], showed 1.8 m of medium-grained planar-bedded sand with scattered pebbles and some cross-bedding. The hummocky topography of the moraine suggests that it was deposited in and around stagnant ice (Boulton and Worsley, 1965) left during retreat of the late-Devensian ice sheet.

The **Foxley Moraine** probably resulted from a still-stand during the retreat of the late-Devenian ice sheet. It forms a morainic belt, first recognised by Wedd (*in* Gibson, 1925), which is best developed in the ridge on High Foxley Farm [793 528]. A series of topographic features and corrugations in the landscape, trending west-south-west, are traversed by a bifurcating glacial drainage channel. Much of the terrain is in till, with pebbly fine-grained sand locally visible beneath the till sheet. The largest area of sand [794 540] shows an uneven landscape similar to that of the tills.

The prominent Brockwood Hill [782 525] consists largely of sand, except on its till-covered eastern slopes. During construction of the A500 road a cutting went straight through the hill, revealing 12 m of glacial sand overlain by 1.85 m of till on the flanks of the hill. The lower beds of sand contain a bed of gently inclined laminated clay 0.30 m thick, overlain by planar laminated sands. Stratal disturbance increases to the west, suggesting that the morainic deposits were subject to glacial deformation. A bed of clay in the sand [7829 5255] is folded into a steeply tilted arcuate shape, and ferruginous cemented layers in the adjoining sands appear to be distorted and sheared, resembling the cemented shear planes described by Taylor (1958).

Temporary exposures 1.2 km farther east on the A500 road, and located at the eastern edge of the Foxley Moraine, were in glaciofluvial sands overlain and underlain by till. The lowest deposit, seen to 1.2 m, was a brownish grey compact till with scattered stones, probably a lodgement till. Overlying sands, 7.5 m thick, contained scattered gravel beds towards the top and base, and a few laminae of finely fragmented coal in the middle bed. The overlying till, with its numerous cobbles and pebbles, was at least 6 m thick.

Minor morainic deposits (the **Sheepwash Moraine**, Figure 40) have also been recorded east of the Cheshire Plain, at Sheepwash Farm [953 430] near Caverswall. The ridge, 300 m long and trending north-west, is associated with glaciofluvial deposits and is of probable late-Devensian age.

GLACIOFLUVIAL COMPLEXES

Glaciofluvial complexes are restricted to the Cheshire Plain, where they dominate the areas between the morainic complexes. They represent, in part, the outwash deposits of the retreating late-Devensian ice sheet, and are dominated by sand and gravel. Minor silts, clays and tills also occur. The geometry and lateral relationships of sediment types within the glaciofluvial complexes are irregular, as is shown by the upper parts of the sequence near Stowford Roundabout (Figures 40 and 42). The irregular geometries suggest that the sediments were deposited above, or in contact with, melting bodies of ice. Locally the complexes were affected by aeolian processes, as shown by the presence of facetted pebbles.

Glaciofluvial sand and gravel forms an extensive tract of flat or gently undulating country between Crewe and the Woore Moraine. The area is crossed by Checkley Brook and a brook running past Weston, both of which are flanked for several kilometres by extensive areas of sand with a few pebbles. In a few places the stream sections show indications of an underlying till.

Two extensive workings south-west of Wybunbury provided the best exposures in this area. In 1966 one pit [708 492] exposed a 6 m high eastern face in medium-grained sand containing horizons of pebbles, some of which were facetted. In the western face a bed of reddish brown clay, 0.6 m in thickness, was visible. Other beds of clay, less well exposed, occurred within and below the sand body. The pit, now worked out, has been landscaped and is partly flooded.

The currently active Hough Mill Quarry, 600 m west of the disused pit, is developed in the sandy bluffs flanking Checkley Brook [702 490]. The component pits contain good exposures in some 14 m of sand and subordinate gravel. The base of the central pit, which is flooded, is reputed to be underlain by clay. Workings in the southern pit exposed planar-stratified sands, whereas other parts are cross-bedded and contain sharply defined channels filled with cross-bedded fine gravel. The best exposure was on the promontory dividing the southern central pits [7030 4898]:

	Thickness m
Sand, medium- and fine-grained, cross-bedded; channelling to 0.5 m depth	10.0
Sand, medium-grained with beds of gravel up to 0.5 m thick, clasts to 10 cm, averaging 2 to 3 cm , some cross-bedding. Pebbles include granite, basic igneous rocks, greywacke, vein quartz, flint, Carboniferous limestone with coral, laminated cupriferous siltstone, reddish brown Mercia Mudstone and a few bivalve shell fragments	1.5
Gap to water level in pool	1.2

Marine shells from within gravel in the pond have been retrieved and identified as *Turritella communis*, cf. *Cerastoderma* sp. and *Mya* sp. (R C Preece, personal communication). A fuller list of marine shells recorded from glaciofluvial sand and gravel in the district is given by Wedd (*in* Gibson, 1925).

Part of the expanse of glaciofluvial deposits south of Crewe has been crossed by recent road construction lines (A52 and A5020), where the glaciofluvial material is demonstrably overlain by till. Fine-grained, locally medium-grained, sand occurs in numerous small exposures. Small pebbles are present locally and include scattered facetted pebbles near Meremoor Farm [7400 5287]. This locality is only 500 m from the eastern end of the section at Stowford roundabout (Figure 42). It is probable that this is the same sand body as that at surface along the section line, illustrating the intricate relationship between tills and glaciofluvial deposits on the Cheshire Plain.

Scattered facetted quartzite pebbles 2 cm in length were also observed in fine-grained sand in burrows close to Birchall Brook [6870 4509]. Facetted pebbles of this type in the Cheshire Plain drift have been taken to mark periods of aeolian deflation (Boulton and Worsley, 1967).

To the south of the Woore Moraine an extensive gently southward shelving area of poorly exposed sand and gravel is located at Knighton. This tract probably represents an outwash plain from ice standing at the moraine. The eastern edge of the deposit forms the entrance to the Maer gap. North of Knighton the deposit appears to pass under the till sheet draped on the slopes of the moraine.

GLACIOLACUSTRINE COMPLEXES

The principal areas occupied by these deposits occur at the Cheshire Plain end of the drift-filled valleys (glacial overflow channels; see below) that cross the Irish Sea/North Sea watershed. They form fairly flat-lying, roughly triangular areas filled largely by sheets of parallel-bedded glaciofluvial sand and gravel (including lacustrine delta deposits), and laminated stoneless clays of lacustrine origin (Gibson and Wedd, 1905; Wedd, *in* Gibson, 1925; Jowett and Charlesworth, 1929; Yates, 1955; Yates and Moseley, 1958; Knowles, 1985b).

These deposits have been attributed to proglacial lakes on the edge of the Cheshire Plain, ponded between the ice sheet and ice-free higher ground of the Pennine margins. Although the lakes were originally seen as developing during ice retreat (Jowett and Charlesworth, 1929), Knowles (1985a) suggests that they may also have formed during ice advance. Drift-filled channels link the areas of glaciofluvial sheet deposit, which suggests that spillways existed between the lakes. The apparent absence of strandlines associated with the lakes probably reflects their ephemeral nature, with fluctuating water levels controlled by seasonal factors and ice movements. It is possible that even larger proglacial lakes have existed in the district. For instance, Poole and Whiteman (1961) postulated that a large lake, 'Lake Lapworth', abutted the western margin of the uplands along their entire length in the present district, with a strandline at about 100 m above OD. Local evidence for this is poor, amounting to a few very tentative lake edge features at about 100 m above OD south of Alsager [804 541, 807 543].

Areas occupied by glaciofluvial complexes are shown in Figure 40. At **Kidsgrove** a gently inclined area around Woodshutts [833 543] is underlain by glaciofluvial sand and gravel, commonly 2 m or more in thickness, though locally up to 9.5 m. Exposures are in clayey sandy gravel. This sand sheet terminates southwards at the mouths of three glacial drainage channels feeding into the main Kidsgrove Channel. At **Miles Green** a gently inclined sheet of medium-grained sand is located close to the mouth of the Apedale Channel [805 495]. Boreholes show up to 15 m of sand resting on bedrock. The shelving areas round the peat-filled depression of **Craddocks Moss** [775 482] are in fine-grained sand likely to have been laid down in a periglacial lake. Peat commonly directly overlies the sand, though clay intervenes locally between these deposits (Yates and Moseley, 1958). Surrounding **Madeley** is a depression with a flat-lying floor, composed largely of level-bedded sands. This is likely to mark the site of a former lake at the northern end of the Whitmore Channel (Yates and Moseley, 1958). Boreholes at the former Madeley Training College situated close to the edge of the sand sheet proved up to 11.75 m of sand and gravel, overlying a probable till that was proved to 15.25 m [7739 4542]. More recently, Al Saigh (1977) carried out research on the Madeley depression involving a comparison of the efficacy of seismic refraction and electrical resistivity techniques in assessing the thickness of sand and gravel and underlying till in the depression and the Madeley Channel. His view was that on the former college campus, immediately south of the borehole quoted above, there is further southward thickening of the sand and gravel to between 16 and 28 m. This suggests that a major rockhead depression lies beneath Madeley, infilled with sand, gravel, and subordinate beds of clay. Stoneless plastic clay forms thick beds south of Madeley [780 430] (Yates and Moseley, 1958).

GLACIAL VALLEY-FILL COMPLEXES

East of the Cheshire Plain, on the Pennine margins, several drift-filled valleys cut across the North Sea/Irish Sea watershed, between the Trent and Mersey drainage basins. The main channels have been described by several workers, including Jowett and Charlesworth (1929), Yates (1955) and Yates and Moseley (1958), and are well summarised by Knowles (1985a). Recently, the distribution of these has become much better understood, particularly in the Stoke-on-Trent conurbation where borehole data have enabled the drift thickness to be plotted (Wilson et al., 1992).

The valleys (Figure 40) all probably acted as glacial meltwater drainage channels, transporting water derived from wasting ice on the Cheshire Plain into the Trent drainage system. The channels were originally thought to have formed solely by meltwaters from a retreating ice sheet, ponded between the ice and the high ground of the Pennine margins (see glaciolacustrine complexes, above). According to this view, rising meltwaters eventually breached the high ground, escaping eastwards along the valleys of pre-existing west-flowing streams. The development of the channels is now thought to have been more complex (Knowles, 1985a), in that it may also have involved drainage of meltwater during ice advance, and meltwater flow beneath ice cover. Subglacial erosion of the channels is suggested by the existence of topographical highs along their lengths (see below) which could only have been overcome by water flowing under pressure beneath ice cover. Knowles points out that as the district is likely to have been subjected to more than one glaciation, the channels probably were utilised, and expanded, by meltwaters on several occasions during the Pleistocene. However, it is likely that any deposits preserved within them are of late-Devensian age only. The channels are commonly floored by glaciofluvial sand and gravel, with pockets of laminated clay and till, which are commonly overlain by postglacial river terrace deposits and alluvium.

The channels are described from north to south (Figure 40). The prominent valley east of **Stockton Brook** was probably cut in pre-late-Devensian times by headwaters of the Trent flowing westwards from the Rudyard area to the north (King, 1960; Johnson, 1965). However, during deglaciation it probably acted as a conduit for glacial meltwaters to flow eastwards into the Churnet valley. Well-defined channels at Biddulph, in the adjoining Macclesfield district, feed into the highest tributaries of the River Trent (Walton, 1964), as do those on the higher slopes of Brown Edge, in the present district. One of several channels recognised around Chell Heath and Goldenhill breaches the escarpment of the sandstone above the Bungilow Coal [872 522] on the interfluve east of Ford Brook at 190 m above OD.

The **Kidsgrove Channels** (Wedd, *in* Gibson, 1925; Jowett and Charlesworth, 1929; and detailed by Yates,

1955) are feeder channels north of a 40 m deep valley which extends from Kidsgrove, past Bath Pool, into the valley of Fowlea Brook, and on to the River Trent. One of the feeders at Harding's Wood bifurcates before joining the main channel. The main channel, followed by the Kidsgrove to Stoke-on-Trent railway, has a high point at surface close to Bath Pool, 2 km south of the intake, with a downhill gradient of the channel bed to both north and south. Boreholes have been drilled on the route of the railway line in recent years, proving a 20 m deep valley fill, mostly of sand and gravel. Farther south, this, the Fowlea Brook channel, is joined by others following the Trent valley and its tributaries; at Sideway the combined channel splits and rejoins. The surface depressions commonly contain some 4 to 8 m of glaciofluvial sand and gravel capped by up to 9 m of alluvium, as shown in numerous site investigation boreholes.

The **Apedale Channel** is represented by a valley 100 m deep that cuts across the Western Anticline. The channel has a discontinuous fill near its highest point. Farther south, in the valley of Lyme Brook, boreholes indicate a fill of glaciofluvial sand and gravel and alluvium, up to 14 m thick, extending southwards to Trentham. Two other valleys west of Silverdale [801 471, 800 460] may also have carried meltwater, but apparently contain no glacial drift; westwards they pass into an extensive sand sheet at Silverdale.

The valley of the **Whitmore Gap** contains a col near Whitmore Station (Yates and Moseley, 1958). East of this the valley extends into that of Meece Brook. The valley of the **River Tern**, extending into Swinchurch Brook, probably marks the site of another glacial meltwater channel (Knowles, 1985a). This is floored by sands and gravels, though they are probably of Flandrian age (see below). South of the district the Sow–Coal Brook system was described by Whitehead et al. (1927).

On the margin of the Cheshire Plain a network of channels appears to have carried glacial meltwaters flowing southwards along the margin of the plain (Figure 40). They were first described by Gibson and Wedd (1905), who suggested that water escaped between the ice margin and higher ground east of the plain as the ice retreated. A general model for the region was proposed by Jowett and Charlesworth (1929), and expanded by Yates (1955) who suggested that the channels were incised by waters emptying from a sequence of proglacial lakes, which successively emptied from south to north, in parallel with ice retreat. Of these channels two are notable. The **Audley Channel** (Wedd, *in* Gibson, 1925; Jowett and Charlesworth, 1929; Yates, 1955), located beneath the escarpment of the Newcastle formation at Audley, is fed by several channels, one of which crosses another [810 525]. These may have been active at different times. The triangular shaped basin at Craddocks Moss received this drainage, with exits to the west and south. The southerly exit is the deeper **Madeley Channel**, which makes a deep incision at Walton's Wood (Wedd, *in* Gibson, 1925; Jowett and Charlesworth, 1929; Yates, 1955; Yates and Moseley, 1958). Site investigations for the M6 motorway proved 4.5 m of peat and soft clay

on 14.0 m of glaciofluvial sand resting on 4.5 m varved clay at the base of the channel. The channel terminates in the Madeley 'Flat', the probable site of an ice-marginal lake (Yates and Moseley, 1958).

It is likely that some of the channels on the margin of the Cheshire Plain have had other origins, apart from those described above. For instance, a subglacial tunnel valley origin was suggested for the series of closed drift-filled depressions proved to run along the eastern edge of the Cheshire Plain between Audley and Kidsgrove by Wilson et al. (1992). Whether these are filled solely with Devensian deposits is unknown.

PERIGLACIAL AND POSTGLACIAL DEPOSITS

During deglaciation, periglacial conditions prevailed in the district. This environment saw the genesis of deposits such as head, alluvial fan deposits and landslips, which were derived from landforms made unstable by the retreat of the ice. Several erosional landforms in the district are likely to have been formed in a periglacial setting, such as dry valleys and nivation patches (reviewed by Knowles, 1985a). Periglacial conditions also left smaller-scale structures, such as frost wedges, frost cracks and convolute strata (Knowles, 1985b). Most of the post-glacial deposits formed under temperate conditions and are associated with rivers and lakes, and include layers and patches of peat.

HEAD

Head commonly consists of a heterogeneous mixture of rock fragments set in a finer-grained matrix, and is thus not dissimilar to glacial till. It is the main product of solifluction, the downslope flow of rock and soil material induced under periglacial conditions by repeated freezing and thawing. Head is widespread in the district, forming thin but extensive blankets over bedrock and drift formations. Commonly it is too thin to map, and in places it is difficult to distinguish from underlying deposits, particularly till; the latter generally appears more consolidated, however. Erratics are common in many head deposits in the area.

In the Stoke-on-Trent district, head is most common downslope of outcrops of rocks of the Sherwood Sandstone Group, where it consists mainly of pebbly sands, as may be seen in Trentham Park [858 406]. Head also is common below sandstone scarps, such as those of the Millstone Grit and the Newcastle and Keele formations.

ALLUVIAL FAN DEPOSITS

Alluvial fans consist of delta-shaped deposits of rock and drift debris, mainly in valleys at the bottom of sandstone scarps. Compositionally they consist of material indistinguishable from head.

LACUSTRINE DEPOSITS

Lacustrine deposits dominantly comprise stone-free, laminated, clays and silts. They are restricted to the Cheshire Plain where many fill subcircular and linear subsidence hollows probably formed by solution of salt formations in the bedrock. They include large depressions like that around Betley Mere [745 481, 745 490, 738 500]. Other occurrences, such as those in depressions along the A52 Barthomley Link to the M6 motorway, occur in large kettleholes.

PEAT

Peat commonly fills glacial drainage channels and sites of glacial lakes. For instance, the peat at Craddocks Moss [777 483] was formed after a former lake, fed by glacial drainage channels, had silted up. A borehole close to the edge of the deposit proved 1 m of peat, and greater thickness is likely near the centre of the moss. A further patch of peat occurs in the former glacial drainage channel at Walton's Wood [777 463]. Much of it was removed to accommodate the weight of the motorway embankment. An area of peat 3 km long, and at least 0.9 m thick, is located in the glacial drainage channel of Whitmore Gap between Baldwin's Gate and Madeley [785 420]. The valley north of Maer and that of the River Tern have several areas of peat, like Dorrington Bogs [737 394], Willoughbridge Bogs [752 395] and The Bogs, Blackbrook [775 385]. Thicknesses of peat up to 1.2 m have been observed. Peats are preserved within the sand sheet flanking The Bogs. G S Boulton (personal communication) reports that the palynology of these indicates the presence of the pollen zones III and IV. This shows that the sand deposit which forms the valley floor is of Flandrian age and not the product of an earlier proglacial lake (cf. Yates and Moseley, 1958, p.420). A section of the deposit has been observed in a tributary valley [7690 3839]:

	Thickness m
Topsoil	0.15
Sand	0.15
Clay, very light grey	0.01
Sand, medium-grained, light yellow	0.20
Peat, silty	0.01
Clay, grey	0.04
Peat, silty in top 2 cm	0.26
Clay, grey	0.02
Sand, orange with a few small sandstone fragments	—

Peat has been formed in a number of semicircular and linear depressions, probably salt solution subsidence hollows. They include large depressions like that around Betley Mere [747 481] and many smaller hollows such as Wybunbury Moss [697 502], Pepperstreet Moss [703 465], Buerton Moss [684 447] and Blakenhall Moss [722 483] where peat reputedly 3.6 m thick overlies white sand. The best documented peat-filled hollows are on the A52 Barthomley Link to the M6 motorway, where well-defined subsidence hollows [755 527] are underlain by 4 m of peat. A similar thickness is recorded in a nearby hollow 40 m wide [7527 5259]. Towards the west the road crosses a large irregularly shaped hollow in which peat and lacustrine deposits occur. Boreholes reveal up to 9 m of peat buried beneath lacustrine silt. In an arm of the main hollow, peat up to 2.8 m thick is exposed and proved in boreholes.

Locally, on the Woore Moraine, peat fills kettleholes. Scattered hollows from 30 to 100 m across in the moraine contain an infill of peat up to at least 1.5 m thick [7123 4206, 7136 4208, 7459 4289, 7450 4274]. Many of these are too small to show on the 1:50 000 map.

Peat also occurs as lenses in valley alluvium, as shown by boreholes in the Stoke-on-Trent urban area in the valley of the River Trent and Fowlea Brook.

RIVER TERRACE DEPOSITS

Deposits in the river terraces are dominated by pebbly clayey sands, and probably are all of late-Devensian or early Flandrian age. The most extensive system is that in the valley of Checkley Brook on the Cheshire Plain. Of the terraces in this system, the third, which stands 11 m above the present stream in the area near Bowsey Wood [758 464], and the second, may be of glaciofluvial origin. The terraces mapped in the Blithe valley, west of Longton, probably represent outwash from the late-Devensian ice sheet (Chisholm et al., 1988).

ALLUVIUM

Alluvium consists of clay, silt, sand and gravel, locally with lenses of peat. Most deposits have gravel at the base. Alluvium commonly has a high organic content, giving it a dark grey colour. It occurs at the bottom of most modern valleys, though in many is too thin to show on the map.

The most extensive spreads of alluvium lie in the valley of the River Trent and its tributaries, Fowlea Brook and Lyme Brook, where the deposit commonly rests on earlier glaciofluvial sands and gravels. The valley of Horton Brook [921 531] contains an extensive spread of silty alluvium which is likely to rest on glaciofluvial deposits. On the Cheshire Plain alluvial spreads are best developed in the valley of Checkley Brook [723 470] and Basford Brook [726 520].

LANDSLIPS

Landslips are developed in a variety of locations and geological settings, but the largest occur in the vicinity of glacial drainage channels where there has been oversteepening of slopes. At **Walton's Wood** [782 464] five separate landslips have been mapped on the sides of the glacial drainage channel, which contains an overdeepened sand-filled buried valley as well as the present gorge (Early and Skempton, 1972). Slips on the east side are multiple rotational slips with arcuate back scars, developed in the top of the Coal Measures, which here contains some reddish brown mudstones transitional to the overlying Etruria Formation. Early and Skempton believe that initial movement of the landslip occurred in the late-Devensian. During motorway construction, loading on the landslip caused a further failure of the slipped area (Chapter 9). Landslips on the west side of the valley are in the topmost part of the Etruria Formation. West of **Moddershall** landslips occur in mudstones and sandstones of the Keele formation on the margins of glacial overflow channels [918 367, 925 368]. Near **Bath Pool**, a tributary channel of the Fowlea Brook glacial drainage channel has a landslip

[835 528], developed in Coal Measures, on its southern side. At **Hollybush**, the western slopes of a glacial drainage channel [896 535] are affected by three small rotational landslips in mudstones and sandstones of the Etruria Formation.

In general the strata most prone to landslips are those in the upper part of the Etruria Formation. Below the escarpment of the Newcastle formation on the west side of Fowlea Brook near **Bradwell Wood**, extensive landslips affect the highest beds of the Etruria Formation [850 503] (Searle, 1973). Locally, beds of the Newcastle formation form the backscar. The upper parts of the landslips consist of rotational segments, whilst the lower limit of the slip is in places hard to define due to flowage of the soft mudstone.

Some landslips in the Etruria Formation have been active in recent years. A small landslip [8450 5057] which was exclusively in Etruria Formation in 1961 has since tripled in size and migrated back to the boundary between the Etruria and Newcastle formations. Over the same period the backscar of a second, larger landslip [8485 5045], has migrated 30 m farther to the west.

Oversteepening of slopes in quarries in the Etruria Formation has led to rotational failures with associated flowage of the soft mudstone. Examples are seen in disused portions of the active quarries at Knutton [8268 4675] and Walleys [831 460]. The Etruria Formation is also prone to solifluction on slopes, with evidence of solifluction lobes, as for instance in fields 300 m south of Walleys Quarry [829 456].

Landslips have also occurred locally in formations other than the Etruria Formation and away from glacial drainage channels. For example, mudstones of the Edale Shale Group at Hollinhouse Wood, Brown Edge [914 544] are extensively affected by rotational slips 300 m in length, and mudstones of the Keele formation and some of the overlying Sherwood Sandstone form a slip 500 m long on the face of the escarpment of Sherwood Sandstone at Heighley Castle [774 469]. Landslips affecting only drift deposits are rare.

MAN-MADE DEPOSITS

Made ground is extensive in the urban areas. It is described in Chapter 9.

EIGHT

Structure

One of the most notable features of the geological map of the district is the contrast in structural complexity between the Carboniferous rocks and the Triassic rocks that unconformably overlie them. The major folds and faults trend north-east to south-west on the western side of the coalfield and broadly north–south in the area to the east of here (Figure 43). A subordinate set of folds, of east-north-easterly trend, occurs towards the south of the coalfield. The two main structural trends converge just north of the district and this, combined with an overall southward plunge beneath the Triassic rocks, gives the Carboniferous outcrop its distinctive triangular shape.

Most structures in the Carboniferous rocks resulted from compressional events during the Variscan Orogeny and subsequent extensional events. All of the major Variscan and younger structures of the district display the influence of deep-seated basement faults, which, on regional evidence, have a history of activity that dates back to the late Proterozoic. It is important to note that many major structures observed in the Stoke-on-Trent district continue into adjoining districts.

This chapter falls into two parts. The first defines and describes the pre-Variscan, Variscan, and post-Variscan structures of the district. Knowledge of these is based on data derived from many sources, including surface and underground mapping, gravity, magnetic and seismic reflection data, and boreholes. The second part of the chapter gives a synthesis of the tectonic history of the region, incorporating basin evolution and the effects of the Variscan Orogeny. This account identifies episodes of crustal extension in the early parts of the Carboniferous, Permian, Triassic and Jurassic periods, with intervening episodes of thermal subsidence, interrupted in the earliest Permian by the Variscan Orogeny.

PRE-VARISCAN STRUCTURES

The trends of the major faults in the Midlands coalfields and flanking Permo-Triassic basins are determined by the presence of long-lived basement faults. The form of these at depth is poorly understood, but at surface they are represented by zones containing many faults of similar orientation. The main basement feature in Central England is a triangular-shaped crustal block of late Proterozoic rocks, the Midlands Microcraton (Pharaoh et al., 1987) (Chapter 2). This was bounded to the north-west by major north-westerly dipping faults (Coward and Siddans, 1979; Smith, 1987; Brewer et al., 1983). Locally, the **Pontesford Lineament** (Figure 3; Woodcock, 1984), a zone of steep gravity gradients, probably marks the position of boundary faults in the

Lower Palaeozoic and Proterozoic rocks. The eastern margin of the microcraton is bounded by faults trending north-west and downthrowing to the north-east. Bisecting the triangular block is a set of north–south-trending faults, the **Malvern Lineament**, which includes the Malvern Fault and the western boundary faults of the Worcester Graben, Warwickshire Coalfield and South Staffordshire Coalfield.

North Staffordshire lies just inside the northern apex of the microcraton. The influence of the north-west-trending structures on the eastern margin of the microcraton is best seen at the margin of the stable Derbyshire block in the adjacent Ashbourne district. Such structures formed the north-eastern margin of the North Staffordshire Basin in Dinantian and Namurian times (Chisholm et al., 1988) (Chapter 3), and are evident in the easternmost part of the Stoke-on-Trent district. In much of this district, however, structures associated with the north-east-trending Pontesford Lineament and north–south-trending Malvern Lineament dominate the structural pattern.

The expression of the Pontesford Lineament in North Staffordshire comprises the Wem, Hodnet and Red Rock faults (Woodcock, 1984). Some confirmation that the faults of this system are reactivated boundary faults of the Lower Palaeozoic Welsh Basin (Chapter 2) is provided in the district to the south, where the Edgemond Borehole (Wills, 1956) proved shallow Proterozoic rocks in the footwall of the Wem Fault, and further to the west where the Prees Borehole (Colter, 1978) encountered Ordovician rocks beneath the Permo-Triassic sequence at a depth of over 4 km. According to Soper and Hutton (1984), the faults within the lineament probably moved in a sinistral sense during the Caledonian Orogeny.

The tectonic influence of the Malvern Lineament is indicated by the dominant northerly to north-north-westerly trend of structures in the central and eastern parts of the coalfield, a trend that continues north into Lancashire, where it is recognisable in the Pennine Monocline.

The major expression of the Malvern Lineament in the subsurface of the district is a structure inferred mainly from gravity modelling, the Lask Edge Fault. This trends north-north-west (Lee, 1988) and forms the western margin of the Dinantian and Namurian North Staffordshire Basin (Figure 4; Chapter 3). It is thought to have been reactivated during Variscan compression. In the Stoke-on-Trent district the location of the fault can be inferred with varying degrees of accuracy. On borehole and outcrop evidence, a fault, assumed to be the Lask Edge Fault, must lie between the thick Namurian basinal sequence encountered in the Werrington Borehole (Chapter 3) and the thin, onlapping Mars-

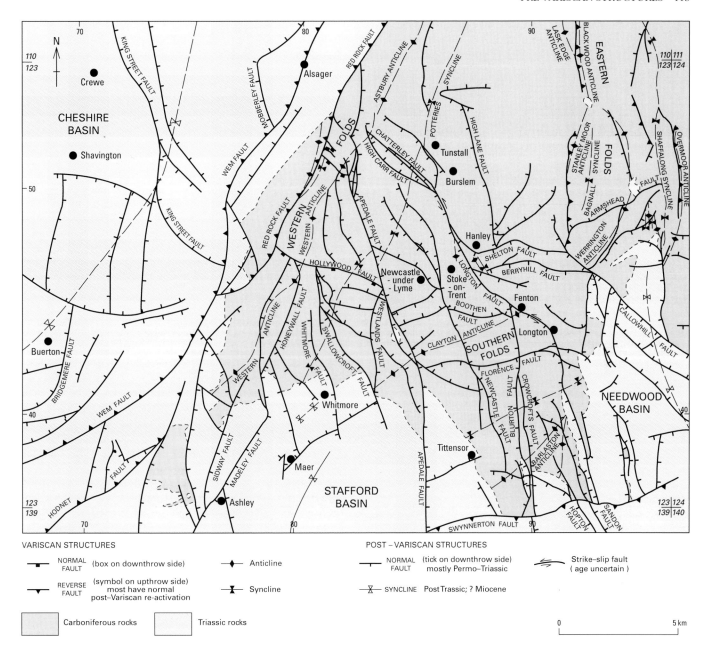

Figure 43   Map showing the principal structures of the district.

denian and Yeadonian sequences proved in the Bowsey Wood and Apedale No.2 boreholes (Chapter 3, Figure 6). Farther north, in the Macclesfield district, the presence of basinal Dinantian strata proved in Nook's Farm Borehole suggests that the fault lies to the west of the Eastern Folds (Figure 43) of the present district.

The evidence from structural mapping using mine plans and seismic reflection profiles provides a more accurate location. This suggests that there is an easterly dipping fault with a westerly thickening wedge of sediments in its hanging wall, on the western margin of the Eastern Folds. Extrapolated towards the surface, the fault lies beneath the outcrop of the Lower Coal Measures, in a tract where the westward dip at outcrop

increases locally to over 75°. Nowhere does the Lask Edge Fault reach the surface and the region of increased dip is inferred to be a monoclinal flexure located above the limit of reversed fault displacement during Variscan inversion.

Farther south, in the late Carboniferous sequence of the Barlaston area, a westerly facing monocline trending nearly north–south appears to be the product of reversed movement on an easterly dipping fault at depth. The deep fault forms the western margin of a thick westerly thickening wedge of sedimentary rocks inferred, on regional evidence, to be of Dinantian (and perhaps late Devonian) age. By extrapolation, the surface expression of this fault may be the Crowcrofts Fault (see below). At depth it extends to the north-north-west as far as Bucknall and to the south it may be represented by the Sandon Fault, which now has a post-Triassic throw to the east. The fault is of similar trend to the Lask Edge Fault and probably forms part of the same fault system.

VARISCAN STRUCTURES

Folds

The main structures formed during the Variscan compression are the Western Folds (the Western Anticline and Astbury Anticline) and the Eastern Folds (which include the Werrington Anticline and Shaffalong Syncline). The Southern Folds, a more open set located in the south-eastern part of the coalfield, include the Clayton Anticline and Barlaston Anticline. There is a strong contrast in geometry between the Western and the Eastern folds. The Eastern Folds are upright, concentric periclines that generally lack a sense of vergence. In contrast, the Western Folds are complex structures with steeply dipping north-western limbs. The different trends of these sets of folds are related to the different orientations of the underlying basement lineaments during Variscan reactivation. The major syncline in the district, the Potteries Syncline, developed on a more stable block between the Western and Eastern folds.

WESTERN FOLDS

The Western and Astbury anticlines trend dominantly north-north-east, approximately parallel to the Wem and Red Rock faults of the Pontesford Lineament. The folds are periclinal in form and the culmination of the northernmost fold, the Astbury Anticline, occurs just north of the district where Dinantian rocks crop in the core of the fold at Astbury (Chapter 3). The western limb of the anticline here is truncated by the post-Triassic throw of the Red Rock Fault system. The anticlines plunge towards each other in the Talke area, forming a saddle. The Western Anticline can be traced to the south-west as least as far as Knighton. The presence of the complete Coal Measures sequence at relatively shallow depths has enabled the preparation of detailed structure maps of the folds from mine plans (Figure 44).

The anticlines have an unusual geometry: the southeastern limbs have dips of 30 to 45° (over 60° in the north of the district), whereas the north-western limbs, on the 'Cheshire dip' (Cope, 1954), have average dips of over 60° and are locally overturned, as at Podmore Colliery, south of Miles Green. The difficulty of mining coal in 'the Rearers', near-vertical strata near the Rearers Fault (Figure 44), is recorded by Grimshaw (1878). At the base of the north-western limb, there is a tight syncline which has a thickened hinge zone indicating local flexural flow during folding. In the north of the district this syncline is flanked to the north-west by a south-westerly plunging anticline of low amplitude, the north-western limb of which has been truncated by the Red Rock Fault.

The coals in the axis of the syncline to the north-west of the Astbury and Western anticlines are consistently elevated

by about 500 m relative to their position in the axis of the Potteries Syncline on the south-eastern limbs of the anticlines. The anticlines thus have an apparent south-easterly vergence even though the north-western limb is steeper.

The Potteries Syncline (Figures 43 and 44) trends north-north-east parallel to the Western and Astbury anticlines, and plunges to the south-south-west at 8°. In the north the structure is a tight asymmetrical fold, with dips on its western limb often over 60° and on its eastern limb generally between 10 and 15°. The syncline maintains its asymmetry to the south, and the axis of the fold remains within 5 km of the Wem–Red Rock fault system as far as it can be traced within the Carboniferous outcrop. However, dips on the western limb in the Newcastle-under-Lyme area are commonly less than in the north, typically between 30 and 45°. The axial zone is flat-lying and the transition to the western limb is sharp, forming a distinct kink, as observed in the Brown Lees opencast site (Figure 7a), which straddled the synclinal axis.

The parallelism of the Pontesford Lineament and the Western Folds strongly suggests a genetic link. Corfield (1991) proposed that the north-westerly dipping Pontesford Lineament at depth beneath the anticlines was reactivated as a south-easterly directed reverse fault zone during the Variscan Orogeny, with a resulting uplift of basement beneath the anticlines. He argued that the lack of deformation beneath the flat-lying Potteries Syncline suggests that the deformation decreased to a limit of thrust displacement beneath the syncline and that the displacement was transferred to a north-westerly directed backthrust. Reverse movement on the Pontesford Lineament during the Variscan Orogeny is also suggested by the higher structural level of synclines west of the Western and Astbury anticlines relative to the Potteries Syncline (see above), and the lower rates of Westphalian subsidence on the western side of the Potteries Coalfield (Figure 7). Variscan reverse movement can also be demonstrated across the Hodnet Fault (which represents the lineament) in the south-western part of the district (Evans et al., 1993; Figure 45).

EASTERN FOLDS

These comprise the periclinal Lask Edge, Blackwood, Werrington, Stanley Moor and Overmoor anticlines and the Bagnall and Shaffalong (Rudyard) synclines (Figure 43). The Stanley Moor Anticline is a relatively small fold that appears to be the southern continuation of the Blackwood Anticline. The folds have near-vertical axial planes and are orientated nearly north to south, thus differing both in form and trend from the Western Folds. The folds are concentric box folds, with steep limbs, commonly near vertical, and flattish axial zones. This is clearly illustrated by the Stanley Moor Anticline, the Werrington Anticline and the Shaffalong Syncline, as noted by Gibson (1905). Most limbs are underlain by reverse faults, which rarely break surface; exceptions include the fault which displaces the Blackwood Anticline on to the eastern limb of the Lask Edge Anticline. To the west of the westernmost anticlines, where the dip of the limbs decreases westwards, is the supposed subsurface position of the Lask Edge Fault, an important deep-seated eastward-dipping structure

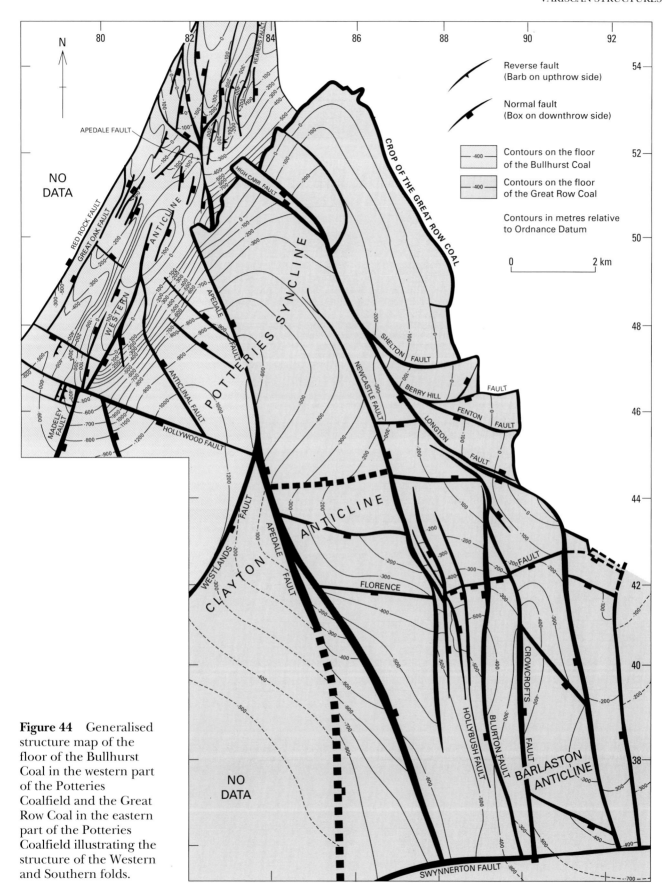

Figure 44 Generalised structure map of the floor of the Bullhurst Coal in the western part of the Potteries Coalfield and the Great Row Coal in the eastern part of the Potteries Coalfield illustrating the structure of the Western and Southern folds.

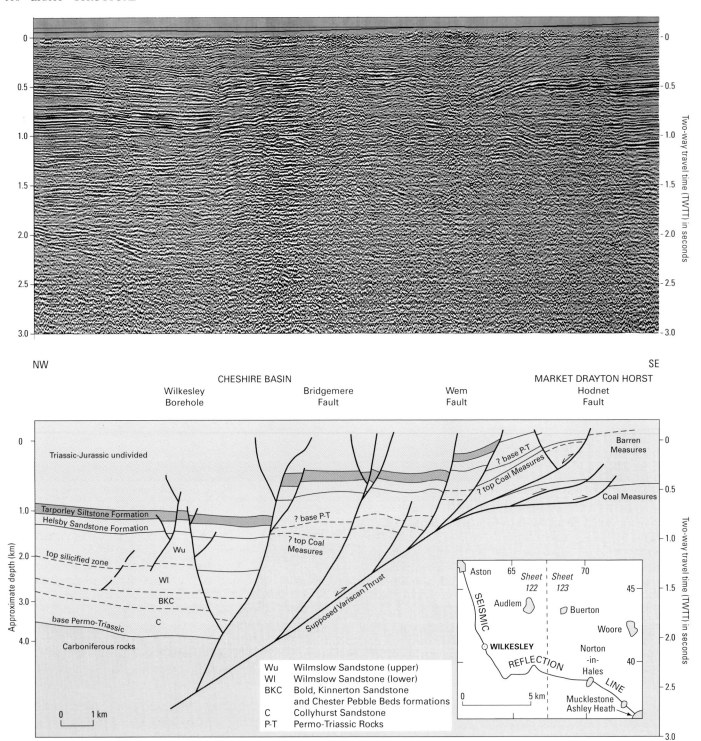

Figure 45 Seismic reflection profile and drawing illustrating the nature of the eastern margin of the Cheshire Basin across the south-eastern part of the Nantwich distict and the south-western part of the Stoke-on-Trent district. Note how synsedimentary normal faults of Permian and Triassic age detach on to the plane of a Variscan thrust representing the Pontesford Lineament. Modified from Evans et al. (1993).

inferred from gravity modelling (see above). It is thought likely that the Eastern Folds were formed in the hanging wall of the Lask Edge Fault during the Variscan inversion. If so, the *en-échelon* disposition of the folds, and their periclinal form, suggest that during inversion the fault had a component of sinistral strike-slip.

SOUTHERN FOLDS

These comprise the Clayton Anticline, the Barlaston Anticline and an intervening syncline, which trend between east-north-east and north-east. The limbs of the folds generally dip at between 6 and 8°, and the axes plunge west-south-west at about 6° (Figures 43 and 44). Although of similar trend to the Western Folds (see above) they are considered separately here because they appear to have had a different origin. The Clayton Anticline (Figures 43 and 44) seems, on seismic evidence, to have been formed above a north-west-directed reverse fault in the basement. It is possible that the Barlaston Anticline has a similar origin, especially as the Swynnerton Fault, to the south of the anticline (Figure 44), had a reverse displacement during the Variscan Orogeny.

Faults

Reverse faults related to the Variscan inversion are relatively uncommon within the Western Folds, although thrusts were recorded by Cope (1954). The only prominent example is the Rearers Fault on the steep north-west limb of the Astbury Anticline (Figure 44). This trends north-north-east, dips 50° to the south-east and has a north-westerly directed reverse displacement of between 20 and 30 m. Minor reverse faults have been recorded on mine plans of the north-western limb of the Western Anticline in the area where it is traversed by the Hollywood Fault (Figure 44). These have a north-westerly directed reverse displacement in the Bullhurst and Banbury seams. However, they are not apparent in the overlying Ten Feet and Moss seams and a small, north-west-facing monocline has formed above the upper limit of displacement of the faults. Reverse faults also occur on the Eastern Folds (see above). Most of the reverse faults are broadly parallel to the fold axes and probably formed by accommodating some of the strain during folding; where reverse faults are absent the strain was taken up by bedding plane detachment, as was observed in High Lane and Brown Lees opencast sites (Figure 7a).

Normal faults associated with folds have a west-north-westerly trend on the Western and Astbury anticlines and an easterly trend on the Werrington Anticline. Slickenside striations measured on faults on the south-eastern limb of the Western Anticline at High Lane opencast site (Figure 7a) indicate that these faults have a consistent extension direction that parallels the axis of the anticline. It is probable that these faults formed at a late stage in the folding.

Minor structures

The excavations at High Lane and Brown Lees opencast sites (Figure 7a) revealed the presence of abundant minor thrusts and folds in thinly bedded (0.1 to 0.5 m thick) sandstone and mudstone sequences. The thrusts have small (0.1 to 0.5 m) displacements, are directed both to south-east and north-west, and pass into bedding-parallel detachment surfaces. Striations on these show that the transport direction of the thrusts was broadly perpendicular to the axis of the Western Anticline. The strain in the mudstones commonly gave rise to folds and localised cleavage. The thrusts are folded, and are postdated by both south-easterly and north-westerly verging large-scale kink folds, which have wavelengths of less than 1 m, and are vertically persistent in 10 m-high exposures. The structures associated with the thrusts include small-scale imbricate stacks, pop-up structures and triangle zones (areas below the intersection of thrusts and backthrusts; Butler, 1982), which are common in both sandstones and ironstones. Shortening in coals is commonly taken up by thrusts; they are rarely folded, and the mudstones between leaves of coal seams are commonly sheared. Cleat orientations are discussed by Corfield (1991).

As interpreted by Corfield (1991), the south-easterly directed minor thrusts were probably formed during an early stage in the growth of the anticline, by a process of layer-parallel shortening. He suggested that the kink folds resulted from a later stage of the same process, and that the growth of the Western Anticline subsequently rotated these early-formed structures to their present orientation.

POST-VARISCAN STRUCTURES

Faults

Few of the faults in the district have a proven Variscan, or earlier, history; the great majority are post-Variscan in age. Those that affect Carboniferous rocks, but not the Triassic cover, are mainly early Permian in age and thus probably were formed during collapse of the Variscan Orogen. Most Permian faults were reactivated during Triassic, and possibly Jurassic, extensional phases.

The orientation of post-Variscan faults is broadly north to south (Figure 43), though varies between three parts of the district, as described below. In each of these areas, the trend of secondary faults is broadly orthogonal to the main trend.

Post-Variscan faults in the area west of the Astbury Anticline, the Western Anticline and the Sidway Fault (including part of the Permo-Triassic Cheshire Basin)

The post-Variscan faults in this area generally trend north-north-east, parallel to the Pontesford Lineament. Towards the southern extent of the Western Anticline they include the Madeley, Hodnet and Sidway faults (Figure 43). In the Permo-Triassic rocks of the Cheshire Basin (Chapters 4 and 5) the fault orientations are somewhat more diverse, though some of the faults that parallel the Pontesford Lineament, such as the Wem Fault and the Bridgemere Fault (Figure 45), have downthrows to the west of over 1 km. White (1949) interpreted the gravity gradients on the south-eastern side of the Cheshire Basin (Figures 2, 3) to represent these faults, a view borne out by recent gravity work using data incorporating detailed traverses (Brooks, 1961; Buller-well, in Taylor et al., 1963). These, and other faults of similar trend, such as the Red Rock Fault, controlled subsidence near the eastern margin of the Cheshire Basin in Permian, Triassic and probably Jurassic times. The thickening of the Permo-Triassic sequence towards the centre of the Cheshire Basin occurs progressively across several faults (Figure 45). This is illustrated by the gravity gradient zone associated with them, which lies almost entirely within the outcrop of the Triassic rocks and has almost disappeared at the surface crop of the eastern-most faults. The steep upper parts of several faults, including the Red Rock Fault (discussed by Challinor, 1978b), where they affect Permian and Triassic rocks, are interpreted to be hanging-wall 'short-cuts' (Gibbs, 1983). The upper parts of the faults were formed during Permo-Triassic extension, when Variscan reverse faults of the Pontesford Lineament were normally reactivated and propagated through the Permian and Triassic strata.

Post-Variscan faults between the Western Anticline and Sidway Fault, and the Longton and Sandon faults (including part of the Permo-Triassic Stafford Basin)

The Apedale, Newcastle, and Longton faults, trending north to north-north-west, extend from the Swynnerton Fault in the Stafford Basin virtually across the whole worked area of the coalfield northwards to the 'saddle' area of the Western Anticline. All decrease in throw towards the north-west. The Apedale Fault has the largest displacement, most of which accumulated in Permian times; this reaches a maximum of over 600 m west of Newcastle-under-Lyme. The mining evidence indicates a relatively shallow fault plane dip of about 45° east.

Southwards there is an increase in the number of faults (Figure 44) and an increase in the net amount of east–west fault extension. The Longton Fault undergoes a local change in strike from north-north-west to west-north-west near Longton. The effect of this change in strike is a southward broadening of the distance between the surface traces of the Newcastle Fault, throwing east, and the Longton Fault. Three major north-trending normal faults occur in this zone, the Hollybush, Crowcrofts and Blurton faults (Figure 44). In cross-sections these faults form a conjugate system, with faults dipping both to east and west, and displacing each other.

The west-north-west-trending segment of the Longton Fault near Longton shows an unusual geometry. The segments of the fault to north and south are planar to a depth of over a kilometre, with an average westerly dip of 50°. However, in the Longton segment the fault has a near-surface dip of 45°, which steepens dramatically downwards to 70° at a depth of about a kilometre. The throw also increases from an average of 50 m in the segments to north and south to about 150 m near Longton. The Berry Hill, Fenton and Shelton faults are easterly trending splays of the Longton Fault to the north of the Longton segment (Figure 44). These faults are similarly steeply dipping (85° to the south in the case of the Berry Hill and Shelton faults). The geometry of the Longton segment suggests that there has been a significant component of oblique-slip movement there.

West of the Longton Fault, there are several easterly trending faults, such as the Florence Fault (Figures 43 and 44), that are apparently restricted to individual blocks between north-north-west-trending faults. The age of these faults, relative to the north-north-westerly faults, is uncertain. However, the Swynnerton Fault, also of easterly trend (Figure 43), appears to postdate the north-north-westerly faults, as the southerly throw appears to be similar in both the Carboniferous and Triassic rocks. Evidence from seismic reflection data suggests that the fault is a product of dip-slip reactivation of a northerly directed Variscan reverse fault at depth.

Post-Variscan faults east of the Astbury Anticline, Longton and Sandon faults (including part of the Permo-Triassic Needwood Basin)

Unlike the faults described above, which are relatively well understood because of the available mining and geophysical evidence, the faults in this area are poorly known. The dominant trend is north-north-west, though in the Longton area they trend nearly north–south. The Callow Hill Fault, of north-westerly trend, extends westwards into the Berry Hill and Shelton faults (Figure 43). North-east of Weston Coyney the Callow Hill Fault cuts out the whole of the Hawksmoor Formation, and at the eastern limit of the district, where it has a throw of at least 75 m, it also cuts out the Denstone Formation.

Folds

The post-Carboniferous rocks of the Cheshire, Stafford and Needwood basins have all been gently folded. In the axial region of each basin a broad syncline is developed which approximately parallels the dominant fault trend in that basin. Analogy with other districts suggests that this folding is mostly of Palaeocene and Miocene age (Green, 1989).

TECTONIC HISTORY

The above account describes the distribution and geometry of structures in the district, but provides only a broad indication of their history. Variations in thickness, distribution and facies within several of the formations in the district illustrate the growth of faults and folds during sedimentation. The development of many of these structures can thus be effectively dated, and the dating may be better constrained here than in adjoining districts. The following account gives a brief review of the Phanerozoic structural history of the district, which necessarily takes into account evidence from adjoining areas, especially for pre-Carboniferous and post-Triassic events.

Most Carboniferous and Permo-Triassic sediments were deposited in actively developing basins. According to the widely accepted model of McKenzie (1978), basin subsidence takes place in two phases, with initial extensional rifting of the crust being followed by thermal subsidence.

During periods when the crust is being stretched, or extended, it thins. This causes subsidence at the surface, controlled by faults that separate more rapidly subsiding basins from intervening blocks of slower subsidence. Such tectonic regimes are referred to as extensional and may be recognised in the early parts of the Carboniferous, Permian and Triassic systems in the present district.

When crustal thinning by extension ceases, subsidence often continues in the area of the thinned crust, though the style of subsidence changes. Differences in subsidence rates across faults are largely eliminated as movement on faults that had taken up the crustal thinning during extension declines. The cooling and thickening of the mantle below regions of thinned crust is considered to be responsible for this post-extensional subsidence. Episodes of such 'thermal subsidence' can be recognised in the later parts of the Carboniferous, Permian (tentatively) and Triassic in the present district.

Lower Palaeozoic

By analogy with districts to the south, it is likely that during the Lower Palaeozoic a relatively thin succession was deposited on the Midlands Microcraton (Chapter 2) compared with that in the Welsh Basin to the west of it. It is also unlikely, in view of their absence in other districts on the microcraton (Old et al., 1987), that Ordovician rocks occur on the microcraton in the Stoke-on-Trent district. Also, the sediments deposited on the microcraton were probably deposited in shallow-water environments, contrasting with those of deep-water origin to the west.

Devonian

The sedimentary history of the early part of the Devonian in the region is unknown because of the effects of the Caledonian Orogeny, which culminated in the middle Devonian. The orogeny is believed to have caused reverse reactivation of faults associated with the Pontesford Lineament. The Lower Palaeozoic rocks deposited west of the lineament in the Welsh Basin were highly deformed, in contrast to those on the microcraton (Chapter 2). The marginal area of the microcraton (represented by a zone with relatively shallow magnetic basement in the Apedale area) may also have suffered major deformation.

By analogy with the sequence in the Caldon Low Borehole in the adjoining Ashbourne district (Chisholm et al., 1988) it is possible that, following the Caledonian deformation, sedimentation in the Stoke-on-Trent district may have recommenced in the late Devonian, although there is no supporting evidence.

Carboniferous

During the Carboniferous, northern Britain went through a series of tectonic phases. First it was affected by extension, which created a widespread system of blocks and basins such as the Market Drayton Horst and the North Staffordshire Basin (Chapter 3). This was followed by a phase of thermal subsidence. Finally, Variscan compression caused folding and reactivation of major faults (Leeder, 1988). This sequence of events is best documented in those British coalfields where strata of post-Bolsovian age are preserved, such as the Midlands coalfields (Besly, 1988) and the South Wales Coalfield (Kelling, 1988). Of these, the Potteries Coalfield provides some of the best evidence for the timing of the onset of the Variscan Orogeny.

Rifting in North Staffordshire commenced in late Devonian or early Dinantian times with the formation of a north–south-trending half-graben, the North Staffordshire Basin. The half-graben geometry is indicated by the depocentre being on the western side of the basin (Trewin and Holdsworth, 1973) and by geophysical evidence (Lee, 1988). The thinning of the crust during Dinantian extension was accompanied by basaltic volcanism (Chapter 3).

Some time between the late Brigantian and early Namurian, fault-controlled extensional subsidence gave way to general thermal subsidence. The distinction between the North Staffordshire Basin and the Market Drayton Horst, clear during the Asbian Stage, decreased in the Brigantian with the deposition of dominantly mudstone-rich formations both on and off the Market Drayton Horst (Chapter 3). By mid-Langsettian (Westphalian A) time the pre-Brigantian topography was buried beneath a blanket of sediment and the North Staffordshire Basin had a much reduced control on sedimentation. Some effects may still be discerned in the early Langsettian, though only in the eastern edge of the Potteries Coalfield (Figure 7b).

Regional thermal subsidence persisted into the later Carboniferous, but with decreasing importance as its effects were progressively overwritten by the growth of Variscan structures. The extensive mining data in the district, incorporating information from boreholes, mine plans and seismic profiles, have enabled detailed analysis of structures active during Westphalian sedimentation, and provide some of the best evidence for the existence of growth folds. This evidence is summarised in isopachyte maps (Figures 7b–k), which indicate that the major folds were active as growth folds from the mid-Duckmantian (Westphalian B) onwards; a condensed, but complete, sequence is present over the crests of the folds. These early movements apparently did not result in the formation of unconformities over the fold crests. However, in the Midlands coalfields to the south, a late Bolsovian (Westphalian C) deformation event caused uplift which effectively ended local coal formation. This event correlates with the first major northward advance of red beds into the Coal Measures environments of the present district (Besly, 1988).

Figure 7b shows a pronounced hinge line aligned slightly west of north and close to the inferred position of the Lask Edge Fault. The location of this hinge, with thick strata to the east of it, shows the continued influence of the Market Drayton Horst and North Staffordshire Basin, inherited from the Dinantian, up to the horizon of the King Coal. This influence then declines, and through the rest of the Langsettian and

Duckmantian (Figures 7c–e) is represented only by a generalised eastward thickening. Gentle north-easterly folds superimposed on this trend start to appear in the late Langsettian (Figure 7c) and continue to grow, with small shifts of axis, through the Duckmantian (Figures 7d–e). Larger shifts of axis occur early in the Bolsovian (Figures 7f–g) and the new pattern is fully established in the late Bolsovian (Figure 7h), with accelerated growth of north-easterly folds through into the Stephanian (Figures 7i–k).

Differential subsidence and seam splitting are more common at some stratigraphical horizons than at others during this time; for instance, the interseam intervals between the Hams and Yard coals (Figure 14) and, to a lesser extent, between the Bullhurst and Cockshead coals (Figure 12), between the Rowhurst and Rowhurst Rider, and between the Winghay and Bay coals (Figure 16) are notable for their thickness variations. Large-scale rotational slumps have been noted at several of these horizons: for instance at the Winghay-Bay horizon at Hem Heath Colliery (Figure 7a), and through the Burnwood Coal at Silverdale Colliery (Figure 7a), representing the base of a rotational slump at the horizon of the Rowhurst Coal. These horizons may represent periods when the crust was at least locally in tension, thereby allowing growth of faults. Tensional conditions certainly existed during deposition of the Rowhurst Coal, as is shown by mudstone-filled fissures within it at Silverdale and Florence Collieries (information from J O'Dell, of British Coal).

On the basis of detailed plots of interseam intervals, Corfield (1991) suggested that some thickness variations may relate to growth folds caused by the early Variscan compressive stresses. Some structures, including mudstone intrusions in the Cannel Row Coal at Hem Heath Colliery (Figure 7a), and slumps and floor-rolls common over large areas between Florence and Silverdale collieries (Figure 7a), may relate to widespread horizontal movement on shear-planes within the coals.

However, better evidence for growth folding comes from the Bolsovian (Westphalian C) and Westphalian D successions. Isopachyte maps (Figures 7f–k) all show an elongate north-east-trending depocentre between Hanley and Newcastle-under-Lyme, and a marked thinning towards the north-west and south. The long-lived nature of this depocentre and its consistent trend both indicate that its position was structurally controlled. There is no evidence to suggest that it was controlled by faults; it is more likely that the flanking Western Folds and Southern Folds were actively growing. The anticlines were probably formed in response to the reactivation of deep-seated basement faults (see above); evidence for movement of major reverse faults during the Bolsovian is found elsewhere in the Midlands, for instance in Warwickshire and Coalbrookdale (Besly, 1988).

The forces responsible for the development of these anticlines also caused localised uplift of blocks on the margin of the Wales–Brabant High (Chapter 3) to the south. The erosion of these blocks caused an influx of clastic detritus into the present district. Evidence for the clastic influx in the Bolsovian sequence can be seen in the lower gamma ray response of the sequence compared with that of the underlying Coal Measures (Figure 8). Also, exotic clasts of probable southern origin occur in lags at the bases of sandstone channels in the Upper Coal Measures (lower Bolsovian) (information from J O'Dell, of British Coal), and in espley sandstones in the Etruria Formation. Emergence and associated oxidising conditions occurred near the uplifted blocks and, as a consequence, to the south and west of the coalfield, near the Market Drayton Horst, the base of the red-bed Etruria Formation lies at progressively lower levels in the sequence (Figures 7l and 18) (Chapter 3).

The uplift also gave rise to unconformities. These have been recognised in the south Midlands within, and at the top of, the Etruria Formation (Besly, in press). Such unconformities have not been recognised in this district, though the existence of an unconformity at the top of the Etruria Formation was suggested by Malkin (1961).

Gravity profile models (such as that shown in Figure 3) provide no evidence of large thicknesses of Carboniferous rocks beneath the Permo-Triassic rocks of the Cheshire Basin. This contradicts the conclusions of Abdoh et al. (1990), who interpreted the gravity data as indicating the presence of Carboniferous rocks, extending to considerable depths (6km below OD), particularly towards the flanks of the Cheshire Basin. It is suspected that such an interpretation is not required if appropriate densities, background fields and seismic reflection evidence are used in modelling the gravity data. The model preferred here suggests that only minor thicknesses of Carboniferous rocks remain in the Cheshire Basin, as a result of erosion following Variscan inversion.

Permian

In the early Permian the Carboniferous and older rocks were widely deformed during the Variscan Orogeny (Besly, 1988). It was probably in the main phase of this orogeny that the Eastern Folds developed into their present form.

The relative Carboniferous and Triassic displacements of the Longton Fault west of Meir Heath, and the Florence Fault near Weston Coyney, clearly show that most dip-slip movement of faults that affect Carboniferous rocks predated Triassic sedimentation. Regional evidence (Tonks et al., 1931; Thompson, 1985; Evans et al., 1993) shows that the faults were active during the early Permian, probably during mid to late Rotliegend times. Such movement is demonstrated by the Wem and Bridgemere faults (Figures 43, 45), which were active during sedimentation of the early Permian Collyhurst Sandstone. This fault activity was probably related to a phase of east–west extension (Evans et al., 1993).

The Manchester Marl, the late Permian northern equivalent of the Bold Formation, is partially marine in origin. The transgression of the 'Bakevellia Sea', in which it was deposited, may have been caused by rifting in the Arctic or North Atlantic (Ziegler, 1981). However, these formations lack the coarse-grained sediments that are normally associated with regional extension. This fact led Evans et al. (1993) to suggest that the Manchester

Marl and Bold Formation were deposited during a period of thermal subsidence, perhaps accompanied by a eustatic rise of sea level, following on from the early Permian extension.

Triassic

The Sherwood Sandstone Group was mainly deposited in an extensional setting, as indicated by substantial thickness differences of parts of the group across basin-bounding faults. The main phase of extension began during deposition of the Chester Pebble Beds, Kidderminster and (lower) Hawksmoor formations. The conglomerates in these had their source in uplifted areas well to the south of the district (Steel and Thompson, 1983). Subsidence continued during the deposition of the Wildmoor Formation, the upper part of the Hawksmoor Formation and, in the Cheshire Basin, the Wilmslow Sandstone and ?Bulkeley Hill formations. The greatest differential subsidence occurred across broadly north–south-trending faults, such as the bounding faults of the Cheshire, Stafford and Needwood basins (Figure 45).

Following this phase of extension, much of the Triassic sequence in the district appears to have been uplifted and eroded, so forming an unconformity. The erosion mostly affected uplifted horsts between the basins, thus further reducing the thin early Triassic successions present on these basin margins. Figure 45 shows a seismic section across the eastern boundary faults of the Cheshire Basin (reactivated faults of the Pontesford Lineament) in the south-western part of the district. The substantial eastward thinning of the Wilmslow Sandstone (containing the top silicified zone) across the fault blocks probably reflects both growth faulting during deposition of the formation, and differential erosion during the subsequent uplift. In the Stafford Basin, an unconformity may be indicated by the northward thinning of the Wildmoor Formation in the Beech Cliff area, and by the absence of this formation below the Kibblestone Formation at the eastern margin of the basin. The Wilmslow and ?Bulkeley Hill formations were also subject to erosion within the Cheshire Basin (Evans et al., 1993, fig.8), but the cause of this episode is not understood.

The varied thicknesses and depositional environments represented by the Helsby, Bromsgrove and Kibblestone formations indicate that these beds may have been deposited during an extensional period. However, the overlying formations of the Mercia Mudstone and Penarth groups contain little evidence for widespread fault activity during their deposition, though isolated instances are recorded (Tucker and Tucker, 1981). Evans et al. (1993) suggested that they were deposited during a period of overall thermal subsidence.

Jurassic and Cretaceous

The thickest Hettangian (early Jurassic) sequence in onshore Britain occurs in an outlier in the hanging wall of the Wem Fault at Prees, in the Cheshire Basin to the south-west of the Stoke-on-Trent district (Poole and Whiteman, 1966). Evans et al. (1993) showed that the rate of deposition was higher in the early Jurassic than it was through the late Triassic and suggested that the basin was subject to a further episode of extension in the early Jurassic. The pre-existing major faults were probably reactivated at this time. For instance, the Wem–Red Rock fault system displays a significant post-Triassic normal throw. Seismic reflection data, and gravity evidence (based on zones of steep gravity gradients, Figures 2 and 3) indicate that, in the south of the district, most of this throw is taken up by the Wem and Hodnet Faults but in the north of the district it is taken up by the Red Rock Fault. The throw of the latter rapidly increases northwards, from about 200 m near Madeley to about 3 km in the region of Congleton, in the Macclesfield district.

The later Mesozoic history of the district can only be assessed by indirect means. An indication of the depth of Mesozoic burial is provided by the densities of mudstones in the Mercia Mudstone Group which suggest about 2.2 km of burial by Jurassic and Cretaceous rocks (Evans et al., 1993). This figure is endorsed by apatite fission track studies (Green, 1989); samples of Sherwood Sandstone from the horst between the Needwood and Stafford basins, about 30 km to the south of the district, revealed elevated palaeotemperatures due to about 2 km of burial by Jurassic and Cretaceous rocks (see below).

Palaeogene and Neogene

Apatite fission track analysis of the Sherwood Sandstone in the Stafford district (see above) showed that at approximately 60 Ma a phase of cooling, interpreted to represent regional inversion, probably commenced in the Palaeocene and was succeeded by a second phase at between 20 to 30 ± 5 Ma, probably in the Miocene. Green's (1989) estimate of uplift of over 2 km may be an overestimate (Holliday, 1993) though it is in broad agreement with that of Thompson (1985), Colter (1978), and Evans et al. (1993) who estimated similar figures for the Stafford Basin, East Irish Sea Basin and Cheshire Basin respectively.

The orientation of the Butterton–Swynnerton dykes (Chapter 6) indicates that during their emplacement, in the Palaeogene, the crust was in a state of east-north-east to west-south-west tension. It is possible that the magmas used pre-existing faults as conduits: the close proximity of the dykes to the Apedale Fault suggests that this may be the case.

NINE

Applied geology

MADE GROUND

Man-made deposits cover large areas of the district, particularly in Stoke-on-Trent, Newcastle-under-Lyme and Crewe. Details of the distribution and character of made ground in most of the Potteries Coalfield are given by Wilson et al. (1992). Made ground can be broadly subdivided into two types: that which has been built up on a pre-existing land surface; and that which infills previously formed excavations. Backfilled excavations are most common in rocks of the Upper Coal Measures and the Etruria Formation. The main categories of made ground material identified by Wilson et al. (1992) are domestic and industrial waste, colliery waste, ironworks slag, ceramic rejects, hardcore, and clay, bricks and tiles. Made ground, particularly colliery waste, commonly forms large mound-like topographic features, such as the tips at Florence [917 415], Heron's Cross [894 443], Forest Park [885 488] and Sneyd [882 497]. Other made ground deposits are topographically subdued, for instance the extensive, virtually flat-lying tracts of made ground surrounding Hem Heath Colliery [887 415]. The potential resources and hazards that made ground deposits present are described below in the sections on economic geology and hazards.

ECONOMIC GEOLOGY

Stoke-on-Trent owes its origins to the geological resources that it overlies, many of which have been readily available for surface extraction because of the structure of the district. However, different resources were sought at different times. For instance, in the Middle Ages clays for the pottery industry were the most important resource, though their value was superseded in the industrial revolution by ironstone, coal and brick clay. Other resources, such as sand, gravel and oil have only become important in the present century (Rees, 1993). The geological resources of the district are still of economic importance, and there is a need to avoid sterilising them by urban development (Shryane, 1990). In the Potteries Coalfield, identification of potential resource areas through applied mapping has gone some way to combating this threat (Wilson et al., 1992).

Coal

The Potteries Coalfield contains one of the thickest sequences of Coal Measures, and arguably the greatest cumulative thickness of workable seams, in Britain. During most of its history (summarised by Taylor, 1981) the coal produced in the North Staffordshire Coalfield was consumed locally. The industry was initially stimulated by the expansion of the local pottery industry and the growth of the Nantwich salt industry. However, the main impetus was provided by developments in the iron and steel industries, particularly the increasing use of blast furnaces in the middle of the 19th century, and the iron and steel boom of the 1870s. The coal industry continued to grow until near the end of the 19th century when production reached a plateau of between 6 and 7 million tonnes per year; this figure was maintained until the recent decline (Besly, 1993). During this century the economic history of coal mining in the district has generally mirrored national trends. Since nationalisation in 1947, coal has been increasingly used for electricity generation, both inside and outside the district. The long history of deep mining is now coming to an end, although several licensed adit mines continue to work coals on a limited scale (Jones, 1969). These include Acres Nook Mine (Rowhurst Coal), Apedale Mine (Great Row to Bassey Mine coals), Bank Top Mine (Rowhurst Coal), Haying Wood No.2 Mine (Rowhurst Coal), Little Sherriff Mine (Winghay Coal); Lycett Mine and Parklands Mine (Bullhurst and Cockshead coals). Coal is still mined by surface excavations at Brown Lees and High Lane opencast sites. However, the environmental impact of large opencast workings (Trigg and Dubourg, 1993) may cause new sites to be limited in number and size. A review of possible opencast coal resources is given by Wilson et al. (1992).

Throughout most of the history of the industry the depth at which coal was mined increased gradually. At the beginning of the last century very little coal was mined at depth, but by the middle of the century coal was commonly mined at several hundred metres below the surface. As a result, most seams were mined near the outcrop in the early 19th century, and deeper areas, towards the axis of the Potteries Syncline for example, were mined much later. Pits, as well as becoming deeper, also tended to get larger. The earliest were usually bell pits, which were simple, shallow unsupported shafts, mostly working a single seam. Later, adits and supported shafts allowed deeper mining. The roofs of workings were left supported by unworked pillars of coal, though latterly longwall mining, in which the roof is allowed to collapse behind the working face, has been almost exclusively used. Engineering problems of the coalfield have been discussed by Scurfield (1958). Economies of scale and uneconomic working practices have led to a steady decline in the number of collieries.

The most important coalfield of the district is the Potteries Coalfield, though the small Shaffalong Coalfield has been a target for opencasting since the last war.

The character of the coals largely depends on their initial composition, proximity to surface and rank. Details of the character of most seams as low as the Winpenny Coal, in the Pie Rough (Keele No.1) Borehole have been published (Millot et al., 1946). Maps showing seam character (Fenton and Rumsby, 1962) have been published by the National Coal Board (1961) for the Moss, Bowling Alley and Banbury coals and by Millot (1941) for the Cockshead Coal. Sections across the coalfield, showing variations in rank, have been published by Crofts (1952).

The composition of coal seams reflects their inferred depositional environment (Figure 8). Those deposited in lower delta plain environments tend to be below mineable thickness, to have high sulphur contents and to have abundant clay partings. Those deposited on the upper delta plain are commonly thick and have low ash contents. Coals near surface generally have lower chlorine and sulphur contents than seams at depth.

The rank of the coals appears to be a function of increased geothermal gradients associated with Permian, rather than Mesozoic, burial (Chapter 8), as coal rank is displaced across faults that are seen to have had pre-Triassic extensional displacements (Crofts, 1952). All coals in the Potteries Coalfield are bituminous and most are in the high volatile category, making them suitable for general industrial use. The more deeply buried coals towards the centre of the Potteries Syncline, such as those formerly mined at Holditch Colliery, are also strongly caking, and so suitable for gas production and coking. The volatility of all coals increases in the area of the Western Folds.

Until nationalisation, seams of upper delta plain facies were generally most in demand. Those between the King Coal and Maltby Marine Band were favoured because of their coking properties and low sulphur content (needed in iron and gas manufacture), and the cannels and associated coals of the Upper Coal Measures were valued for their long flame (needed for kiln firing). After nationalisation, since when the demand for coal has mainly come from the electricity industry, the high sulphur, poor quality, but cheaply worked Rowhurst and Winghay coals were increasingly mined from the south of the coalfield. However, with the need to reduce acid emissions from power stations, interest in low sulphur seams, such as the Cockshead and Banbury, has been rekindled. The low sulphur and chlorine content of opencast coal has increased its popularity also with electricity generators.

Ceramic clays

The history of the local **pottery industry**, for which the district is famous, is reviewed by Haggar et al. (1967) and by Seckers (1981). The industry was originally based on local clays, which were generally used to manufacture coarse earthenware. It is no coincidence that the towns of Longton, Fenton, Hanley, Burslem and Tunstall are all sited on the outcrop of the Upper Coal Measures, which was able to supply large quantities both of suitable clays and of coals for firing. Weathered mudstones from the Etruria Formation were also used to make red-bodied wares. A fashion for whiteware developed during the latter part of the 18th century, and from the early 19th century most pottery was made from white-firing ball clays and china clays imported into the district, rather than from indigenous clays. However, the industry stayed in the area because of the ready availability of coal, the supply of local clay for the manufacture of kilnware, and the existence of a skilled workforce. Local clays from the Coal Measures, particularly those above the Bassey Mine and Peacock coals (Gibson, 1925), continued to be used in the manufacture of sanitary ware for much of this century.

Local mudstones have also been extensively used for **brick and tile manufacture**. The industry has had a long history and some products, such as the Staffordshire 'Blue Brick', a high-strength engineering brick obtained by firing local clays under reducing conditions, have become well known (Celoria, 1971). The clay, fired at a high temperature, vitrifies to produce a strong brick with low water absorbency.

The Etruria Formation is the prime source of clay for the important local brick and tile industry. The bulk mineralogy of the clay (i.e. the relative proportions of kaolinite, illite, quartz and haematite), and the lack of impurities such as carbon, sulphur, soluble salts and, except locally, calcite, make it very suitable for the manufacture of high quality and high strength facing bricks, paving blocks, and roofing and floor tiles. The Etruria Formation is extensively worked in quarries at Bankeyfields (Holly Wall) [849 519], Chatterley [845 507], Apedale South [827 488], Gorsty Farm [822 475], Walleys [832 460], Fenton Manor [885 455], Hanford [856 430], Keele [788 450], Knutton [826 468] and Lightwood [920 410]. Wilson et al. (1992) divided the Etruria Formation into three stratigraphical units (Chapter 3), based broadly on their suitability for brick manufacture. The best clays occur in the lower and middle divisions, although these are increasingly being sterilised by urban development (Wilson et al., 1992). The lower-grade clays of the upper division could be improved by blending with other clays, perhaps from the Radwood Formation; correlatives of the latter formation have been extensively used for brickmaking elsewhere in the Midlands.

The mudstones in the Coal Measures were at one time widely worked for brick manufacture (especially those between the Hoo Cannel and Spencroft coals and between the Twist and Bungilow coals), but are now only periodically used at Birchenwood Brickpit [855 540].

Working of clays from formations younger than the Etruria Formation has been only on a very small scale. Clays from the Radwood Formation were the most commonly worked, and a pit near Willoughbridge Wells [744 394] currently produces clay for brick making. There may be scope for mudstones of the Radwood Formation to be worked more widely in the future. Mudstones from the Mercia Mudstone Group have been used on a small scale to make bricks and pipes, such as those from the Denstone Formation near Blythe Bridge [938 426]. Tills were commonly worked for brick manufacture in the Crewe area (Gibson, 1925; Thompson, 1980).

Ironstone

Blackband and clayband ironstones occur in the Coal Measures, and are common in the Upper Coal Measures where they occur both as continuous layers and as nodules. They are no longer of economic interest, but their usage is described here as they have been of major economic importance to the district in the past. Despite the ready availability of these iron ores, and the abundant supply of local coal suitable for coke manufacture, the iron and steel industry in the district was slow to develop, and it was not until about the middle of the 19th century, with the development of blast furnaces and the construction of the railways, that local iron production grew rapidly.

Analyses indicate ranges for blackband ironstones of 29 to 39 per cent Fe, 0.19 to 0.44 per cent P, 0.64 to 11.0 per cent CaO; and ranges for clayband ironstones of 25 to 46 per cent Fe, 0.17 to 1.20 per cent P and 1.24 to 5.07 per cent CaO (Slater and Highley, 1978).

The blackband ironstones were most popular for the production of iron as they are highly carbonaceous, making them in part self-calcining. They also are closely associated with coal seams, which were usually mined at the same time as the ore. Iron produced from the blackband ironstones was suitable for steel manufacture, particularly by the Siemens-Martin open hearth process, which tended to be used locally. The value of the ironstones was such that they were mined underground as well as at surface; several collieries started by working ironstones. By the mid-1920s the horizons worked were mostly confined to the Gubbin, Cannel Row and Pennystone ironstones, and the Burnwood Ironstone which was usually mined with the coal. Descriptions of most of the ironstone horizons are given by Homer (1875), Hallimond (1925) and, in more detail, by Gibson (1905) and Dewey (1920).

The importance of Coal Measures ironstones declined during the latter part of the 19th century and into the 20th century, due to the increasing use of Jurassic ores and ultimately to the availability of high-grade, low-priced imported iron ores. Production in north Staffordshire finally ceased in the 1950s, the district being perhaps the last in Britain to see Coal Measures ironstones exploited.

Iron and steel production became centred on coastal sites from the 1960s onwards, the local industry went into decline, and by the late 1970s had effectively ended (Harrison, 1983).

Sand and gravel

Sand and gravel are worked principally from the Sherwood Sandstone Group and from Quaternary glaciofluvial deposits.

Large quantities of sand and gravel have been produced from the Sherwood Sandstone Group in the past. Most of this has been from the Chester Pebble Beds, Kidderminster and Hawksmoor formations which contain a wide range of sand and gravel grades by virtue of the conglomeratic horizons they contain; it is these horizons which have been exploited primarily. Only the quarries at Lightwood [924 420] appear to have worked the sand-dominated parts of the Hawksmoor Formation. Other disused quarries, many of which remain unfilled, include those at Hulme [930 445], Trentham Park [855 403], Acton [820 413], and Tadgedale [733 366] (Figure 32). The quarries at Willoughbridge [750 385] are currently active. Here most of the sand and gravel is mechanically ripped because the Kidderminster Formation near-surface is weathered and friable. The demand for further local resources has led in recent years to the publication of reports outlining reserves to the east (Piper, 1982) and south (Malkin, 1985) of Stoke-on-Trent. Planned quarry sites near Moddershall and south of Hanchurch (Shryane, 1990) are being reviewed.

The Wildmoor Formation, also part of the Sherwood Sandstone Group, was formerly used as a source of naturally bonded foundry sand. Its usage for this purpose was facilitated by the consistency of sand grade within most excavations; notable quarries include those near Baldwin's Gate [799 395] and at Hatton [831 367].

Glaciofluvial sands and gravels have only been worked on a small scale in the past, as at Rye Hill [802 501], Betley [752 487] and Chorlton's Moss [779 385]. These deposits often form broad spreads at surface, as at Silverdale [813 466], Miles Green [803 498], Mosshouse [807 534], Alsager [802 549], Kidsgrove [833 543], near Madeley [778 434] and Maer [770 385]. Alternatively, they fill glacial overflow channels (Chapter 7), such as that which runs from Kidsgrove [841 540] through Bathpool Park [837 524] and Longport [857 500] to Stoke [879 453], and that which follows the present course of the Trent from Milton [904 500]. Sands and gravels are currently produced at Hough Mill, Wybunbury [702 492], and sands at White Moss, Alsager [775 550], for use by the construction industry. The prospects of more extensive use of glaciofluvial sand and gravel deposits near Newcastle-under-Lyme and Stoke-on-Trent are restricted by urban expansion (Wilson et al., 1992). Outside these areas some of the spreads probably have economic potential.

Building stone

The working of building stone in the district has been on a limited scale. Stone has been quarried from the Cheddleton Sandstones near Endon [9198 5277], Stanley Moor [9240 5125] and Stanley Pool [9290 5202]; from the Chatsworth Grit at Armshead, near Werrington [935 483], Bagnall [9299 5080] and from south of Brown Edge [914 527, 916 524]; and from the Rough Rock north-east of Deep Hayes Reservoir [964 534], at Washerwall Quarry, Werrington [935 477], and at Baddeley Green [9140 5126]. Sandstones from the younger Carboniferous formations have been very little utilised for building. In the Coal Measures, the Ten-Feet Rock was worked at Coalpit Hill, Talke [8252 5355], and at Alsagers Bank [808 485]; the sandstone above the Bungilow Coal was worked at Tunstall [871 522]. The Hanchurch Sandstone was worked in Job's Wood Quarry [823 460] from the Newcastle formation and at Quarry

Bank Quarry [807 461] from the Keele formation; a higher sandstone in the Keele formation was quarried near Keele [8203 4565]. Sandstones in the Radwood Formation have been quarried at Willoughbridge Lodge [740 386].

Much building stone has been worked from rocks of the Sherwood Sandstone Group. The most popular formations for provision of quality building stones are the Bromsgrove and Helsby formations, especially where cemented with barytes. The former has been extensively quarried at Beech [855 382] (Middleton, 1986b).

Rock salt

Two major halite formations, the Northwich Halite and the Wilkesley Halite (Chapter 5) crop out in the present district. The outcrop of these, particularly the latter, is very extensive, though attempts to prove or work the halite have been minimal. The only area where the Wilkesley Halite, the purer of the two halite formations, is potentially extractable by modern controlled pumping methods is in a small area in the western part of the district, where it is overlain by mudstones (Brooks Mill Mudstone Formation, Blue Anchor Formation and Penarth Group). Elsewhere the Wilkesley Halite is too thin, due to near-surface solution, for controlled pumping to be used. The Northwich Halite is present beneath a large part of the Mercia Mudstone outcrop in the district, and offers considerable potential for future brine extraction using the technique of controlled brine pumping by creation of artificial cavities.

Base metals

Occurrences of base metal ores in the district are of academic interest only; none is potentially workable. Vein-hosted malachite is present in the Helsby Sandstones in the south-western part of the district, and malachite was worked at one time at Bearstone [7217 3924]. Locally, the malachite occurs as a cement, as in sandstones to the north-east of Norton-in-Hales [7228 3944] (Appendix 2). The malachite is intimately associated with baryte, which is widespread as a cement in the Sherwood Sandstone Group on the margins of the Cheshire Basin (as in samples E32808a,b; Appendix 2) (Wedd, 1899). Isotope work on the baryte (Naylor et al., 1989) shows that most of the sulphur was derived from evaporites in the Mercia Mudstone Group. This implies that the minerals were deposited from formation waters that had migrated out of the Cheshire Basin. The galena and sphalerite commonly recorded from the Coal Measures (Cadman, 1901; Gibson, 1905) (and as may be presently found at Birchenwood Brickpit [855 540]) are likely to have a different origin, probably by circulation of meteoric waters during early Permian extension (Chapter 8) (Besly, 1993).

Oil and gas

Oil, from organic rich mudstones between coals, has often been reported in mines in the Potteries Coalfield (Torrens, 1994). Reports are mainly of seepages, usually from coals of the Upper Coal Measures and the Bowling Alley and Holly Lane seams (Gibson, 1905), though shafts and roadways have on occasion been flooded (Challinor, 1990; Wilson et al., 1992). Indeed, Century Oils Ltd. started out by selling oil from where its present headquarters are sited in Hanley [878 477]. In many cases oil has migrated into local sandstone reservoirs, as in the Bullhurst Rock, and sandstones above the Burnwood, Winghay and Great Row coals (J O'Dell, British Coal).

Exploration for oil and gas in the district started when exploratory boreholes were drilled soon after the First World War (Giffard, 1923). Two were on the Western Anticline (Apedale Nos. 1 and 2 boreholes, the latter being a re-drill of the first) and the third on the Werrington Anticline (Werrington Borehole). These did not prove oil or gas. However, a small gas field, producing gas at potentially economic rates, has been discovered beneath the Lask Edge Anticline in the adjoining Macclesfield district. The reservoir is in the Onecote Sandstone (Chapter 3); this has a low permeability, but has been enhanced by the fracture system in the anticline (Besly, 1993). Other exploratory boreholes have been drilled within, or just outside, the district; most have targeted potential reservoirs in the Sherwood Sandstone or Millstone Grit. Some have contained shows of oil and gas, but none has proved a commercial discovery.

Coal-bed methane

The value of methane in coal seams has long been recognised, indeed the Pie Rough (Keele No. 1) Borehole (Millot et al., 1946) was an early attempt at obtaining methane from this source. The method of producing gas by drilling boreholes into seams and extracting the gas by fracturing the coal under pressure, though commercially viable in the USA, has not been much practiced in Britain. The methane potential of coals in the Potteries Coalfield is demonstrated by the common occurrence of methane emissions in mines (Rhydderch and Yates, 1964), and a methane drainage system, the North Staffordshire Grid, connects Silverdale, Hem Heath and Florence collieries (Burton, 1984). Most of the exposed part of the coalfield has little potential because of the extent of mine workings; however, the concealed coalfield to the south can be expected to contain large reserves. Methane data for coals in the Hobgoblin Borehole, in the concealed coalfield, are given by Creedy (1983).

Geothermal energy

The Crewe Heat Flow Borehole was drilled to explore the potential for hot water production from the Sherwood Sandstone Group in the area between Crewe and Nantwich. The Sherwood Sandstone was never reached, and though potentially exploitable waters at near 60°C were discovered, these have yet to be utilised (Gale et al., 1984; Downing and Gray, 1986).

Made ground

Man-made deposits provide potentially economic resources on a small scale in the urbanised parts of the district. Colliery waste can locally be used for bulk fill material, or may be washed to produce coal if it contains a sufficiently high coal content. Landfill deposits containing domestic waste have the potential to produce large volumes of methane, which may be utilised commercially; however, this can usually only be done successfully if the operation is planned prior to dumping. Most areas of made ground have the potential to be redeveloped for industrial, recreational or even agricultural use (Wilson et al., 1992).

HYDROGEOLOGY

The principal river system is that of the Trent, whose headwaters drain the greater part of the district flowing in a general southerly and easterly direction. The extreme south-west corner of the district drains south and west into the River Tern system, while the north-west drains to the River Weaver system. The average annual rainfall over the district varies between 740 and 860 mm, and the average annual evapotranspiration is in the order of 480 to 500 mm. Licensed abstractions of groundwater and surface water together amount to some 72 million cubic metres per annum (million m^3/a). Groundwater forms 52 per cent of the licensed total, the greater part of this being for public water supply and taken from a single aquifer, the Sherwood Sandstone. The distribution of the licensed abstractions is shown in Table 3.

The **Sherwood Sandstone Group** forms the major aquifer in the district. The most important sandstones are those in the Stafford and Needwood basins where large areas of sandstone crop out, and are relatively free of drift cover. The sandstones bordering the Cheshire Basin have a relatively small outcrop and are dissected by faults. The aquifer is not wholly homogeneous since beds of mudstone and siltstone provide local horizontal divisions, so that the horizontal permeability is frequently greater than the vertical by one or more orders of magnitude. Although there is commonly a significant intergranular component in the groundwater flow in these strata, it has been shown that fissure flow is normally predominant (Williams et al., 1972; Lovelock, 1972; Brereton and Skinner, 1974). Transmissivities determined from borehole pumping tests typically vary from 20 to 2000 cubic metres per metre per day (m^2/day), and specific yield from 10 to 20 per cent. Yields, although generally very good, are variable and are a function of aquifer thickness, degree of cementation and intensity of fissuring around the borehole. The sandstones generally stand without support, but occasionally broken rock or a seam of running sand necessitate the use of a sand screen.

The natural replenishment to the aquifers is from rainfall, plus some influent streams such as the Meece Brook. The mean annual infiltration into the Sherwood Sandstone Group in this district varies from about 380 mm/year over drift-free outcrop to less than 120 mm/year where drift hinders percolation. As the Stoke-on-Trent district does not contain any single, complete aquifer unit, it is difficult to estimate annual aquifer recharge purely within the district. Little recharge takes place during the period April through September because the potential evapotranspiration exceeds the available rainfall, and groundwater levels usually fall through these months. The range of fluctuation in the sandstones is rarely more than a few metres and generally less than 3 m. Prior to the Water Resources Act (1963), groundwater development was haphazard, resulting in full development of resources. Since 1963, development has been reduced to some extent, but there is little, if any, groundwater to supply new demands.

Groundwater in the Sherwood Sandstones is generally soft, with a total hardness less than 300 milligrammes per litre (mg/l), and often less than 200 mg/l (as $CaCO_3$); carbonate (temporary) hardness predominates, and the water type is usually calcium-bicarbonate. The concentration of nitrate can be high in the outcrop areas although it does not generally exceed the recommended limit of 50 mg/l (as NO_3). The chloride ion concentration appears for the most part to be less than 30 mg/l (as Cl), while sulphate seems to be less than 50 mg/l (as SO_4). Low concentrations of fluoride are present, usually in

Table 3
Licensed water abstraction in the Stoke-on-Trent district.

The category 'Other aquifers' includes the Mercia Mudstone and strata of Carboniferous age. Units are millions of cubic metres per annum.

Source	Spray irrigation	Other agricultural	Industrial	Energy generation	Public water supply	Other uses	TOTALS
Surface water	0.670	3.669	8.871	21.402	0.076	0.069	34.757
Superficial deposits	0.031	0.171	0.188	0	0	0	0.390
Sherwood Sandstone	0.256	0.153	1.252	0	32.569	0.014	34.244
Other aquifers	0.000	0.134	1.079	0	1.227	0	2.440
TOTAL	0.957	4.127	11.390	21.402	33.872	0.083	71.831
Groundwater as percentage of total	30%	11%	22%	0	99%	17%	52%

the range 0.04 to 0.10 mg/l (as F). Iron is also present only in low concentrations of less than 0.20 mg/l (as Fe). Typical analyses are shown in Table 4.

The main producing boreholes and pumping stations (often extracting from several boreholes) in the Sherwood Sandstone Group occur at Bearstone, Blacklake, Cresswell, Fulford, Hatton, Madeley, Mill End, Moddershall, Mossgate, Swynnerton, The Spot, The Wellings, Wallmires, and Whitmore (Appendix 1).

The **Carboniferous strata** form a multilayered aquifer with groundwater restricted to fissures in the sandstones. As the fissures appear to diminish with depth it is rarely worth drilling to more than 60 m. The situation is complicated by extensive faulting that tends to divide aquifers into discrete, contained, blocks and which limits natural replenishment of the water-bearing strata. Consequently, initial yields may diminish with time as the amount of groundwater in storage is reduced by pumping. Sand screens are frequently required to support the borehole as the sandstones are often broken and the mudstones may show a tendency to 'squeeze'.

The strata in the **Millstone Grit Group** generally tend to give higher yields to boreholes than those in the Coal Measures, and water quality is normally good at depths of less than 60 m. The Chatsworth Grit and the Rough Rock provide the best aquifers, and adits were sunk into the former, for the purposes of extraction, at Stockton Brook [9129 5208]. At depths of less than 60 m and where mining has not taken place, groundwater in the **Coal Measures** and **Barren Measures** is usually of moderately good quality. The total hardness is generally less than 300 mg/l, mostly carbonate hardness. Neither the sulphate ion nor the chloride ion concentrations normally exceed 100 mg/l. Locally, the groundwater may be ferruginous, with iron concentrations approaching 1.0 mg/l.

In the vicinity of old coal workings (and with increasing depth), groundwater quality tends to deteriorate markedly, with large increases in the concentration of sulphate, chloride and iron; such waters are rarely potable. Due to the presence of hydrocarbons in the rocks, methane may be a particular hazard in boreholes and even in fairly shallow pits. Consequently, wellhead works should be constructed above ground level and well ventilated. Methane tends to collect in subsurface voids, and a potential hazard can exist in, for example, underground pump-houses, the more so when these are excavated through thin till cover into the bedrock beneath. High yields are obtainable from the Coal Measures where boring takes place close to, or through, old coal workings. In the past, heavy pumping was necessary just to keep workings in some areas dewatered, and this led to large falls in the local water tables.

The presence of spoil tips and coal processing plant, even where now disused, may produce mineralised leachate from percolating rainfall, and that may pass down into the subsoil where sufficiently permeable. The construction of water wells in the vicinity of such sites is best avoided. The compartmentalisation by faulting of the sandstones does, however, reduce the vulnerability of the bulk of these formations to surface sources of pollution.

Groundwater in the **Mercia Mudstone Group** is essentially restricted to the few sandstones. Yields tend to be small (of use only to domestic and small farming requirements), and groundwater quality is variable. In sandstones close to the surface, the total hardness may be less than 300 mg/l and the chloride and sulphate ion concentrations less than 50 mg/l. However, at depth the presence of gypsum and anhydrite can lead to concentrations of the sulphate ion in excess of 400 mg/l. Although the chloride ion concentration is usually less than 100 mg/l, it may increase very markedly in the vicinity of halite beds. The sandstones are largely protected by overlying mudstones from sources of surface pollution.

Among the **Drift** deposits, only those comprising sands and gravels hold any potential as aquifers. River terrace deposits tend to drain rather rapidly, so that only the sands and gravels of the floodplains and those of glacial

Table 4 Typical chemical analyses of groundwater in the Sherwood Sandstone Group. Values are means for recent years, provided by the National Rivers Authority.

Determinand	Units	The Wellings No. 1	Whitmore	Swynnerton (combined)	Mossgate	Sheepwash No. 1
National Grid reference		768 384	812 395	864 359	953 372	951 452
pH	pH units	7.2	7.5	7.6	7.7	7.5
Electrical conductivity	μS/cm	447	409	440	327	346
Calcium (Ca)	mg/l	84	63	55	60	67
Magnesium (Mg)	mg/l	8	15	33	7	5
Sodium (Na)	mg/l	9	10	8	6	6
Potassium (K)	mg/l	3.2	4.1	2.4	2.3	2.9
Bicarbonate (HCO_3)	mg/l	195	189	243	174	163
Sulphate (SO_4)	mg/l	38	35	36	18	33
Chloride (CI)	mg/l	24	21	21	14	15
Nitrate (NO_3)	mg/l	41	31	27	19	18
Iron (Fe)	mg/l	0.04	0.03	0.01	0.02	0.01
Fluoride (F)	mg/l	0.05	0.04	0.08	0.04	0.04
Carbonate Hardness (as $CaCO_3$)	mg/l	160	155	199	143	134
Non-Carbonate Hardness (as $CaCO_3$)	mg/l	83	65	74	34	53

origin provide possibilites. Work carried out by the Trent River Authority (Anon, 1973) in areas along the Trent valley from King's Bromley to Gainsborough suggested that hydraulic conductivities in the river gravels varied from 20 to more than 200 cubic metres per square metre per day (m^2/day), and similar values may be expected in the Stoke-on-Trent district. These gravels are generally in hydraulic continuity with their associated rivers, so yields from boreholes and shallow shafts tend to be more dependent upon the rate of recharge induced from the river by pumping than on the available storage in the gravels. Nonetheless, a useful function is performed by the gravels in these circumstances by filtering the river water, with a consequent improvement in quality.

In general, glacial sands and gravels have lower hydraulic conductivities than their floodplain counterparts. However, their distribution is wider, they are often of greater thickness, and the amount of groundwater held in storage is consequently greater. Where they are laterally extensive they may be significant aquifers. Because many rivers, streams, lakes and springs are supported by baseflow from these deposits, as with the floodplain gravels, groundwater development often has to be limited to avoid damage to fishery, amenity and conservation interests. This is often a problem where dewatering for working of sand and gravel pits is proposed. Small local supplies may be obtained from shallow wells and boreholes both in the exposed sand and gravel and where covered by till.

Groundwater in both the floodplain gravels and in the glacial sands and gravels tends to be surprisingly hard. An average total hardness of 400 mg/l or more may be increased to more than 1000 mg/l under till. The natural chloride ion concentration is almost always less than 100 mg/l, and values greater than this usually indicate pollution. Iron presents potentially the most serious problem, especially in the floodplain gravels, where the concentration may locally exceed 3.0 mg/l.

Where sands and gravels have no impermeable cover such as till, they are particularly at risk from surface sources of pollution. The deposits are generally highly permeable and the water table is close to the ground surface so that pollutants can quickly reach the saturated zone and can also travel rapidly towards pumped water sources. In agricultural areas, the concentration of nitrate frequently exceeds the recommended limit of 50 mg/l (as NO_3). Considerable care has to be taken in treating water that is to be used for drinking.

HAZARDS

Foundation conditions

A characteristic feature of the engineering geology of the district is the high degree of lithological variability of the outcropping formations. The conglomerates, sandstones, siltstones, mudstones and halite all fall within the engineering category of weak rock (Meigh, 1981), as they have intrinsic unconfined compression strengths of less than 12.5 meganewtons per square metre (MN/m^2).

They are susceptible to weathering, and in their highly weathered state can either become cohesionless soils of variable relative density, or stiff to very stiff cohesive soils, with unconfined compressive strengths of less than 1.25 MN/m^2. When weathered, their physical behaviour in the rock-mass depends on local structure, on the frequency and type of discontinuities, on the distribution and proportion of cohesionless/cohesive sediments, on the presence or absence of halite beds and, perhaps most importantly, on groundwater conditions. A summary of geotechnical properties of bedrock and drift formations at outcrop in the urban area is given by Waine et al. (1991) and Wilson et al. (1992).

Because most rocks in the Carboniferous outcrop dip steeply, variations in lithology and rock mass quality can be expected over small distances. Where such rocks are drift-free they tend to be more heavily weathered, and thus lie in the weaker part of the compressive strength range, with lower average effective shear strengths, drained modulus values, and poor rock-mass characteristics (see below). Firman and Lovell (1988) note that **sandstones** in the Sherwood Sandstone Group tend to be geotechnically variable because of differences in the nature of cementing materials. This observation probably also applies, though perhaps to a lesser degree, to sandstones of Carboniferous age in the district. Local solution by groundwater may in some places have created leached and widened fault zones, or conduits, which have become infilled subsequently with loose weathered sands, or glacial materials of greater compressibility than the surrounding rock. Groundwater, moving in such sandstones, may also have tended to become channelled locally, as a result of lateral changes in topography, and the presence of lithological boundaries. Groundwater tends to flow downwards under gravity in the aquifers towards underlying mudstones, eventually discharging through zones of surface saturation. These zones, which follow seepage lines, may be found either on slopes or near river valleys, and need to be investigated carefully prior to excavation or construction, especially if confined groundwater conditions have developed, or are suspected, near the area of discharge. These are the areas which are also more likely to be affected by slope instability and shallow ground movements such as swelling and shrinkage, following vegetation change, drought, heavy rainfall, man-made alterations to natural infiltration, and runoff processes in the areas of recharge and discharge.

Weathering has a variable but important effect on the engineering behaviour of **Carboniferous mudrocks**, and Taylor (1988) suggested that sedimentary structures, slaking, type of diagenetic minerals, and expansion of mixed-layer clay minerals control their susceptibility to physical and chemical breakdown. According to Taylor (1988), the more durable mudrock lithologies have unconfined compressive strengths greater than 3.6 MN/m^2 with a slake durability value of over 60 per cent in a three-cycle test. However, good quality engineering results are not available from the Stoke-on-Trent district, and individual beds showing these characteristics have not yet been separately identified.

Lord and Nash (1974), Meigh (1976) and Marsland (1977) have discussed the engineering geology of weak **Triassic mudrocks** (Mercia Mudstone Group), with reference to predicted and observed performance of some major building foundations. These materials are susceptible to weathering (Chandler, 1969), and break down to cohesionless and cohesive soils with plasticity indices varying from 10 to 35 per cent. A review of case histories covering a wide geographical area has suggested that lithological factors and weathering profiles are important controls on foundation performance, and that the conventional practice, which uses a constant compressibility modulus and assumes an infinite depth of foundation material, is unrealistic for settlement predictions. The case records illustrated that the modulus values were broadly related to the geological weathering scheme suggested by Skempton and Davis (1966), and quoted by Chandler (1969), with a significant change in stiffness values normally taking place between the zone II and zone III grades. However, Meigh (1976) discussed the difficulties of applying this scheme to all Triassic lithologies, and Bacciarelli (1990) has since suggested a modification which takes into account the presence in the Mercia Mudstone of alternating weathered and unweathered mudstones.

The glacial sediments consist largely of stiff cohesive silty **tills**. Index properties and undrained strengths, summarised by Waine et al. (1991), do not distinguish between till types, but show that they are heterogeneous, with a low to medium plasticity and with variable strength and stiffness. These spatial variations in stiffness have not been studied in the Stoke-on-Trent district. However, Gostelow and Browne (1986) have summarised numerous vertical borehole profiles recording variations with depth of Standard Penetration Test (SPT) values and coefficient of volume compressibility (m_v) from site investigations in other tills. They found that SPT increased and m_v decreased with depth, and that a significant change in stiffness occurred at about 8 m, with m_v reducing from 0.16 m^2/MN to 0.08 m^2/MN; this change was attributed to weathering. In a similar study, Eyles and Sladen (1981) described weathering profiles, also of between 3 and 8 m, which they suggested were influenced locally by the presence of intraformational fluvial sediments, and the relationships of the till to underlying solid lithologies, topography, and the movement of groundwater. They concluded that the average depth of alteration was related to current mean annual precipitation, the areas of deep weathering being found where this is less than 75 mm. They suggested that low precipitation allowed a large summer soil moisture deficit to develop, which encouraged surface cracking, the development of discontinuities, volume change, and shallow groundwater movements. It seems likely that the change in stiffness described above may also reflect these processes, and that the spatial variations in weathering profile characteristics and till structure in the Stoke-on-Trent district may likewise influence engineering behaviour. These factors should be considered in site investigation planning.

The **glaciofluvial sand and gravel** also has a wide range in apparent relative density (from 0.15 to 1.0), with SPT values (N) ranging from 4 to 77, that is from very loose to very dense (Waine et al., 1991). However, Skempton (1986) has suggested that for comparative purposes, SPT values should, ideally, be normalised with respect to effective overburden pressure. He also discussed the influence of test procedures, ageing, particle size, and overconsolidation on measured N blow counts. These factors should be taken into account when using tabulated values, and carrying out investigations into cohesionless glacial deposits.

The deposits of **alluvium** and **peat** in the district generally have not been compacted as much as other drift formations and may be considered as soft and compressible soils which can sometimes lead to structural damage to buildings.

Glaciofluvial and alluvial deposits may also be prone to **'running conditions'** caused by flow of sand and gravel under high water pressures, especially in excavations. Problems caused by **differential subsidence** may locally arise over areas where loosely tipped made ground infills excavations. Made ground may sometimes have a greater compressibility towards the centre of an excavation than towards the margins. Construction across these margins may thus experience variable degrees of unpredictable settlement.

Slope stability

The formation most prone to landslips (Chapter 7) is the Etruria Formation (Chapter 3), particularly in its upper part, as at Bradwell Wood [850 500] where a large landslip affected the construction of the A 500 road (Searle, 1973). The description of a landslip in this formation (at Bury Hill, Wolverhampton) by Hutchinson et al. (1973) illustrates well some of the geological problems that can arise in weak rocks with mixed lithologies.

The best recorded landslip in the district is that at Walton's Wood [782 464] (Early and Skempton, 1972). A layer of head and colluvium 10 m thick, which developed in the late- to postglacial period, was present on a 35 m-high, 11° valley slope in weak rocks of the Upper Coal Measures. It was reactivated by construction of the M6 motorway in 1962, when downslope movements of an embankment crossing the slope deposits were recorded as 50 mm per month. The in-situ geological succession within the hill consists of alternating mudstones and sandstones with a coal (probably the Red Shagg seam). A perched water table, developed in the slope deposits, was apparently drained by sandstones in the Coal Measures. However, since only a limited number of piezometers was used during the investigation, the source of the groundwater recharge feeding the unstable area was not clear. An important observation was that groundwater was seeping into trial pits along exposed, discoloured, and chemically altered slip planes within the colluvium, which appeared to be acting as zones of high hydraulic conductivity and discharge. The investigation of the Walton's Wood slide has become historically significant, because it led to an understanding of the important concept of residual shear strength (Skempton, 1964). This discovery revolutionised soil mechanics and civil engineering design and practice.

These case records suggest that all areas with natural slope angles greater than 11° (highlighted by Wilson et al., 1992) should be carefully investigated before building or other forms of landscape development take place.

Seismic risk

Whilst this area appears to be of low seismicity, several small earthquakes have been felt in the Stoke district in recent years. A larger shock, locally of intensity 7 on the MSK scale, was recorded in 1916 centred on Chebsey [860 287], near Stafford, only 7 km south of the district (Neilson et al., 1984), illustrating that there is a minor seismic risk. However, in recent times most local recorded earthquakes have been the direct, or indirect, result of underground coal mining (Westbrook et al., 1980). The recent seismicity of the district has been reviewed by Lovell et al. (1996).

Salt subsidence

The occurrence of readily soluble halite at rockhead in the Cheshire Basin has led to a long history of subsidence in the district and further north in Cheshire. Such solution continues today, though is likely, in the absence of human intervention, to be very slow. Subsidence caused by mining of salt, or more commonly by 'wild brining' (uncontrolled pumping of brines) has caused many problems in Cheshire in the past (Calvert, 1915; Poole and Whiteman, 1966; Evans et al., 1968) though in the Stoke-on-Trent district these activities have been minimal. The only known 'wild brining' has been well outside the district at Nantwich and Lawton; subsidence effects (well illustrated by Calvert, 1915) in the nearest part of the present district are small, and may be partly remedied (Ege, 1984; Waltham, 1989).

Mining subsidence

The Potteries and Shaffalong coalfields have had a long history of coal and ironstone mining which have left a legacy of dereliction and mining subsidence. Early mining methods left open or incompletely filled shafts, and underground voids between pillars of coal in the old workings. The cavities have since collapsed, or remain liable to collapse in the future. These early workings are numerous and poorly documented; several thousand shafts and shallow excavations have already been recorded in the district and many more are found each year. A statutory duty to lodge plans of workings has only existed since 1872, so it is not possible to discount the possibility that old workings may be present at shallow depths anywhere in the Coal Measures. The areas most at risk, particularly those in central Hanley and Longton, are indicated by Wilson et al. (1992).

Modern longwall mining methods allow the coal roof to collapse behind the area of current working, so that most subsidence takes place within a few years of undermining. Differential subsidence can cause problems, nevertheless; near Barlaston, for example, undermining in recent years has reactivated pre-existing faults such as the Hollybush Fault (Chapter 8). As a consequence, many faults in the area between Barlaston, Meir Heath and Moddershall have formed active scarps at the surface, up to 2 m high.

Leachates

Domestic waste commonly produces leachates which may contaminate surface or groundwater supplies. Leachates may be preferentially transported via underground cavities, mine workings, faults, and permeable lithologies such as sandstones, which can facilitate lateral migration. Colliery waste commonly produces sulphate-rich leachates which may also contaminate groundwater, making it aggressive towards concrete. Sulphate-resistant concrete therefore needs to be used in any constructions associated with colliery waste; aggressive groundwaters may also be produced by slag from ironworks. Made ground deposits that consist dominantly of ceramic waste are very variable in the leachate hazard they present, as they may contain chemically active pottery 'slip' (Wilson et al. 1992).

Methane

Domestic waste tips and disused mine workings both have the potential to produce large volumes of methane gas. In certain circumstances, this may form an explosive mixture with air. Methane may be preferentially transported via cavities, faults and permeable lithologies (Hooker and Bannon, 1993), and can collect in subsurface voids. Consequently, where a development exists, or is proposed, for an area adjacent to a domestic waste tip or old workings, the possible presence of gases should be investigated.

Spontaneous combustion

The sulphides in mudstones and siltstones mined in association with coal may cause spontaneous combustion within colliery waste tips, particularly in the presence of sufficient volumes of coal (commonly over 25 per cent) and air. Tips established since 1971 are better compacted and have flatter profiles than older tips and are less liable to spontaneous combustion.

REFERENCES

Most of the references listed below are held in the Library of the British Geological Survey at Keyworth, Nottingham. Copies of the references can be purchased subject to the current copyright legislation.

ABDOH, A, COWAN, D, and PILKINGTON, M. 1990. 3D gravity inversion of the Cheshire Basin. *Geophysical Prospecting*, Vol. 38, 999–1011.

AITKENHEAD, N, CHISHOLM, J I, and STEVENSON, I P. 1985. Geology of the country around Buxton, Leek and Bakewell. *Memoir of the British Geological Survey*, Sheet 111 (England and Wales).

AITKENHEAD, N, and CHISHOLM, J I. 1982. A standard nomenclature for the Dinantian formations of the Peak District of Derbyshire and Staffordshire. *Report of the Institute of Geological Sciences*, No. 82/8.

ALI, A D. 1982. Triassic stratigraphy and sedimentology in central England. Unpublished PhD thesis, University of Aston in Birmingham.

AL SAIGH, N H. 1977. Geophysical investigations of glacial sediments in the region of Madeley, Staffordshire. Unpublished MSc thesis, University of Keele.

ALLPORT, S. 1874. On the microscopic structure and compositions of British Carboniferous dolerites. *Quarterly Journal of the Geological Society of London*, Vol. 30, 137–138.

ANDERSON, F W. 1964. Rhaetic ostracoda. *Bulletin of the Geological Survey of Great Britain*, No. 21, 133–174.

ANON. 1947. The north west county boundary. *Transactions of the North Staffordshire Field Club*, Vol. 81, 152–154.

ANON. 1973. Artificial recharge: river gravels. *Trent Research Programme*, Vol. 8. (Reading: Water Resources Board.)

ARTHURTON, R S. 1973. Experimentally produced halite compared with Triassic layered halite-rock from Cheshire, England. *Sedimentology*, Vol. 20, 145–160.

ARTHURTON, R S. 1980. Rhythmic sedimentary sequences in the Triassic Keuper Marl (Mercia Mudstone Group) of Cheshire, northwest England. *Geological Journal*, Vol. 15, 43–50.

ASHTON, C A. 1974. Palaeontology, stratigraphy and sedimentology of Kinderscoutian and lower Marsdenian (Namurian) of North Staffordshire and adjacent areas. Unpublished PhD thesis, University of Keele.

AUDLEY-CHARLES, M G. 1970a. Stratigraphical correlation of the Triassic rocks of the British Isles. *Quarterly Journal of the Geological Society of London*, Vol. 126, 19–47.

AUDLEY-CHARLES, M G. 1970b. Triassic palaeogeography of the British Isles. *Quarterly Journal of the Geological Society of London*, Vol. 126, 49–89.

BACCIARELLI, R. 1990. Keuper Marl (Mercia Mudstone Group): a revised weathering classification. 139–199 in *Preprints of the Symposium on the engineering geology of weak rock*. (Leeds: Engineering Group of the Geological Society of London.)

BARCLAY, W J, AMBROSE, K, CHADWICK, R A, and PHAROAH, T C. In press. Geology of the country around Worcester. *Memoir of the British Geological Survey*, Sheet 199.

BARKE, F. 1920. The evolution of river valleys. *Transactions of the North Staffordshire Field Club*, Vol. 54, 17–27.

BARKE, F. 1929. The old course of the River Churnet. *Transactions of the North Staffordshire Field Club*, Vol. 63, 90–97.

BARNSLEY, G B, CLOWES, J M, and FOWLER, W. 1966. Kaolin tonsteins in the Westphalian of North Staffordshire. *Geological Magazine*, Vol. 103, 508–521.

BENTLEY, K. 1983. Geologists. (Stoke-on-Trent: City Museum and Art Gallery.)

BENTON, M J, WARRINGTON, G, NEWELL, A J, and SPENCER, P S. 1994. A review of the British Middle-Triassic tetrapod assemblages. 131–160 in *In the shadow of the dinosaurs: Early Mesozoic tetrapods*. FRASER, N C, and SUES, H D (editors). (New York: Cambridge University Press.)

BESLY, B M. 1983. The sedimentology and stratigraphy of red beds in the Westphalian A to C of Central England. Unpublished PhD thesis, University of Keele.

BESLY, B M. 1988. Palaeogeographical implications of late Westphalian to early Permian red-beds, Central England. 200–221 in *Sedimentation in a synorogenic basin complex: the Upper Carboniferous of Northwest Europe*. BESLY, B M, and KELLING, G (editors). (London and Glasgow: Blackie.)

BESLY, B M. 1993. Geology. 17–36 in *The Potteries region: coninuity and change in a Staffordshire conurbation*. PHILLIPS, A D M (editor). (Stroud: Alan Sutton Publishing Ltd.)

BESLY, B M, BURLEY, S D, and TURNER, P. 1993. The late Carboniferous 'Barren Red-Bed' play of the Silver Pit area, southern North Sea. 727–740 in *Petroleum geology of North west Europe*. Proceedings of the 4th International Conference on the Petroleum Geology of North West Europe. PARKER, J R (editor). (London: Geological Society.)

BESLY, B M, and CLEAL, C J. 1995. Lithostratigraphy and macrofloral biostratigraphy of the Late Carboniferous of the Midlands and Oxfordshire Coalfields. *British Geological Survey Technical Report*, WH/95/81R.

BESLY, B M, and CLEAL, C J. 1997. Upper Carboniferous lithostratigraphy of the West Midlands (UK) revised in the light of borehole geophysical logs and detrital compositional suites. *Geological Journal*, Vol. 32, 85–118.

BESLY, B M, and FIELDING, C. 1989. Palaeosols in Westphalian coal-bearing and red-bed sequences, central and northern England. *Palaeogeography, Palaeoclimatology, Palaeoecology*, Vol 70, 303–330.

BESLY, B M, and TURNER, P. 1983. Origin of red beds in a moist tropical climate (Etruria Formation, Upper Carboniferous, UK.). 131–147 *in* Residual deposits. WILSON, R C L (editor). *Special Publication of the Geological Society of London*, No. 11.

BIGGS, A. 1987. The potential of clays from Keele Quarry for tile manufacture. Unpublished MSc thesis, University of Hull.

BIRMINGHAM UNIVERSITY, GEOLOGICAL SCIENCES and CIVIL ENGINEERING DEPARTMENTS. 1981. Saline Groundwater Investigation Phase 1 — Lower Mersey Basin, Final Report to the North West Water Authority. Appendix Vol. 1 Geology.

BOARDMAN, E L. 1978.   The blackband ironstones of the North Staffordshire coalfield.   *North Staffordshire Journal of Field Studies*, Vol. 18, 1–13.

BOARDMAN, E L. 1989.   Coal measures (Namurian and Westphalian) Blackband Iron Formations: fossil bog iron ores. *Sedimentology*, Vol. 36, 621–633.

BOARDMAN, E L, EXLEY, C S, RANKILOR, P R, and WILSON, A A. 1972.   A note on the stratigraphy of the uppermost Carboniferous of North Staffordshire.   *North Staffordshire Journal of Field Studies*, Vol. 12, 39–45.

BONNY, A P, MATHERS, S J, and HAWORTH, E Y. 1986.   Interstadial deposits with Chelford affinities from Burland, Cheshire. *Mercian Geologist*, Vol. 10, 151–160.

BOLTON, T. 1978.   The palaeontology, sedimentology and stratigraphy of the upper Arnsbergian, Chokierian and Alportian of the North Staffordshire Basin.   Unpublished PhD thesis, University of Keele.

BOULTON, G S, and PAUL, M A. 1976.   The influence of genetic processes on some geotechnical properties of glacial tills. *Quarterly Journal of Engineering Geology*, Vol. 9, 159–194.

BOULTON, G S, and WORSLEY, P. 1965.   Late Weichselian glaciation in the Cheshire–Shropshire Basin.   *Nature, London*, Vol. 207, 704-706.

BOWEN, D Q, and SYKES, G A. 1988.   Correlations of marine events and glaciations on the northeast Atlantic margin. *Philosophical Transactions of the Royal Society of London*, B318, 619–635.

BRERETON, N R, and SKINNER, A C. 1974.   Groundwater flow characteristics in the Triassic sandstone of the Fylde area of Lancashire.   *Water Services*, August issue, 3–7.

BREWER, J A, MATTHEWS, D H, WARNER, M R, HALL, J, SMYTHE, D K, and WHITTINGTON, R J. 1983.   BIRPS deep seismic reflection studies of the British Caledonides.   *Nature, London*, Vol. 305, 206–210.

BRISTOW, C S. 1988.   Controls on the sedimentation of the Rough Rock Group (Namurian) from the Pennine Basin of northern England.   114–131 in *Sedimentation in a synorogenic basin complex: the Upper Carboniferous of Northwest Europe*. BESLY, B M, and KELLING, G (editors). (London and Glasgow: Blackie.)

BROOKFIELD, M E. 1977.   The origin of bounding surfaces in ancient aeolian sandstones.   *Sedimentology*, Vol. 24, 303–332.

BROOKS, M. 1961.   Geophysical surveys in the Crewe area, 3rd–29th July, 1961.   *British Geological Survey Geophysics Department Report*, GD/16/29.

BRUGMAN, W A. 1986.   Late Scythian and Middle Triassic palynostratigraphy in the Alpine realm.   *Albertiana*, Vol. 5, 19–20.

BRUGMAN, W A, VELD, H, VAN BUGGENUM, J M, HOLSHUIJSEN, R P, BOEKELMAN, W A, VAN DEN BERGH, J J, ALMEKINDERS, M P, POORT, R J, ABBINK, O A, and D'ENGELBRONNER, E R. 1988.   Palynological investigations within the Triassic of the Germanic Basin of southern Germany.   *Stuifmail*, Vol. 6, 52–54.

BUIST, D S, and THOMPSON, D B. 1982.   Sedimentology, engineering properties and exploitation of the Pebble Beds in the Sherwood Sandstone Group (?Lower Trias) of North Staffordshire, with particular reference to highway schemes. *Mercian Geologist*, Vol. 8, 241–268.

BURTON, J B. 1984.   Colliery methane utilisation schemes within the Western Area.   *Mining Engineer*, Vol. 144, No. 276, 175–180.

BUSBY, J P. 1987.   An interactive Fortran 77 program using GKS graphics for 2.5D modelling of gravity and magnetic data. *Computers and Geosciences*, Vol. 13, 639–644.

BUTLER, R H. 1982.   The terminology of structures in thrust belts.   *Journal of Structural Geology*, Vol. 4, 239–245.

BUTTERWORTH, M A, and SMITH, A H. 1976.   The age of the British Upper Coal Measures with reference to their miospore content.   *Review of Palaeobotany and Palynology*, Vol. 22, 281–306.

CADMAN, J. 1901.   The occurrence, mode of working and treatment of the ironstones found in the North Staffordshire Coalfield.   *Transactions of the Institute of Mining Engineers*, Vol. 22, 89–112.

CALVER, M A. 1956.   Die stratigraphische Verbreitung der nicht-marinen Muscheln in den penninischen Kohlenfeldern Englands.   *Zeitschrift der Deutschen Geologischen Gesellschaft*, Vol. 107, 26–39.

CALVER, M A. 1968a.   Coal Measures invertebrate faunas. 147–177 in *Coal and coal-bearing strata*. MURCHISON, D G, and WESTOLL, T S (editors). (Edinburgh and London: Oliver and Boyd.)

CALVER, M A. 1968b.   Distribution of Westphalian marine faunas in northern England and adjoining areas.   *Proceedings of the Yorkshire Geological Society*, Vol. 37, 1–72.

CALVERT, A F. 1915.   *Salt in Cheshire*.   (London: Spon.)

CAMPBELL-SMITH, W. 1963.   Description of the igneous rocks represented among pebbles from the Bunter Pebble Beds of the Midlands of England.   *Bulletin of the British Museum of Natural History*, Vol. 2, 1–17.

CELORIA, F (editor). 1971.   Edward Dobson's "A rudimentary treatise on the manufacture of bricks and tiles" (1850). *Journal of Ceramic History*, No. 5.

CHADWICK, R A. 1986.   Extension tectonics in the Wessex Basin, southern England.   *Journal of the Geological Society of London*, Vol. 143, 465–488.

CHALLINOR, J. 1921.   Notes on the geology of the Roaches district.   *Transactions of the North Staffordshire Field Club*, Vol. 55, 76–87.

CHALLINOR, J. 1978a.   Literature relating to the geology, mineralogy and palaeontology of North Staffordshire.   *North Staffordshire Journal of Field Studies*, Vol. 18, 14–20.

CHALLINOR, J. 1978b.   The "Red Rock Fault", Cheshire: a critical review.   *Geological Journal*, Vol. 13, 1–10.

CHALLINOR, P J. 1990.   Oil ingress into mine workings.   *The Mining Engineer*, Vol. 150, No. 347, 68–74.

CHANDLER, R J. 1969.   The effect of weathering on the shear srength properties of Keuper Marl.   *Geotechnique*, Vol. 19, 321–334.

CHARSLEY, T J. 1982.   A standard nomenclature for the Triassic formations of the Ashbourne district.   *Report of the Institute of Geological Sciences*, No. 81/14.

CHERRY, J L. 1877.   Sectional Reports. No. 1: Geology.   *North Staffordshire Naturalists' Field Club*, 23–25.

CHISHOLM, J I. 1990.   The Upper Band–Better Bed sequence (Lower Coal Measures, Westphalian A) in the central and south Pennine area of England.   *Geological Magazine*, Vol. 127, 55–74.

CHISHOLM, J I, CHARSLEY, T J, and AITKENHEAD, N. 1988. Geology of the country around Ashbourne and Cheadle. *Memoir of the British Geological Survey*, Sheet 124 (England and Wales).

CLAOUÉ-LONG, J C, ZHANG ZICHAO, MA GUOGAN and DU SHASHUA. 1991.   The age of the Permian–Triassic boundary.   *Earth and Planetary Science Letters*, Vol. 105, 182–190.

CLEMMENSEN, L B, and ABRAHAMSEN, K. 1983. Aeolian stratigraphy and facies associations in desert sediments, Arran Basin (Permian), Scotland. *Sedimentology*, Vol. 30, 31–39.

COLLINSON, J D. 1988. Controls on Namurian sedimentation in the Central Province basins of northern England. 85–101 in *Sedimentation in a synorogenic basin complex: the Upper Carboniferous of Northwest Europe*. BESLY, B M, and KELLING, G (editors). (London and Glasgow: Blackie.)

COLLINSON, J D, HOLDSWORTH, B K, JONES, C M, and MARTINSEN, O J. 1992. Discussion of: 'The Millstone Grit (Namurian) of the southern Pennines viewed in the light of eustatically controlled sequence stratigraphy' by W A READ. *Geological Journal*, Vol. 27, 173–180.

COLTER, V S. 1978. Exploration for gas in the Irish Sea. *Geologie en Mijnbouw*, Vol. 57, 503–516.

COLTER, V S, and BARR, K W. 1975. Recent developments in the geology of the Irish Sea and Cheshire Basins. 61–75 in *Petroleum and the Continental Shelf of North West Europe*. WOODLAND, A W (editor). (London: Applied Science Publishers.)

COPE, F W. 1946. The correlation of the coal measures of the Cheadle Coalfield, North Staffordshire. *Transactions of the Institute of Mining Engineers*, Vol. 105, Pt 2, 75–91.

COPE, F W. 1948. A boring in the Millstone Grit at Timbersbrook, near Congleton. *Journal of the Manchester Geological Association*, Vol. 2, 7–16.

COPE, F W. 1954. The North Staffordshire Coalfields. 219–243 in *The coalfields of Great Britain*. TRUEMAN, A (editor). (London: Edward Arnold.)

Cope, F W. 1966. The Butterton Dyke near Keele and Butterton, Staffordshire. *North Staffordshire Journal of Field Studies*, Vol. 5, 25-37.

COPE, W S. 1852. Mineral map and section of the North Staffordshire Coalfield. (HANLEY.)

CORFIELD, S M. 1991. The Upper Palaeozoic to Mesozoic structural evolution of the North Staffordshire Coalfield and adjoining areas. Unpublished PhD thesis, University of Keele.

CORNWELL, J D, and DABEK, Z K. 1992. Reprocessing detailed gravity data from the 1961 Crewe Survey. *British Geological Survey Regional Geophysics Project Note*, PN/92/08.

CORNWELL, J D, and DABEK, Z K. 1994. Geophysical investigations in the Stoke-on-Trent district. *British Geological Survey Technical Report*, WK/94/04.

COWARD, M P, and SIDDANS, A W B. 1979. The tectonic evolution of the Welsh Caledonides. 187–198 *in* The Caledonides of the British Isles — reviewed. HARRIS, A L, HOLLAND, C H, and LEAKE, B E (editors). *Special Publication of the Geological Society of London*, No. 8.

CREEDY, D P. 1983. Seam gas-content data-base aids firedamp prediction. *The Mining Engineer*, Vol. 143, No. 263, 79–82.

CROFTS, H J. 1953. The coking coals of North Staffordshire. *Transactions of the Institute of Mining Engineers*, Vol. 112, 719–739.

CROFTS, R G. 1990a. Geology of the Trentham district. *British Geological Survey Technical Report*, WA/90/06.

CROFTS, R G. 1990b. Geology of the Kidsgrove district. *British Geological Survey Technical Report*, WA/90/07.

CROOKALL, R. 1955–1976. Fossil plants of the Carboniferous rocks of Great Britain. *Palaeontological Memoir of the Geological Survey of Great Britain*, Vol. 4, Pts 1–7, 1–1004.

CURTIS, C D, PEARSON, M D, and SOMOGYI, V A. 1975. Mineralogy, chemistry and origin of a concretionary siderite sheet (clay-ironstone band) in the Westphalian of Yorkshire. *Mineralogical Magazine*, Vol. 40, 385–395.

DAGLEY, P. 1969. Palaeomagnetic results from some British Tertiary dykes. *Earth and Planetary Science Letters*, Vol. 6, 349–354.

DE RANCE, C E. 1898. 107–110 in Geological Survey of Great Britain. *Summary of Progress for 1897.* (London: HMSO.)

DEWEY, H. 1920. Carboniferous bedded ores, North Staffordshire. 75–81 in *Memoir of the Geological Survey, Special Reports on the Mineral Resources of Great Britain*, Vol. 13, STRAHAN, A, GIBSON, W, CANTRILL, T C, SHERLOCK, R L, and DEWEY, H.

DIX, E. 1931. The flora of the upper portion of the Coal Measures of North Staffordshire. *Quarterly Journal of the Geological Society of London*, Vol. 87, 160–179.

DIX, E, and TRUEMAN, A E. 1931. Some non-marine lamellibranchs from the upper part of the Coal Measures. *Quarterly Journal of the Geological Society of London*, Vol. 87, 180–211.

DOWNING, R A, and GRAY, D A (editors). 1986. *Geothermal energy — the potential in the United Kingdom*. (London: HMSO.)

EAGAR, R M C. 1956. Additions to the non-marine fauna of the Lower Coal Measures of the North Midlands coalfields. *Liverpool and Manchester Geological Journal*, Vol. 1, 328–369.

EAGAR, R M C. 1960. A summary of the results of recent work on the palaeoecology of Carboniferous non-marine lamellibranchs. *Compte Rendu du 4me Congrès International de Stratigraphie et de Géologie du Carbonifère*, Heerlen 1958, Vol. 1, 137–149.

EARLY, K R, and SKEMPTON, A W. 1972. Investigations of the landslide at Walton's Wood, Staffordshire. *Quarterly Journal of Engineering Geology*, Vol. 5, 19–41.

EARP, J R, and CALVER, M A. 1961. Exploratory boreholes in the North Staffordshire Coalfield. *Bulletin of the Geological Survey of Great Britain*, No. 17, 153–190.

EARP, J R, and TAYLOR, B J. 1986. Geology of the country around Chester and Winsford. *Memoir of the British Geological Survey*, Sheet 109.

EDEN, R A. 1954. The Coal Measures of the *Anthraconaia lenisulcata* Zone in the East Midlands Coalfield. *Bulletin of the Geological Survey of Great Britain*, No. 5, 81–106.

EDWARDS, W. 1951. The concealed coalfield of Yorkshire and Nottinghamshire. 3rd edition. *Memoir of the Geological Survey of England and Wales.*

EGE, J R. 1984. Mechanisms of surface subsidence resulting from extraction of salt. *Geological Society of America Review of Engineering Geology*, Vol. 6, 203–221.

ETHERIDGE, R. 1865. On the Rhaetic or *Avicula contorta* Beds at Garden Cliff, Westbury-upon-Severn, Gloucestershire. *Proceedings of the Cotteswold Naturalists' Field Club*, Vol. 3, 218–235.

EVANS, A L. 1969. On dating the British Tertiary Igneous Province. Unpublished PhD thesis, University of Cambridge.

EVANS, W B, WILSON, A A, TAYLOR, B J, and PRICE, D. 1968. Geology of the country around Macclesfield, Congleton, Crewe and Middlewich. *Memoir of the British Geological Survey*, Sheet 110 (England and Wales).

EVANS, D J, REES, J G, and HOLLOWAY, S. 1993. The Permian to Jurassic stratigraphy and structural evolution of the central Cheshire Basin. *Journal of the Geological Society of London*, Vol. 150, 857–870.

EXLEY, C S. 1970. Observations on the geology of the Keele district, Staffordshire. *North Staffordshire Journal of Field Studies*, Vol. 10, 49–63.

EYLES, N, and McCABE, A M. 1989. The late Devensian (< 22,000 BP) Irish Sea Basin: the sedimentary record of a collapsed ice sheet margin.   *Quaternary Science Reviews*, Vol. 8, 307–351.

EYLES, N, and SLADEN, J A. 1988.   Stratigraphy and geotechnical properties of weathered lodgement till in Northumberland, England.   *Quarterly Journal of Engineering Geology*, Vol. 14, 129–141.

FENTON, G W. and RUMSBY, P L. 1962.   The mapping and appraisal of the characteristics of British coal seams.   *The Mining Engineer*, Vol. 121, No. 19, 454–467.

FIELDING, C R. 1984a.   Upper delta plain lacustrine and fluvio-lacustrine facies from the Westphalian of the Durham Coalfield, NE England.   *Sedimentology*, Vol. 31, 547–567.

FIELDING, C R. 1984b.   A coal depositional model for the Durham Coal Measures of NE England.   *Journal of the Geological Society of London*, Vol. 141, 919–931.

FIELDING, C R. 1986.   The anatomy of a coal seam split, Durham Coalfield, northeast England.   *Geological Journal*, Vol. 21, 45–57.

FIRMAN, R J, and LOVELL, M A. 1988.   The geology of the Nottingham region: A review of some engineering and environmental aspects.   *Geological Society Engineering Geology Special Publication*, No. 5, 33–51.

FITCH, F J, MILLER, J A, and THOMPSON, D B. 1966.   The palaeogeographic significance of isotopic age determinations on detrital micas from the Triassic of the Stockport–Macclesfield district, Cheshire, England. *Palaeogeography, Palaeoclimatology and Palaeoecology*, Vol. 2, 281–312.

FITCH, F J, MILLER, J A, GRASTY, A L and MENEISY, M Y. 1969. Isotopic age determinations on rocks from Wales and the Welsh borders.   23–46 in *The Pre-Cambrian and Lower Palaeozoic Rocks of Wales*. WOOD, A (editor). (Cardiff: University of Wales Press.)

FLOYD, P. 1964.   A note on fossil trees found in situ at Stanfield, Staffordshire.   *North Staffordshire Journal of Field Studies*, Vol. 4, 103–107.

FORSTER, S C, and WARRINGTON, G. 1985.   Geochronology of the Carboniferous, Permian and Triassic.   99–113 in *The chronology of the geological record*. SNELLING, N J (editor). *Memoir of the Geological Society of London*, No. 10.

FREYTET P. 1984.   Les sédiments lacustres carbonatés et leurs transformations par émersion et pédogenèse. Importance de leur identification pour les reconstructions paléogéographiques.   *Bulletin Centre de recherche exploration et production Elf-Aquitaine*, Vol. 8, pt. 1, 223–247.

FRYBERGER, S G, and SCHENK, C J. 1988.   Pin stripe lamination: A distinctive feature of modern and ancient eolian sediments. *Sedimentary Geology*, Vol. 55, 1–15.

FULTON, I M. 1987.   Genesis of the Warwickshire Thick Coal: a group of long residence histosols.   201–218 in Coal and coal-bearing strata: Recent Advances. SCOTT, A C (editor). *Special Publication of the Geological Society of London*, No. 32.

GALE, I N, EVANS, C J, EVANS, R B, SMITH, I F, HOUGHTON, M T, and BURGESS, W G. 1984.   *Investigation of the geothermal potential of the UK. The Permo-Triassic aquifers of the Cheshire and West Lancashire Basins*. (Keyworth: British Geological Survey.)

GARNER, R. 1844.   *The natural history of the county of Staffordshire*. (London: J. Van Voorst.)

GEIGER, M E, and HOPPING, C A. 1968.   Triassic stratigraphy of the southern North Sea Basin.   *Philosophical Transactions of the Royal Society of London*, B254, 1–36.

GEMMELL, A M D, and GEORGE, P K. 1972.   The glaciation of the West Midlands: a review of recent research.   *North Staffordshire Journal of Field Studies*, Vol. 12, 1–20.

GEOLOGICAL SURVEY OF GREAT BRITAIN. 1857.   Old series sheet 72SW.   Solid. 1: 63 360. (London: HMSO.)

GIBBS, A D. 1983.   Balanced cross-section construction from seismic sections in areas of extensional tectonics.   *Journal of Structural Geology*, Vol. 5, 153–160.

GIBSON, W. 1899a.   122–129 in Geological Survey of Great Britain.   *Summary of Progress for 1898*. (London: HMSO.)

GIBSON, W. 1899b.   Some recent work among the Upper Carboniferous Rocks of North Staffordshire, and its bearing on concealed coalfields.   *Geological Magazine*, Dec. IV, Vol. 6, 505–506.

GIBSON, W. 1900.   108 in Geological Survey of Great Britain. *Summary of Progress for 1899*. (London: HMSO.)

GIBSON, W. 1901.   On the character of the Upper Coal Measures of North Staffordshire, South Staffordshire, Denbighshire and Nottinghamshire, and their relation to the Productive Series.   *Quarterly Journal of the Geological Society of London*, Vol. 57, 251–265.

GIBSON, W. 1905.   The geology of the North Staffordshire Coalfields.   *Memoir of the Geological Survey of Great Britain*.

GIBSON, W. 1925.   The geology of the country around Stoke-upon-Trent (3rd edition).   *Memoir of the Geological Survey of England and Wales*.

GIBSON, W, and HIND, W. 1899.   On the agglomerates and tuffs in the Carboniferous Limestone Series of Congleton Edge. *Quarterly Journal of the Geological Society of London*, Vol. 55, 548–559.

GIBSON, W, and WEDD, C B. 1902.   The geology of the country around Stoke-on-Trent.   *Memoir of the Geological Survey of England and Wales*.

GIBSON, W, and WEDD, C B. 1905.   The geology of the country around Stoke-upon-Trent (2nd edition).   *Memoir of the Geological Survey of England and Wales*.

GIFFARD, H P W. 1923.   The recent search for oil in Great Britain.   *Transactions of the Institute of Mining Engineers*, Vol. 65, 221–250.

GOLDTHWAIT, R P, and MATSCH, C L (editors). 1988.   *Genetic classification of glacigenic deposits*. (Rotterdam: Balkema.)

GOSTELOW, T P, and BROWNE, M A E. 1986.   The engineering geology of the Upper Forth Estuary.   *Report of the British Geological Survey*, Vol. 16, No. 8.

GREEN, P F. 1989.   Thermal and tectonic history of the East Midlands shelf (onshore UK) and surrounding regions assessed by apatite fission track analysis.   *Journal of the Geological Society of London*, Vol. 146, 755–773.

GREENWOOD, H W. 1918.   Trias of the Macclesfield district with notes on its relation to the adjacent Carboniferous rocks and to the Trias of the Midlands.   *Proceedings of the Liverpool Geological Society*, Vol. 12, 325–338.

GRIMSHAW, W J. 1878.   On the method of working "rearing" mines at Leycett, Staffordshire.   *Transactions of the Manchester Geological Society*, Vol. 14, 155–168.

GUION, P D, and FIELDING, C R. 1988.   Westphalian A and B sedimentation in the Pennine Basin, U.K.   153–177 in *Sedimentation in a synorogenic basin complex: the Upper Carboniferous of Northwest Europe*. BESLY, B M, and KELLING, G (editors). (London and Glasgow: Blackie.)

HAGGAR, R G, MOUNTFORD, A R, and THOMAS, J. 1981.   *The Staffordshire pottery industry*. (Reprinted from Greenslade, M W, and Jenkins, J G. 1967.   *The Victoria history of the county of Stafford*, Vol. II). (Stafford: Staffordshire County Library.)

HALLIMOND, A F. 1925. Iron ores: The bedded ores of England and Wales. Petrography and chemistry. *Memoir of the Geological Survey. Special Report on the Mineral Resources of Great Britain*, No. 29.

HALLIMOND, A F. 1929. Magnetic observations on the Swynnerton Dyke. *Mining Magazine*, Vol. 41, December 1929, 16–22.

HARLAND, W B, ARMSTRONG, R L, COX, A V, CRAIG, L E, SMITH, A G, and SMITH, D G. 1990. *A geological time scale 1989.* (Cambridge: Cambridge University Press.)

HASLAM, H W. 1993. Geochemistry of Carboniferous sediments from the Sidway Mill borehole, Staffordshire. *British Geological Survey Technical Report*, WP/93/04.

HESS, J C, and LIPPOLT, H J. 1986. $^{40}Ar/^{39}Ar$ ages of tonstein and tuff sanidines: new calibration points for the improvement of the Upper Carboniferous time scale. *Chemical Geology*, Vol. 59, 143–154.

HESTER, S W. 1932. The Millstone Grit succession in North Staffordshire. 34–48 in Geological Survey of Great Britain. *Summary of Progress for 1931.* (London: HMSO.)

HIND, W. 1893. On the affinities of *Anthracoptera* and *Anthracomya*. *Quarterly Journal of the Geological Society of London*, Vol. 49, 249–275.

HIND, W. 1894. A monograph of *Carbonicola, Anthracomya* and *Naiadites*, Pt 1. *Palaeontographical Society Monograph*, 1–80.

HIND, W. 1895. A monograph of *Carbonicola, Anthracomya* and *Naiadites*, Pt 2. *Palaeontographical Society Monograph*, 81–170.

HIND, W. 1896. A monograph of *Carbonicola, Anthracomya* and *Naiadites*, Pt 3. *Palaeontographical Society Monograph*, 171–182.

HIND, W. 1906. Speculations on the evolution of the River Trent. *Transactions of the North Staffordshire Field Club*, Vol. 41, 93–100.

HIND, W. 1910. Staffordshire. 546–591 in *Geology in the field.* MONCKTON, H W, and HERRIES, R S (editors). Jubilee Volume of the Geologists' Association (1858–1908). (London: Stanford.)

HIND, W, and HOWE, J A. 1901. The geological succession and palaeontology of the beds between the Millstone Grit and the limestone-massif at Pendle Hill and their equivalents in certain other parts of Britain. *Quarterly Journal of the Geological Society of London*, Vol. 57, 347–404.

HIND, W, and STOBBS, J T. 1903. Chart of fossil shells found in connection with the seams of coal and ironstone of North Staffordshire. *Transactions of the North Staffordshire Institute of Mining and Mechanical Engineering.*

HODGKINSON, D. 1986. Silverdale Colliery. *Mining Magazine*, Vol. 155, September, 212–219.

HOLDRIDGE, D A. 1956. Compositional variation in Etruria Marls. *Transactions of the British Ceramic Society*, Vol. 58, 301–328.

HOLDSWORTH, B K. 1964. The 'Crowstones' of Staffordshire, Derbyshire and Cheshire. *North Staffordshire Journal of Field Studies*, Vol. 4, 89–102.

HOLDSWORTH, B K. 1966. A preliminary study of the palaeontology and palaeoenvironment of some Namurian limestone 'bullions'. *Mercian Geologist*, Vol. 1, 315–317.

HOLDSWORTH, B K, and COLLINSON, J D. 1988. Millstone Grit cyclicity revisited. 132–152 in *Sedimentation in a synorogenic basin complex: the Upper Carboniferous of Northwest Europe.* BESLY, B M, and KELLING, G (editors). (London and Glasgow: Blackie.)

HOLLAND, T H. 1912. The origin of desert salt deposits. *Proceedings of the Liverpool Geological Society*, Vol. 11, 227–250.

HOLLIDAY, D W. 1993. Mesozoic cover over northern England: interpretation of apatite fission track data. *Journal of the Geological Society, London*, Vol. 150, 657–660.

HOMER, C J. 1875. The North Staffordshire coalfield with the ironstones contained therein. *Journal of the Iron and Steel Institute*, No. 2, 540–573.

HOOKER, P J, and BANNON, M P. 1993. Methane: its occurrence and hazards in construction. *CIRIA report* 130.

HOWARD, A S. 1990. Geology of the Werrington district. *British Geological Technical Report*, WA/90/09.

HUDSON, R G S, and COTTON, G. 1945. The Lower Carboniferous in a boring at Alport, Derbyshire. *Proceedings of the Yorkshire Geological Society*, Vol. 25, 254–330.

HULL, E. 1860. On the new sub-divisions of the Triassic rocks of the central counties. *Transactions of the Manchester Geological Society*, Vol. 2, 22–34.

HULL, E. 1869. The Triassic and Permian rocks of the Midland Counties of England. *Memoir of the Geological Survey of Great Britain.*

HUNTER, R E. 1977. Basic types of stratification in small eolian dunes. *Sedimentology*, Vol. 24, 361–387.

HUTCHINSON, J N, SOMERVILLE, S H, and PETLEY, D J. 1973. A landslide in periglacially disturbed Etruria Marl at Bury Hill, Staffordshire. *Quarterly Journal of Engineering Geology*, Vol. 6, 377–404.

IRELAND, R J, POLLARD, J E, STEEL, R J, and THOMPSON, D B. 1978. Intertidal sediments and trace fossils from the Waterstones (Scythian–Anisian?) at Daresbury, Cheshire. *Proceedings of the Yorkshire Geological Society*, Vol. 41, 399–436.

JOHNSON, R H. 1965. The origin of the Churnet and Rudyard Valleys. *North Staffordshire Journal of Field Studies*, Vol. 5, 95–105.

JONES, J I. 1969. Licensed coal mining in North Staffordshire. *North Staffordshire Journal of Field Studies*, Vol. 9, 75–91.

JONES, C M. 1980. Deltaic sedimentation in the Roaches Grit and associated sediments (Namurian R_2b) in the south-west Pennines. *Proceedings of the Yorkshire Geological Society*, Vol. 43, 39–67.

KEELING, P S. 1961. Geochemistry of the common clay minerals. *Transactions of the British Ceramic Society*, Vol. 60, 678–689.

KEELING, P S, and HOLDRIDGE, D A. 1958. Mineralogical and chemical variation in a short section of the Etruria Marl. *British Ceramic Research Association*, Research Paper 399.

KELLING, G. 1988. Silesian sedimentation and tectonics in the South Wales Basin: a brief review. 38–42 in *Sedimentation in a synorogenic basin complex: the Upper Carboniferous of Northwest Europe.* BESLY, B M, and KELLING, G (editors). (London and Glasgow: Blackie.)

KEREY, I E. 1978. Sedimentology of the Chatsworth Grit sandstone in the Goyt- Chapel en le Frith area. Unpublished MSc thesis, University of Keele.

KIDSTON, R. 1892. The fossil flora of the Staffordshire Coalfields, Part II. The fossil flora of the coalfield of the Potteries. *Transactions of the Royal Society of Edinburgh*, Vol. 36, 63–98.

KIDSTON, R. 1905. On the divisions and correlation of the upper portions of the Coal Measures, with special reference to their development in the Midland counties of England. *Quarterly Journal of the Geological Society of London*, Vol. 61, 308–323.

KING, C A M. 1960. The Churnet Valley. *East Midlands Geographer*, Vol. 14, 33–40.

KIRKBY, J. 1894.   On the trap dykes in the Hanchurch Hills. *Transactions of the North Staffordshire Naturalists' Field Club*, Vol. 28, 129–140.

KIRTON, S R, and DONATO, J A. 1985.   Some buried Tertiary dykes of Britain and surrounding waters deduced by magnetic modelling and seismic reflection studies.   *Journal of the Geological Society of London*, Vol. 142, 1047–1057.

KLEIN, G DE V. 1962.   Sedimentary structures in the Keuper Marl (Upper Triassic).   *Geological Magazine*, Vol. 99, 137–144.

KNOWLES, A J. 1985a.   The Quaternary history of North Staffordshire.   222–236 in *The geomorphology of north-west England.* JOHNSON, R H (editor). (Manchester: Manchester University Press.)

KNOWLES, A J. 1985b.   The relationship of periglacially disturbed ground phenomena and associated sediments to supposed former pro-glacial lakes near Baldwins Gate, north-west Staffordshire, England.   *North Staffordshire Journal of Field Studies*, Vol. 21, 37–55.

LEE, A G. 1988.   Carboniferous basin configuration of central and northern England modelled using gravity data.   69–84 in *Sedimentation in a synorogenic basin complex: the Upper Carboniferous of Northwest Europe.* BESLY, B M, and KELLING, G (editors). (London and Glasgow: Blackie.)

LEE, M K, PHAROAH, T C, and SOPER, N J. 1990.   Structural trends in central Britain from images of gravity and magnetic fields. *Journal of the Geological Society of London*, Vol. 147, 241–258.

LEEDER, M R. 1988.   Recent developments in Carboniferous geology: a critical review with implications for the British Isles and northwest Europe.   *Proceedings of the Geologists' Association*, Vol. 99, 73–100.

LEEDER, M R, and MCMAHON, A H. 1988.   Upper Carboniferous (Silesian) basin subsidence in northern Britain. 43–52 in *Sedimentation in a synorogenic basin complex: the Upper Carboniferous of Northwest Europe.* BESLY, B M, and KELLING, G (editors). (London and Glasgow: Blackie.)

LEWIS, H C. 1894.   Papers and notes on the glacial geology of Great Britain and Ireland. (London: Longman.)

LEWIS, C L E, GREEN, P F, CARTER, A, and HURFORD, A J. 1992. Elevated K/T palaeotemperatures throughout Northwest England: Three kilometres of Tertiary erosion? *Earth and Planetary Science Letters*, Vol. 112, 131–145.

LISTER, J H, and STOBBS, J T. 1917.   Erratics in coal seams, with special reference to new discoveries in North Staffordshire. *Transactions of the North Staffordshire Field Club*, Vol. 51, 33–47.

LISTER, J H, and STOBBS, J T. 1918.   Additional erratics from the Woodhead Coal of Cheadle, North Staffordshire. *Transactions of the North Staffordshire Field Club*, Vol. 52, 93–95.

LORD, J A, and NASH, D F T. 1974.   Settlement studies of two structures on Keuper Marl.   292–310 in *Proceedings of the Conference on settlement of structures.* (London: British Geotechnical Society.)

LOVELL, J H, FORD, G D, HENNI, P H O, BAKER, C, SIMPSON, I, and PETTITT, W. 1996.   Recent seismicity in the Stoke-on-Trent area, Staffordshire.   *British Geological Survey Technical Report*, WL/96/20.

LOVELOCK, P E R. 1972.   Aquifer properties of the Permo-Triassic sandstones of the United Kingdom.   Unpublished PhD thesis, University of London.

MACHETTE, M N. 1985.   Calcic soils of the southwestern United States.   1–21 in WIEDER, D L (editor). Soils and Quaternary Geology of the Southwestern United States.   *Geological Society of America Special Paper*, No. 203.

MAGRAW, D. 1957.   New boreholes into the Lower Coal Measures below the Arley Mine of Lancashire and adjacent areas.   *Bulletin of the Geological Survey of Great Britain*, No. 13, 14–38.

MALKIN, A B. 1961.   Contributions to the study of the stratigraphy of the North Staffordshire Coalfield.   Unpublished MSc thesis, University of Manchester.

MALKIN, A B. 1985.   *The conglomerate resources of the Sherwood Sandstone Formation between Stoke-on-Trent and Stone, Staffordshire.* (Newcastle-under-Lyme: Wardell and Partners.)

MARSLAND, A. 1977.   In-situ measurement of the large scale properties of Keuper Marl.   335–344 in *Proceedings of the conference on the Geotechnics of structurally complex formations.* Vol 1. (Capri: Associazone Geotecnica Italiana.)

MAYHEW, R W. 1966.   A sedimentological investigation of the Marsdenian grits and associated measures in north-east Derbyshire.   Unpublished PhD thesis, University of Sheffield.

MAYNARD, J R. 1992.   Sequence stratigraphy of the Upper Yeadonian of northern England.   *Marine and Petroleum Geology*, Vol. 9, 197–207.

MAYNARD, J R, and LEEDER, M R. 1992.   On the periodicity and magnitude of the late Carboniferous glacio-eustatic sea-level changes.   *Journal of the Geological Society of London*, Vol. 149, 303–311.

MCCLINTOCK, W F P, and PHEMISTER, J. 1928.   A gravitational survey over the Swynnerton Dyke, Yarnfield, Staffordshire. 1–14 in Geological Survey of Great Britain.   *Summary of Progress for 1927.* (London: HMSO.)

MCKENZIE, D. 1978.   Some remarks on the development of sedimentary basins.   *Earth and Planetary Science Letters*, Vol. 40, 25–32.

MCQUILLIN, R. 1964.   Geophysical investigations of seismic shot-holes in the Cheshire Basin.   *Bulletin of the Geological Survey of Great Britain*, No. 21, 197–203.

MEIGH, A C. 1976.   The Triassic rocks, with particular reference to predicted and observed performance of some major foundations.   *Geotechnique*, Vol. 26, Pt 3, 391–452.

MEIGH, A C, and WOLKSI, W. 1981.   Design parameters for weak rocks.   55–79 in *Proceedings of the 7th European Conference on Soil Mechanics Foundation Engineering.* Vol 5. (London: British Geotechnical Society.)

MELVILLE, R V. 1947.   The non-marine lamellibranchs of the North Staffordshire Coalfield.   *Annals and Magazine of Natural History*, Series 11, Vol. 12, No.101, 289–337.

MIDDLETON, T. 1986a.   Kent Hill Gravel Mine and Quarry, Audley, Staffordshire.   *Bulletin of the Peak District Mines Historical Society*, Vol. 9, No.6, 393–399.

MIDDLETON, T. 1986b.   A survey of Beech Cave, Staffordshire. *Bulletin of the Peak District Mines Historical Society*, Vol. 9, No.6, 401–403.

MILLOT, J O'N. 1939.   The microspores in the coal-seams of North Staffordshire. Part 1.   The Millstone Grit — Ten Foot Coals.   *Transactions of the Institute of Mining Engineers*, Vol. 96, 317–353.

MILLOT, J O'N. 1941.   Regional variations in properties of the Eight Foot Banbury or Cockshead seam in the North Staffordshire Coalfield.   *Transactions of the Institute of Mining Engineers*, Vol. 101, 2–24.

MILLOT, J O'N, COPE, F W, and BERRY, J. 1946.   The seams encountered in a deep boring at Pie Rough, near Keele, North Staffordshire.   *Transactions of the Institute of Mining Engineers*, Vol. 105, 528–586.

MOLYNEUX, W. 1864. Report of the committee on the distribution of the organic remains of the North Staffordshire Coalfield. *Report of the meeting of the British Association for the Advancement of Science, Bath*, 1864, 342–344.

MORGAN, A V. 1973. The Pleistocene geology of the area north and west of Wolverhampton, Staffordshire, England. *Philosophical Transactions of the Royal Society of London*, B265, 233–297.

MORGAN, A V, DUTHIE, H C, MORGAN, A, FRITZ, P, and REARDON, E J. 1977. The Stafford project; a multi-disciplinary analysis of a late-Glacial sequence in the West Midlands. 309 in *Abstracts of the 10th conference of the international union for Quaternary research*. (Birmingham: INQUA.)

MOSTLER, H, and SCHEURING, B W. 1974. Mikrofloren aus dem Langobard und Cordevol der nördlichen Kalkalpen und das Problem des Beginns der Keupersedimentation in Germanischen Raum. *Geologisches-Paläontologisches Mitteilungen, Innsbruck*, Vol. 4, No. 4, 1–35.

MUSSET, A E, DAGLEY, P, and SKELHORN, R R. 1988. Time and duration of activity in the British Tertiary Igneous Province. 337–348 *in* Early Tertiary Volcanism and the opening of the NE Atlantic. MORTON, A C, and PARSON, L M (editors). *Special Publication of the Geological Society of London*, No. 39.

MYERS, J. 1954. On the occurrence of *Anthraconauta tenuis* (Davies and Trueman) in North Staffordshire. *Geological Magazine*, Vol. 91, 171–173.

MYKURA, W. 1960. The replacement of coal by limestone and the reddening of Coal Measures in the Ayrshire Coalfield. *Bulletin of the Geological Survey of Great Britain*, No. 16, 69–109.

NATIONAL COAL BOARD. 1960. *North Staffordshire Coalfield Seam Maps*. (London: National Coal Board.)

NAYLOR, H, TURNER, P, VAUGHAN, D J, BOYCE, A J, and FALLICK, A E. 1989. Genetic studies of red bed mineralization in the Triasssic of the Cheshire Basin, northwest England. *Journal of the Geological Society of London*, Vol. 146, 685–699.

NEILSON, G, MUSSON, R M W, and BURTON, P W. 1984. Macroseismic reports on historical British earthquakes. V: Midlands. *Report of the Global Seismology Unit, British Geological Survey*, No. 228.

OLD, R A, SUMBLER, M G, and AMBROSE, K. 1987. Geology of the country around Warwick. *Memoir of the British Geolological Survey*, Sheet 184 (England and Wales).

OLD, R A, HAMBLIN, R J O, AMBROSE, K, and WARRINGTON, G. 1991. Geology of the country around Redditch. *Memoir of the British Geological Survey*, Sheet 183 (England and Wales).

OWENS, B, NEVES, R, GUEINN, K J, MISHELL, D R F, SABRY, H S M Z, and WILLIAMS, J E. 1977. Palynological division of the Namurian of Northern England and Scotland. *Proceedings of the Yorkshire Geological Society*, Vol. 41, 381–398.

PATTISON, J. 1970. A review of the marine fossils from the Upper Permian rocks of northern Ireland and north-west England. *Bulletin of the Geological Survey of Great Britain*, No. 32, 123–165.

PATTISON, J, SMITH, D B, and WARRINGTON, G. 1973. A review of late Permian and early Triassic biostratigraphy in the British Isles. *Memoirs of the Canadian Society of Petroleum Geologists*, No. 2, 220–260.

PHAROAH, T C, MERRIMAN, R J, WEBB, P C, and BECKINSALE, R D. 1987. The concealed Caledonides of eastern England: preliminary results of a multidisciplinary study. *Proceedings of the Yorkshire Geological Society*, Vol. 46, 355–369.

PIPER, D P. 1982. The conglomerate resources of the Sherwood Sandstone Group of the country east of Stoke-on-Trent, Staffordshire. *Mineral Assessment Report of the Institute of Geological Sciences*, No. 91.

POLLARD, J E. 1981. A comparison between the Triassic trace fossils of Cheshire and South Germany. *Palaeontology*, Vol. 24, 555–588.

POLLARD, J E, and WISEMAN, J F. 1971. Algal limestone in the Upper Coal Measures (Westphalian D) at Chesterton, North Staffordshire. *Proceedings of the Yorkshire Geological Society*, Vol. 38, 329–342.

POOLE, E G. 1969. The stratigraphy of the Geological Survey Apley Barn borehole, Witney, Oxfordshire. *Bulletin of the Geological Survey of Great Britain*, No. 29, 1–104.

POOLE, E G, and WHITEMAN, A J. 1955. Variations in thickness of the Collyhurst Sandstone in the Manchester area. *Bulletin of the Geological Survey of Great Britain*, No. 9, 33–41.

POOLE, E G, and WHITEMAN, A J. 1961. The glacial drifts of the southern part of the Shropshire–Cheshire Basin. *Quarterly Journal of the Geological Society of London*, Vol. 117, 91–130.

POOLE, E G, and WHITEMAN, A J. 1966. Geology of the country around Nantwich and Whitchurch. *Memoir of the Geological Survey of Great Britain*, Sheet 122 (England and Wales).

PUGH, W. 1960. Triassic salt: discoveries in the Cheshire–Shropshire Basin. *Nature, London*, Vol. 187, 278–279.

RAMSBOTTOM, W H C. 1962. Boreholes in the Carboniferous rocks of the Ashover district. *Bulletin of the Geological Survey of Great Britain*, No. 19, 114–117.

RAMSBOTTOM, W H C. 1969. The Namurian of Britain. *Compte Rendu du 6eme Congrès International de Stratigraphie et de Géologie du Carbonifere, Sheffield 1967*, Vol. 1, 219–232.

RAMSBOTTOM, W H C. 1977. Major cycles of transgression and regression (mesothems) in the Namurian. *Proceedings of the Yorkshire Geological Society*, Vol. 41, 261–291.

RAMSBOTTOM, W H C. 1980. Eustatic control in Carboniferous ammonoid biostratigraphy. 369–387 *in* The Ammonoideae. HOUSE, K R, and SENIOR, J R (editors). *Special Publication of the Systematics Association*, No. 18.

RAMSBOTTOM, W H C, CALVER, M A, EAGAR, R M C, HODSON, F, HOLLIDAY, D W, STUBBLEFIELD, C J, and WILSON, R B. 1978. A correlation of Silesian rocks in the British Isles. *Special Report of the Geological Society of London*, No. 10.

READ, W A. 1991. The Millstone Grit (Namurian) of the southern Pennines viewed in the light of eustatically controlled sequence stratigraphy. *Geological Journal*, Vol. 27, 173–180.

REES, J G. 1990a. Geology of the Hanley district. *British Geological Survey Technical Report*, WA/90/05.

REES, J G. 1990b. Geology of the Longton district. *British Geological Survey Technical Report*, WA/90/10.

REES, J G. 1993. Stoke-on-Trent: a brief view of its economic geology. *Mercian Geologist*, Vol. 13, 138–144.

REES, J G, and CLARK, M C. 1992. Geology of the Tunstall district. *British Geological Survey Technical Report*, WA/90/08.

REES, J G, CORNWELL, J D, DABEK, Z K, and MERRIMAN, R J. 1996. The Apedale tuffs, North Staffordshire: probable remnants of a late Asbian/Brigantian (P_1a) volcanic centre. 345–357 *in* Recent advances in Lower Carboniferous geology. STROGEN, P, SOMERVILLE, I D, and JONES, G U (editors). *Special Publication of the Geological Society of London*, No. 107.

RHYDDERCH, L D, and YATES, D C. 1964. Outbursts of firedamp in the North Staffordshire Coalfield. *The Mining Engineer*, Vol. 124, No. 51, 168–184.

ROWLAND, J, and CADMAN, B. 1960. *Ambassador for oil. The life of John, First Baron Cadman.* (London: Herbert Jenkins.)

SCHILLER, H J. 1980. X-ray textural analysis of clays and compacted shales. 17–22 in *Abstracts of the International Association of Sedimentologists 1st European Meeting.* (Bochum: International Association of Sedimentologists.)

SCOTT, A. 1920. Notes on the petrography of the Butterton Dyke. *Transactions of the North Staffordshire Field Club,* Vol. 54, 36–43.

SCOTT, A. 1925. The intrusive igneous rocks of North Staffordshire. 86–99 in The geology of the country around Stoke-upon-Trent. GIBSON, W. *Memoir of the Geological Survey of England and Wales.*

SCOTT, A. 1927. A marine band in the Shaffalong Coalfield. *Transactions of the North Staffordshire Field Club,* Vol. 61, 87–88.

SCURFIELD, R W. 1958. Reconstruction in the North Staffordshire Coalfield. *Transactions of the Institution of Mining Engineers,* Vol. 117, 248–262.

SEARLE, I W. 1973. An investigation into the movement of an embankment on the Potteries 'D' road at Stoke-on-Trent. Unpublished MSc thesis, University of Keele.

SECKERS, D. 1981. *The Potteries.* (Aylesbury: Shire Publications.)

SHEARMAN, D J. 1970. Recent halite rock, Baja California, Mexico. *Transactions of the Institution of Mining and Metallurgy,* Vol. 79, B155–162.

SHOTTON, F W. 1966. The problems and contributions of methods of absolute dating within the Pleistocene period. *Quarterly Journal of the Geological Society of London,* Vol. 122, 356–383.

SHRAYNE, J. 1990. *Aggregates local (subject) plan 1989–2001. Draft written statement including report of Survey proposals and policies.* (Stafford: Stafford County Council Planning Department.)

SKEMPTON, A W. 1964. Long term stability of clay slopes *Geotechnique,* Vol. 14, 77–101.

SKEMPTON, A W. 1986. Standard penetration test procedures and the effects in sands of overburden pressure, relative density, particle size, ageing, and overconsolidation. *Geotechnique,* Vol. 36, 425–447.

SLATER, D, and HIGHLEY, D E. 1978. United Kingdom. 283–299 in *The iron ore deposits of Europe and adjacent areas. Vol. 2.* ZITZMANN, A (editor). (Hannover: Bundesanst.)

SMITH, D B, BRUNSTROM, R G W, MANNING, P I, SIMPSON, S and SHOTTON, F W. 1974. A correlation of Permian rocks in the British Isles. *Special Report of the Geological Society of London,* No. 5.

SMITH, N J P. 1987. The deep geology of central England: the prospectivity of the Palaeozoic rocks. 217–224 in *Petroleum geology of northwest Europe.* BROOKS, J, and GLENNIE, K (editors). (London: Graham and Trotman.)

SMITH, A H V, and BUTTERWORTH, M A. 1967. Miospores in the coal seams of the Carboniferous of Great Britain. *Special Papers in Palaeontology,* No. 1, 1–324.

SMITH, I F, and ROYLES, C P R. 1989. The digital aeromagnetic survey of the United Kingdom. *British Geological Survey Technical Report,* WK/89/05.

SMITH, I F, HOUGHTON, M T, and BURGESS, W G. 1980. *Investigation of the geothermal potential of the UK. The Permo-Triassic Cheshire and West Lancashire Basins.* (London: Institute of Geological Sciences.)

SMITH, K, SMITH, N J P, and HOLLIDAY, D W. 1985. The deep structure of Derbyshire. *Geological Journal,* Vol. 20, 215–225.

SMITHSON, F. 1947. Geology. *Transactions of the North Staffordshire Field Club,* Vol. 81, 124–126.

SMYTH, W W. 1861. Iron ores of Great Britain. Part IV. Iron ores of North Staffordshire. *Memoir of the Geological Survey.*

SOPER, N J, and HUTTON, D H W. 1984. Late Caledonian sinistral displacements in Britain: implications for a three-plate collision model. *Tectonics,* Vol. 3, 781–794.

SOPER, N J, WEBB, B C, and WOODCOCK, N H. 1987. Late Caledonian (Acadian) transpression in North West England: timings, geometry and geotectonic significance. *Proceedings of the Yorkshire Geological Society,* Vol. 46, 175–192.

SOWERBUTTS, W T C. 1987. Magnetic mapping of the Butterton Dyke: an example of detailed geophysical surveying. *Journal of the Geological Society of London,* Vol. 144, 29–33.

SOWERBUTTS, W T C. 1988. Discussion on the magnetic mapping of the Butterton Dyke: an example of detailed geophysical surveying. *Journal of the Geological Society of London,* Vol. 145, 181–184.

STEEL, R J, and THOMPSON, D B. 1983. Structures and textures in Triassic braided stream conglomerates ('Bunter' Pebble Beds) in the Sherwood Sandstone Group, North Staffordshire, England. *Sedimentology,* Vol. 30, 341–367.

STEPHENS, J V. 1961. North Western District. 37–39 in Geological Survey of Great Britain. *Summary of progress for 1960.* (London: HMSO.)

STEVENSON, I P, and GAUNT, G D. 1971. Geology of the country around Chapel en le Frith. *Memoir of the Geological Survey of Great Britain,* Sheet 99 (England and Wales).

STEVENSON, I P, and MITCHELL, G H. 1955. Geology of the country between Burton upon Trent, Rugeley and Uttoxeter. *Memoir of the Geological Survey of Great Britain,* Sheet 140.

STOBBS, J T. 1902. Recent work in the correlation of the measures of the Potteries Coalfield of North Staffordshire, with suggestions for further development. *Transactions of the Institute of Mining Engineers,* Vol. 22, 229–247.

STOBBS, J T. 1905. The marine beds in the coalfields of North Staffordshire. *Quarterly Journal of the Geological Society of London,* Vol. 61, 495–527.

STOBBS, J T. 1915. A glossary of terms in use in the North Staffordshire Coalfields. *Transactions of the North Staffordshire Field Club,* Vol. 1, 42–62.

STOBBS, J T. 1920. More erratics found in the coal-seams of the North Staffordshire and Leicestershire coalfields. *Transactions of the North Staffordshire Field Club,* Vol. 54, 100–101.

STOBBS, J T. 1922. Further erratics in and near coal-seams of North Staffordshire and India. *Transactions of the North Staffordshire Field Club,* Vol. 56, 97–99.

STUBBLEFIELD, C J, and TROTTER, F M. 1957. Divisions of the Coal Measures on Geological Survey maps of England and Wales. *Bulletin of the Geological Survey of Great Britain,* No. 13, 1–5.

SUTHERLAND, D S (editor). 1982. *Igneous rocks of the British Isles.* (Chichester: Wiley.)

TANNER, P W G. 1989. The flexural-slip mechanism. *Journal of Structural Geology,* Vol. 11, 635–655.

TAYLOR, A J. 1981. *The Staffordshire coal industry.* (Reprinted from Greenslade, M W, and Jenkins, J G. 1967. *The Victoria history of the County of Stafford,* Vol. II). (Stafford: Staffordshire County Library).

TAYLOR, B J. 1958. Cemented shear-planes in the Pleistocene Middle Sands of Lancashire and Cheshire. *Proceedings of the Yorkshire Geological Society,* Vol. 31, 359–365.

TAYLOR, B J, PRICE, R H, and TROTTER, F M. 1963. Geology of the country around Stockport and Knutsford. *Memoir of the Geological Survey of Great Britain*, Sheet 98 (England and Wales).

TAYLOR, R K. 1988. Coal Measures mudrocks: composition, classification and weathering. *Quarterly Journal of Engineering Geology*, Vol. 21, 85–99.

TEALL, J J H. 1888. *British petrography: with special reference to the igneous rocks.* (London: Dulau and Co.)

THOMPSON, D B. 1969. Dome-shaped aeolian dunes in the Frodsham Member of the so-called 'Keuper' Sandstone Formation (Scythian–?Anisian: Triassic), at Frodsham, Cheshire (England). *Sedimentary Geology*, Vol. 3, 263–289.

THOMPSON, D B. 1970a. The stratigraphy of the so-called Keuper Sandstone Formation (Scythian–?Anisian) in the Permo-Triassic Cheshire Basin. *Quarterly Journal of the Geological Society of London*, Vol. 126, 151–181.

THOMPSON, D B. 1970b. Sedimentation of the Triassic (Scythian) Red Pebbly Sandstones in the Cheshire Basin and its margins. *Geological Journal*, Vol. 7, 183–216.

THOMPSON, D B. 1980. The influence of Pre-history on the growth of Betley. 9–32 in *Betley, a village of contrasts*. SPEAKE, R (editor). (Keele: Department of Adult Education).

THOMPSON, D B. 1982. Conservation, planning and other issues relating to the construction of highways across areas underlain by Pebble Beds of the Sherwood Sandstone Group. *Mercian Geologist*, Vol. 8, 271–284.

THOMPSON, D B. 1985. *Field excursion to the Permo-Triassic of the Cheshire, East Irish Sea, Needwood and Stafford basins.* Poroperm Excursion Guide No. 4. (Chester: Poroperm-Geochem Ltd.)

THOMPSON, D B. 1989. The geology of the neighbourhood of Chester — an essay review. *Amateur Geologist*, Vol. 13, Pt 1, 45–54.

THOMPSON, D B, and WINCHESTER, J A. 1995. Chemical and field studies and the tectonic context of the largely Tertiary dyke suites in Staffordshire and Shropshire, Central England. *Proceedings of the Yorkshire Geological Society.*

THOMPSON, D B, and WORSLEY, P. 1967. Periods of ventifact formation in the Permo-Triassic and Quaternary of East Cheshire. *Mercian Geologist*, Vol. 2, 279–298.

TONKS, L H, JONES, R C B, LLOYD, W, and SHERLOCK, R L. 1931. Geology of Manchester and the south-east Lancashire Coalfield. *Memoir of the Geological Survey.*

TORRENS, H S. 1994. 300 years of Oil: mirrored by development in the West Midlands. *Geological Society of London: British Association lectures 1993*, Geological Society of London, 4–8.

TREWIN, N H. 1968. Potassium bentonites in the Namurian of Staffordshire and Derbyshire. *Proceedings of the Yorkshire Geological Society*, Vol. 37, 73–91.

TREWIN, N H, and HOLDSWORTH, B K. 1972. Further K-bentonites from the Namurian of Staffordshire. *Proceedings of the Yorkshire Geological Society*, Vol. 39, 87–88.

TREWIN, N H, and HOLDSWORTH, B K. 1973. Sedimentation in the lower Namurian rocks of the North Staffordshire Basin. *Proceedings of the Yorkshire Geological Society*, Vol. 39, 371–408.

TRIGG, A B, and DUBOURG, W R. 1993. *Valuing the environmental impacts of open cast coal mining; the case of the Trent Valley in North Staffordshire.* (Milton Keynes: Open University.)

TROTTER, F M. 1954. Reddened beds in the Coal Measures of South Lancashire. *Bulletin of the Geological Survey of Great Britain*, No. 5, 61–80.

TROTTER, F M. 1955. North-western district. 37–40 *in* Geological Survey of Great Britain. *Summary of progress for 1954.* (London: HMSO.)

TROTTER, F M. 1960. Upper Carboniferous. Pt. 3 a VIII *in* England, Wales and Scotland. WHITTARD, W F, and SIMPSON, S (editors), in *Lexique Stratigraphique International.* (Paris: Centre National de la Recherche Scientifique for Congrès Géologique International — Commission de Stratigraphie).

TRUEMAN, A, and WEIR, J. 1945–1968. A monograph of British Carboniferous non-marine Lamellibranchia. *Palaeontographical Society Monograph*, Pts 1-13, 1–449.

TUCKER, R M. 1981. Giant polygons in the Triassic salt of Cheshire, England: a thermal contraction model for their origin. *Journal of Sedimentary Petrology*, Vol. 51, 779–786.

TUCKER, R M, and TUCKER, M E. 1981. Evidence of synsedimentary tectonic movements in the Triassic halite of Cheshire. *Nature, London*, Vol. 290, 495–496.

TURNER, N. 1991. The occurrence of *Elaterites triferens* Wilson 1943 in miospore assemblages from coal measures of Westphalian D age, North Staffordshire, England. *Palynology*, Vol. 15, 35–46.

TURNER, N, SPINNER, E, SPODE, F, and WIGNALL, P B. 1994. Palynostratigraphy of a Carboniferous transgressive systems tract from the earliest Alportian (Namurian) of Britain. *Review of Palaeobotany and Palynology*, Vol. 80, 39–54.

VAN DER EEM, J G L A. 1983. Aspects of Middle and Late Triassic palynology. 6. Palynological investigations in the Ladinian and Lower Carnian of the western Dolomites, Italy. *Review of Palaeobotany and Palynology*, Vol. 39, 189–300.

VERNON, R D. 1912. On the geology and palaeontology of the Warwickshire Coalfield. *Quarterly Journal of the Geological Society of London*, Vol. 68, 507–683.

VISSCHER, H, and BRUGMAN, W A. 1981. Ranges of selected palynomorphs in the Alpine Triassic of Europe. *Review of Palaeobotany and Palynology*, Vol. 34, 115–128.

WAGNER, R H. 1984. Megafloral zones of the Carboniferous. *Compte Rendu, 9eme Congrès International de Stratigraphie et de Géologie du Carbonifère, Washington DC, Champaign Illinois*, Vol. 2, 109–134.

WAINE, P J, HALLAM, J R, and CULSHAW, M G. 1991. Engineering Geology of the Stoke-on-Trent area. *British Geological Survey Technical Report*, WN/90/11.

WALKER, T R. 1976. Diagenetic origin of continental red beds. 240–282 in *The continental Permian in Central, West and Southern Europe*. FALKE, N (editor). (Dordrecht: Holland.)

WALTHAM, A C. 1989. *Ground subsidence.* (Glasgow and London: Blackie.)

WALTON, A D. 1964. Meltwater channels near Head of Trent. *North Staffordshire Journal of Field Studies*, Vol. 4, 67–75.

WARD, J. 1890. The geological features of the North Staffordshire Coalfields, their organic remains, their range and distributions, with a catalogue of the fossils of the Carboniferous System of North Staffordshire. *Transactions of the North Staffordshire Institute of Mining Engineers*, Vol. 10, Pt 5, 1–189.

WARRINGTON, G. 1970a. The stratigraphy and palaeontology of the "Keuper" Series of the central Midlands of England. *Quarterly Journal of the Geological Society of London*, Vol. 126, 183–223.

WARRINGTON, G. 1970b. The "Keuper" Series of the British Trias in the northern Irish Sea and neighbouring areas, *Nature, London*, Vol. 226, 254–256.

WARRINGTON, G, AUDLEY-CHARLES, M G, ELLIOTT, R E, EVANS, W B, IVIMEY-COOK, H C, KENT, P E, ROBINSON, P L, SHOTTON, F W, and TAYLOR, F M. 1980.   A correlation of the Triassic rocks in the British Isles.   *Special Report of the Geological Society of London*, No. 13.

WARRINGTON, G, and THOMPSON, D B. 1971.   The Triassic rocks of Alderley Edge, Cheshire.   *Mercian Geologist*, Vol. 4, 69–72.

WEDD, C B. 1899.   On barium sulphate as a cementing material in the Bunter Sandstone of North Staffordshire. *Geological Magazine*, Dec.IV, Vol.6, 508.

WESTBROOK, G K, KUSZNIR, N J, BROWITT, C W A, and HOLDSWORTH, B K. 1980.   Seismicity induced by coal mining in Stoke-on-Trent.   *Engineering Geology*, Vol. 16, 225–241.

WHITE, P H N. 1949.   Gravity data obtained in Great Britain by the Anglo-American Oil Company Ltd.   *Quarterly Journal of the Geological Society of London*, Vol. 104, 339–364.

WHITEHEAD, T H, DIXON, E E L, POCOCK, R W, ROBERTSON, T, and CANTRILL, T C. 1927.   The country between Stafford and Market Drayton.   *Memoir of the Geological Survey of Great Britain.*

WIGNALL, P B. 1987.   A biofacies analysis of the Gastrioceras cumbriense marine band (Namurian) of the central Pennines. *Proceedings of the Yorkshire Geological Society*, Vol. 46, 111–121.

WILD, E K. 1987.   The sedimentology and reservoir quality of the Kinnerton Sandstone Formation, U.K. and the Tirrawarra Sandstone, S. Australia.   Unpublished PhD thesis, University of Bristol.

WILLIAMS, B P J, DOWNING, R A, and LOVELOCK, P E R. 1972. Aquifer properties of the Bunter Sandstones in Nottinghamshire, England.   169–176 in *Proceedings of the 24th International Geological Congress*, Section 11. (Montreal: 24th International Geological Congress.)

WILLIAMSON, W O. 1946.   Some grits and associated rocks in the Etruria Marls of North Staffordshire.   *Geological Magazine*, Vol. 83, 20–32.

WILLS, L J. 1947.   *The palaeogeography of the Midlands.* (Liverpool: Liverpool University Press.)

WILLS, L J. 1956.   *Concealed coalfields.* (London: Blackie.)

WILLS, L J. 1970a.   The Triassic succession in the central Midlands in its regional setting.   *Quarterly Journal of the Geological Society of London*, Vol. 126, 225–285.

WILLS, L J. 1970b.   The Bunter Formation at the Bellington Pumping Station of the East Worcestershire Waterworks Company.   *Mercian Geologist*, Vol. 3, 387–397.

WILLS, L J. 1976.   The trias of Worcestershire and Warwickshire.   *Report of the Institute of Geological Sciences*, No. 76/2.

WILSON, A A. 1962.   40 in Geological Survey of Great Britain. *Summary of Progress for 1961.* (London: HMSO.)

WILSON, A A. 1990.   Geology of the Silverdale district.   *British Geological Survey Technical Report*, WA/90/04.

WILSON, A A. 1993.   The Mercia Mudstone Group (Trias) of the Cheshire Basin.   *Proceedings of the Yorkshire Geological Society*, Vol. 49, 171–188.

WILSON, A A, and EVANS, W B. 1990.   Geology of the country around Blackpool.   *Memoir of the British Geological Survey*, Sheet 66 (England and Wales).

WILSON, A A, SARGEANT, G A, YOUNG, B R, and HARRISON, R K. 1966.   The Rowhurst Tonstein, North Staffordshire, and the occurrence of crandallite.   *Proceedings of the Yorkshire Geological Society*, Vol. 35, 421–427.

WILSON, A A, REES, J G, CROFTS, R G, HOWARD, A S, BUCHANAN, J G, and WAINE, P J. 1992.   Stoke-on-Trent: a geological background for planning and development.   *British Geological Survey Technical Report*, WA/91/01.

WOODCOCK, N H. 1984.   The Pontesford Lineament, Welsh Borderland.   *Journal of the Geological Society of London*, Vol. 141, 1001–1014.

WORSLEY, P. 1980.   Problems of radiocarbon dating the Chelford Interstadial of England.   289–304 in *Timescales in geomorphology.* CULLINGFORD, D A, DAVIDSON, D A, and LEWIN, J (editors). (Chichester: Wiley.)

WORSLEY, P. 1985.   Pleistocene history of the Cheshire–Shropshire Plain.   201–220 in *The geomorphology of North-west England.* JOHNSON, R H (editor). (Manchester: Manchester University Press.)

WORSLEY, P, COOPE, G R, GOOD, T R, HOLYOAK, D T, and ROBINSON, J E. 1983.   A Pleistocene succession from beneath Chelford sands at Woodoak Quarry, Chelford, Cheshire. *Geological Journal*, Vol. 18, 307–324.

YATES, E M. 1955.   Glacial meltwater spillways near Kidsgrove, Staffordshire.   *Geological Magazine*, Vol. 92, 413–418.

YATES, E M. 1956.   The Keele surface and the Upper Trent drainage.   *East Midlands Geographer*, Vol. 1, 10–22.

YATES, E M. 1957.   A contribution to the geomorphology of north-west Staffordshire and adjacent parts of Cheshire. Unpublished PhD thesis, University of London.

YATES, E M, and MOSELEY, F. 1958.   Glacial lakes and spillways in the vicinity of Madeley, North Staffordshire.   *Quarterly Journal of the Geological Society of London*, Vol. 113, 409–428.

YATES, E M, and MOSELEY, F. 1967.   A contribution to the glacial geomorphology of the Cheshire Plain.   *Transactions and Papers of the Institute of British Geographers*, No. 42, 107–125.

ZEIGLER, P A. 1975.   North Sea Basin history in the tectonic framework of North-Western Europe.   131–148 in *Petroleum and the Continental Shelf of north-west Europe, 1, Geology.* WOODLAND, A W (editor). (London: Applied Science Publishers Ltd.)

ZEIGLER, P A. 1981.   Evolution of sedimentary basins in North-West Europe.   3–39 in *Petroleum Geology of the Continental Shelf of North West Europe.* ILLING, L V, and HOBSON, G D (editors). (London: Heydon and Son.)

APPENDIX 1

List of boreholes and shafts

This list includes the name, permanent BGS record number and location of boreholes and shafts that are referred to in this memoir; u/g = underground borehole, SI = site investigation borehole, UDC = Urban District Council, * = borehole not on sheet 123. Copies of non-confidential records may be obtained from the British Geological Survey, Keyworth, Nottingham NG12 5GG at advertised prices.

Allotment No. 1*	SJ92NW/35	[9461 2679]
Alsager UDC Waterworks	SJ75SE/5	[7989 5392]
Alsager Waterworks	SJ75NE/1	[7938 5517]
Apedale No. 2	SJ84NW/23	[8074 4862]
Apedale Works No. 5	SJ84NW/47	[8288 4882]
AU15*	SJ64SE/10	[6694 4055]
AU16	SJ64SE/9	[6795 4063]
AU17*	SJ63NE/21	[6721 3916]
Audley Waterworks	SJ75SE/3	[7769 5029]
Bagot's Park*	SK02NE/1	[0882 2697]
Barker's Wood	SJ84NW/68	[8221 4516]
Bearstone Waterworks	SJ73NW/2	[7243 3899]
Beech Cliff	SJ83NE/152	[8563 3924]
Berry Hill New Shaft	SJ84NE/14	[8934 4597]
Birchenwood Bath (Harecastle) Pit	SJ85SW/12	[8373 5248]
Bittern's Wood	SJ74SE/92	[7662 4286]
Blacklake	SJ93NW/1	[9343 3930]
Blacklake Waterworks	SJ93NW/52	[9378 3892]
Bowsey Wood	SJ74NE/9	[7695 4643]
Bromley	SJ84SW/80	[8035 4364]
Burley Pit	SJ84NW/7	[8198 4926]
Butterton No. 2	SJ84SW/9	[8426 4173]
Butterton No. 5	SJ84SW/20	[8405 4234]
Byley*	SJ76NW/7	[7207 6942]
Caldon Low	SK04NE/36	[0804 4822]
Chatterley Whitfield u/g No. 2	SJ85SE/16	[8703 5278]
Chatterley Whitfield Hesketh Shaft	SJ85SE/2	[8848 5332]
Checkley New Farm	SJ74NW/2	[7428 4578]
City General Hospital No. 25	SJ84NE/168	[8584 4520]
City General Hospital No. 33	SJ84NE/175	[8584 4530]
City General Hospital No. 53	SJ84NE/195	[8569 4514]
City General Hospital No. 58	SJ84NE/199	[8562 4503]
Clayton	SJ84SW/3	[8485 4315]
Clifford''s Wood	SJ83NW/8	[8480 3745]
Coton Field No. 1*	SJ92SW/27	[9276 2444]
Cresswell No. 1*	SJ93NE/1	[9737 3394]
Crewe Gates No. 1	SJ75NW/16	[7219 5536]
Crewe Heat Flow	SJ65SE/6	[6827 5452]
Crowcrofts	SJ94SW/13	[9009 4001]
Darlaston	SJ83NE/61	[8900 3592]
Dog Lane	SJ83NW/20	[8268 3922]
Dunge No. 6	SJ74SE/21	[7940 4472]
Dunge No. 11	SJ74SE/26	[7932 4461]
Fenton Manor No. 1	SJ84SE/76	[8873 4493]
Fenton Manor Homer Shaft	SJ84SE/3	[8824 4332]
Florence u/g No. 4	SJ94SW/29	[9173 4034]
Florence No. 2 Shaft	SJ94SW/2	[9149 4189]
Foley Deep Ash Pit	SJ84SE/413	[8997 4359]
Fulford	SJ93NW/15	[9471 3825]
Grangewood	SJ84SW/88	[8392 4204]
Groundslow	SJ83NE/5	[8666 3771]
Gun Hill*	SJ96SE/18	[9723 6182]
Hatton Waterworks	SJ83NW/2	[8295 3699]
Hanley Deep Pit	SJ84NE/12	[8849 4832]
Hargreaves	SJ84SE/68	[8545 4113]
Harley	SJ83NW/7	[8367 3931]
Harley Thorns	SJ83NW/21	[8457 3873]
Hem Heath No. 1 Shaft	SJ84SE/1	[8865 4142]
Hem Heath No. 2 Shaft	SJ84SE/2	[8855 4149]
Hem Heath u/g No. 8	SJ94SW/21	[9021 4125]
Hesketh Shaft	SJ85SE/2	[8848 5332]
Hey Sprink	SJ74SE/32	[7878 4336]
Highway	SJ84NW/116	[8017 4511]
Hobbergate	SJ93NW/10	[9218 3746]
Hobgoblin	SJ84SW/85	[8340 4052]
Holditch No. 2 Shaft	SJ84NW/3	[8359 4816]
Holditch u/g No. 2	SJ84NW/6	[8316 4720]
Holditch u/g No. 4	SJ84NW/30	[8392 4850]
Holditch u/g No. 27	SJ84NW/129	[8306 4733]
Holditch u/g No. 61	SJ84NW/146	[8349 4579]
Holts Barn	SJ93NW/2	[9005 3640]
Home Farm	SJ84NW/17	[8238 4547]
Keele University Campus SI No. 9	SJ84NW/105	[8176 4522]
Keele University Campus SI No. 14	SJ84NW/108	[8172 4521]
Kibblestone	SJ93NW/60	[9107 3637]
Knenhall	SJ93NW/86	[9153 3822]
Knight's Wood No. 5	SJ84SW/62	[8339 4196]
Knight's Wood No. 8	SJ84SW/12	[8389 4177]
Knowl Wall	SJ83NE/1	[8574 3933]
Knowles Farm	SJ95SW/14	[9258 5487]
Knutsford*	SJ77NE/11	[7560 7770]
Little Paddocks	SJ84SW/150	[8153 4100]
Lower Thornhill	SJ74NE/11	[7552 4613]
Lymes Road	SJ84SW/1	[8257 4366]
Madeley Pumping Station	SJ74NE/10	[7614 4518]
Madeley Training College No.1	SJ74NE/18	[7734 4546]
Madeley Training College No. 2	SJ74NE/19	[7740 4544]
Madeley Training College No. 3	SJ74NE/20	[7745 4543]
Madeley Clarke's Pit	SJ74NE/1	[7910 4675]
Meaford Hall No.1	SJ83NE/3	[8846 3714]
Meaford Farm Bridge	SJ83NE/64	[8860 3746]
Meir Pumping Station	SJ94SW/19	[9375 4206]
Mill End	SJ75SE/11	[7930 5236]
Millbank Colliery	SJ84NW/16	[8309 4585]
Minnie Pit	SJ74NE/5	[7933 4894]
Moddershall	SJ93NW/5	[9239 3632]
Moddershall Pumping Station	SJ93NW/62	[938- 367-]
Mossfield No. 1 Pit	SJ94SW/5	[9155 4498]
Mossgate	SJ93NE/37	[953- 372-]
Nettlebank No. 1 Pit	SJ85SE/10	[8861 5033]
Newstead	SJ84SE/13	[8895 4072]
Nook's Farm*	SJ95NW/12	[9174 5803]
Norton u/g No. 3	SJ85SE/18	[8842 5053]
Onneley	SJ74SE/27	[7505 4365]
Parkhouse No. 3 Shaft	SJ85SW/8	[8381 5020]
Parkhouse u/g No. 1	SJ85SW/19	[8396 5029]
Peacock's Lane	SJ84SW/70	[8445 4113]
Penfields No. 1	SJ84SW/152	[8132 4367]
Penfields No. 2	SJ84SW/156	[8151 4329]
Pewit Lane	SJ74NW/1	[7003 4525]

Pie Rough (Keele No. 1)	SJ84SW/2	[8293 4399]	The Rowe	SJ83NW/34	[8250 3805]
Pipe Gate	SJ74SW/1	[7365 4077]	The Spot	SJ93NW/61	[9390 3770]
Plattlane*	SJ53NW/16	[5140 3645]	The Wellings	SJ73NE/9b	[7692 3852]
Plum Tree	SJ74NE/37	[7662 4788]	Timbersbrook*	SJ86SE/4	[8971 6254]
Prees*	SJ53SE/3	[5572 3394]	Tittensor	SJ83NE/62	[8779 3908]
Racecourse No. 3 Shaft	SJ84NE/8	[8725 4787]	Toft	SJ84SW/84	[8473 4037]
Radwood	SJ74SE/28	[7753 4176]	Town House Farm	SJ85SW/18	[8020 5487]
Ranton No. 1*	SJ82SW/12	[8841 2362]	Trentham Park No. 5	SJ84SE/39	[8589 4065]
Rectory No. 1	SJ84NW/89	[8112 4563]	Trentham Park No. 8	SJ84SE/42	[8609 4057]
Redheath Plantation No. 2	SJ84NW/70	[8151 4568]	Trentham Park No. 10	SJ84SE/44	[8582 4077]
Redheath Plantation No. 3	SJ84NW/71	[8157 4587]	Trentham	SJ84SE/14	[8625 4129]
Ridgeway	SJ85SE/14	[8922 5381]	Turnover	SJ83NE/65	[8937 3657]
Rook Hall	SJ84SW/6	[8354 4167]	Wallmires	SJ94NW/8	[9491 4639]
Rowhurst No. 2 Shaft	SJ84NE/11	[8779 4735]	Werrington	SJ94NW/10	[9434 4856]
Schoolhouse	SJ93NW/54	[9352 3780]	Weston Coyney No. 1	SJ94SW/14	[9361 4349]
Sheepwash	SJ94NE/39	[9510 4510]	Weston Coyney No. 3	SJ94SW/16	[9383 4472]
Shelton Steelworks No. 1	SJ84NE/33	[8659 4796]	Weston Coyney No. 4	SJ94SW/17	[9476 4338]
Sidway Mill	SJ73NE/3	[7603 3934]	Whitchurch No. 1	SJ54SE/24	[5842 4134]
Springpool No. 1	SJ84SW/45	[8264 4396]	Whitchurch No. 2	SJ54SE/35	[5816 4330]
Springpool No. 10	SJ84SW/52	[8209 4459]	Whitmore Pumping Station	SJ83NW/9	[8125 3980]
Stabhill	SJ83NW/10	[8499 3666]	Whitmore No. 4	SJ74SE/9	[7994 4012]
Stallington	SJ94SW/12	[9326 4092]	Whitmore Rectory	SJ84SW/132	[8038 4093]
Stockton Brook	SJ95SW/3	[9129 5208]	Wilkesley*	SJ64SW/7	[6286 4144]
Stony Low	SJ74SE/33	[7905 4429]	Wolstanton No. 2 Shaft	SJ84NE/28	[8606 4811]
Swallowcroft	SJ84SW/79	[8229 4342]	Wolstanton No. 3 Shaft	SJ84NE/29	[8606 4800]
Swynnerton Waterworks	SJ83NE/153	[864- 359-]	Workhouse No. 1	SJ84NW/94	[8048 4528]
Ternhill	SJ63SW/26	[6315 3313]	Workhouse No. 3	SJ84NW/99	[8039 4538]
The Dams	SJ94SW/50	[9499 4309]	Workhouse No. 9	SJ74NE/24	[7963 4524]

APPENDIX 2

List of petrographical samples

E numbers refer to the Petrography Collection of the British Geological Survey. Several hundred petrographical samples from the Stoke-on-Trent district have been registered in these collections, though only those referred to in the text are listed here. A full listing of petrographical samples is available on request.

Each number is followed by locality details; in the case of samples from boreholes, the details can be found in Appendix 1.

E30238   Red Street Foot Rail [8308 5143]. Tonstein, Burnwood Coal, Coal Measures

E3114   Shelton Steelworks No. 1 Borehole, depth 67 m. Sideritic ironstone, Etruria Formation

E31144   Florence Colliery [9154 4184]. Holditch Tonstein, Coal Measures

E31197   Gratton (Sheet 110) [9241 5547]. Minn Sandstones

E31198   Gratton (Sheet 110) [9237 5550]. Minn Sandstones

E32029   Sidway Mill Borehole, depth 651.3 m. Tonstein, Newcastle formation

E32807a   Old quarry west of Maer [7889 3818]. Hematitised sandstone, Bromsgrove Formation

E32807b   North-east of Maer [7984 3950]. Hematitised sandstone, Wildmoor Formation

E32807c   North-east of Norton-in-Hales [7228 3944]. Malachite-impregnated sandstone, Helsby Formation

E32808a   Old quarry near Bearstone [7218 3924]. Sandstone with baryte, Helsby Formation

E32808b   Quarry near Napley Farm [7202 3844]. Baryte-cemented sandstone, Helsby Formation

E32808c   Quarry on Red Hill [7868 3956]. Sandstone, Wildmoor Formation

E32995   Quarry near Stockton Brook Station [9145 5133]. Rough Rock

E32996   Quarry near Stockton Brook Station [9157 5142]. Chatsworth Grit

E32997   Quarry in Broughton Wood [9248 5126]. Cheddleton Sandstones

E32998   Quarry near Manor House [9254 5094]. Kniveden Sandstones

E32999   Old quarry, Bagnall Green [9299 5087]. Chatsworth Grit

E33000   Cutting in Cliff Wood [9348 5118]. Kniveden Sandstones

E33001   Old quarry, The Ashes [9310 5442]. Hurdlow Sandstones

E33002   Escarpment along drive to Henridding Farm [9170 5360]. Hurdlow Sandstones

E33279   Florence Colliery [915 418]. Holditch Tonstein, Coal Measures

E33280   Florence Colliery [915 418]. Holditch Tonstein, Coal Measures

E33282   Florence Colliery [915 418]. Rowhurst Tonstein, Coal Measures

E33283   Florence Colliery [915 418]. Stafford Tonstein, Coal Measures

E33289   Exposure in stream near Heywood Grange [9634 4558]. Sandstone, Edale Shales

E33290   Exposure in stream near Heywood Grange [9631 4555]. Sandstone, Edale Shales

E33292   Excavation near Caverswall [9618 4473]. Chatsworth Grit

E33294   Small quarry near Stansmore Hall [9620 4401]. Chatsworth Grit

E33300   Small quarry near Blakeley Lane Church [9629 4742]. Rough Rock

E33301   Small quarry near Jack Hayes [9221 4981]. Rough Rock

E33302   Exposure near Lark Hall [9310 4967]. Rough Rock

E33303   Exposure in beck near Brookhouse [9390 4967]. Kniveden Sandstones

E33305   Armshead Quarry [9353 4828]. Chatsworth Grit

E33306   Armshead Quarry [9353 4828]. Chatsworth Grit

E33307   Old quarry at Washerwall [9367 4749]. Rough Rock

E33782   Trentham Park No. 8 Borehole. Tonstein, depth 11.2 m. Keele formation

E33783   Trentham Park No. 10 Borehole. Tonstein, depth 14.3 m. Keele formation

E33784   Trentham Park No. 5 Borehole. Tonstein, depth 29.9 m. Keele formation

E35040–   Apedale No. 2 Borehole, depths 466.3–1126.2 m.
E35048   Tuffs

E37674   Wolstanton Colliery No. 3 Shaft, depth 644 m. Stafford Tonstein, Coal Measures

E37675   Wolstanton Colliery No. 3 Shaft, depth 646 m. Stafford Tonstein, Coal Measures

144

AUTHOR CITATIONS FOR FOSSIL SPECIES

To satisfy the rules and recommendations of the international codes of botanical and zoological nomenclature, authors of cited species are listed below.

Chapter 3

Anthracoceratites vanderbeckei (Ludwig, 1863)
Anthraconaia adamsii (Salter, 1861)
Anthraconaia curtata (Brown, 1849)
Anthraconaia dolabrata (J de C Sowerby, 1840)
Anthraconaia ellipsoides Trueman & Weir, 1966
Anthraconaia expansa (Hind, 1893)
Anthraconaia hindi (Wright, 1930)
Anthraconaia insignis Davies and Trueman, 1927
Anthraconaia lanceolata (Hind, 1893)
Anthraconaia librata (Wright, 1929)
Anthraconaia modiolaris (J de C Sowerby, 1840)
Anthraconaia obovata (Hind, 1893)
Anthraconaia prolifera (Waterlot, 1934)
Anthraconaia pruvosti (Chernyshev, 1931)
Anthraconaia pulchella Broadhurst, 1959
Anthraconaia pulchra (Hind, 1895)
Anthraconaia salteri (Leitch, 1940)
Anthraconaia sagittata (Tchernyshev, 1931)
Anthraconaia saravana (Schmidt, 1907)
Anthraconaia stobbsi (Dix and Trueman, 1931)
Anthraconaia varians (Melville, 1947)
Anthraconaia wardi (Hind, 1893)
Anthraconaia warei (Dix and Trueman, 1931)
Anthraconaia williamsoni (Brown, 1849)
Anthraconauta calcifera (Hind, 1899)
Anthraconauta phillipsii (Williamson, 1836)
Anthraconauta tenius (Davies and Trueman, 1927)
Anthraconauta wrighti (Dix and Trueman, 1931)
Anthraconeilo taffiana Girty, 1915
Anthracopupa brittanica Cox, 1926
Anthracosia actuella (Wright, 1929)
Anthracosia aquilina (J de C Sowerby, 1840)
Anthracosia aquilinoides (Tchernyshev, 1931)
Anthracosia atra (Trueman, 1929)
Anthracosia barkeri Leitch, 1947
Anthracosia beaniana King, 1856
Anthracosia caledonica Trueman and Weir, 1951
Anthracosia carissima (Wright, 1929)
Anthracosia concinna (Wright, 1929)
Anthracosia disjuncta Trueman and Weir, 1951

Anthracosia faba (Wright, 1930)
Anthracosia fulva (Davies and Trueman, 1927)
Anthracosia lateralis (Brown, 1843)
Anthracosia nitida (Davies and Trueman, 1929)
Anthracosia ovum (Trueman and Weir, 1951)
Anthracosia phrygiana (Wright, 1929)
Anthracosia planitumida (Trueman, 1929)
Anthracosia regularis (Trueman, 1929)
Anthracosia similis (Brown, 1843)
Anthracosia simulans Trueman & Weir, 1952
Anthracosia subrecta Trueman & Weir, 1952
Anthracosphaerium affine (Davies & Trueman, 1927)
Anthracosphaerium boltoni (Wright, 1929)
Anthracosphaerium cycloquadratum (Wright, 1929)
Anthracosphaerium exiguum (Davies & Trueman, 1927)
Anthracosphaerium gibbosum (Hind, 1894)
Anthracosphaerium propinquum (Melville, 1947)
Anthracosphaerium radiatum (Wright, 1929)
Anthracosphaerium turgidum (Brown, 1843)

Bellisporites nitidus Horst (Sullivan), 1964
Bilinguites bilinguis (Salter, 1864)
Bilinguites gracilis (Bisat, 1924)
Bilinguites metabilinguis (Wright, 1926)
Bilinguites superbilinguis (Bisat, 1924)

Caneyella multirugata (Jackson, 1927)
Caneyella rugata (Jackson, 1927)
Carbonicola aldamae (Brown, 1843)
Carbonicola artifex Eagar, 1954
Carbonicola bipennis (Brown, 1843)
Carbonicola communis Davies & Trueman, 1927
Carbonicola concinna Wright, 1929
Carbonicola crista-galli Wright, 1936
Carbonicola discus Eagar, 1947
Carbonicola fallax Wright, 1934
Carbonicola haberghamensis Wright, 1934
Carbonicola limax Wright, 1934
Carbonicola obliqua Wright, 1934
Carbonicola obtusa Hind, 1894
Carbonicola os-lancis Wright, 1929
Carbonicola polmontensis (Brown, 1849)
Carbonicola protea Wright, 1934
Carbonicola pseudorobusta Trueman, 1929
Carbonicola rectilinearis Trueman & Weir, 1948
Carbonicola rhomboidalis Hind, 1894
Carbonicola subconstricta (J Sowerby), 1813
Carbonicola torus Eagar, 1954
Carbonita bairdioides (Jones & Kirkby, 1879)
Carbonita fabulina (Jones & Kirkby, 1879)
Carbonita humilis (Jones & Kirkby, 1879)
Carbonita inflata (Jones & Kirkby, 1879)

Carbonita pungens (Jones & Kirkby, 1879)
Carbonita salteriana (Jones & Kirkby, 1879)
Carbonita scalpellus (Jones & Kirkby, 1879)
Cirratriradites saturni (Ibrahim) Schopf, Wilson & Bentall, 1944
Cravenoceras subplicatum Bisat, 1932
Cravenoceratoides edalensis Bisat, 1928
Crurithyris carbonaria (Hind, 1905)
Curvirimula candela (Dewar, 1939)
Curvirimula subovata (Dewar, 1939)
Curvirimula trapeziforma (Dewar, 1939)
Cypridina radiata Jones, Kirkby and Bradey, 1874

Dictyoclostus craigmarkensis Muir-Wood, 1937
Donetzoceras aegiranum (H Schmidt, 1925)
Draffania biloba Cummings, 1957
Dunbarella macgregori (Currie, 1937)
Dunbarella papyracea (J Sowerby, 1822)
Dunbarella speciosia (Jackson, 1927)

Eocypridina radiata (Jones & Kirkby, 1874)
Ephippioceras costatum Foord, 1891
Euestheria mathieui (Pruvost, 1911)
Euestheria simoni (Pruvost, 1911)
Euestheria vinti (Kirkby, 1864)
Eumorphoceras erinense Yates, 1962
Eumorphoceras grassingtonense Dunham & Stubblefield, 1944
Euphemites anthracinus (Weir, 1931)
Euphemites urei (Fleming, 1928)

Gastrioceras listeri (J Sowerby, 1812)
Gastrioceras subcrenatum C Schmidt, 1924
Geisina arcuata (Bean, 1836)
Geisina subarcuata (Jones, 1889)

Hiboldtina wardiana (Jones & Kirkby, 1866)
Hollinella bassleri (Knight, 1928)
Hollinella claycrossensis Bless & Calver, 1970
Homoceratoides divaricatus Hind, 1905
Huanghoceras costatum (Hind, 1905)

Ibrahimspores brevispinosus Neves, 1961

Jordanites cristinae (Bless, 1967)

Koninckopora inflata (de Konick, 1842) Lee, 1912

Lingula mytilloides J Sowerby, 1812
Lingula pringlei Currie, 1937 (now Lachrymula)
Lissochonetes minutus Demanet, 1943
Lobatopteris vestita (Les quereux) Wagner, 1958

Megalichthys hibberti Agassiz, 1835
Metacoceras cornutum Girty, 1915
Metacoceras perelegans Girty, 1915
Myalina perlata de Konick, 1885

Naiadites alatus Weir, 1956
Naiadites angustus Trueman & Weir,1956
Naiadites carinata (J de C Sowerby, 1840)
Naiadites daviesi Dix & Trueman, 1932
Naiadites elongatus (Hind, 1893)
Naiadites flexuosus Dix & Trueman, 1932
Naiadites hindi Trueman & Weir, 1956
Naiadites melvillei Trueman & Weir, 1956
Naiadites obliquus Dix & Trueman, 1932
Naiadites productus (Brown, 1849)
Naiadites quadratus (J de C Sowerby, 1840)
Naiadites triangularis (J de C Sowerby, 1840)
Naiadites tumidus (Etheridge jun., 1879)
Neochonetes granulifer (Owen, 1852)
Nuculoceras stellarum (Bisat, 1932)
Nuculopsis gibbosa (Fleming, 1828)

Odontopteris cantabrica Wagner, 1969
Orbiculoidea nitida (Phillips, 1836)
Orthoceras sulcatum (Fleming, 1815)

Paraarchaediscus stilus Grozdilova & Lebedeva, 1953
Paraconularia crustula (White, 1880)
Pernopecten carboniferus (Hind, 1903)
Phestia acuta (J De C Sowerby, 1840)
Phestia attenuata (Fleming, 1828)
Planoarchaediscus concinnus Conil & Lys, 1964
Planolites ophthalmoides (Jessen, 1949)
Platyconcha hindi Longstaff, 1936
Politoceras politum (Schumard, 1858)
Posidonia corrugata Etheridge jun., 1873
Posidonia gibsoni Brown in Salter, 1862
Posidonia sulcata (Hind, 1904)
Productus hibernicus Muirwood, (?1961)
Pseudamussium fibrillosus (Salter, 1864)
Pseudoammodiscus volgensis (Rauser, 1948)

Reticuloceras circumplicatile (Foord, 1903)
Reticuloceras dubium Bisat & Hudson, 1943
Reticuloceras paucicrenulatum Bisat & Hudson, 1943
Reticuloceras reticulatum (Phillips, 1836)
Rugosochonetes skipseyi (Currie, 1937)

Siphonodendron martini (Milne-Edwards & Haime, 1851)
Sphenothallus stubblefieldi Schmidt & Tiechmuller, 1958

Tornquistia diminuta Demanet, 1949
Tornquistia gibbosa (Dorsman, 1945)
Trepospira radians (de Koninck, 1843)

Verneuilites sigma (Wright, 1926)

Chapter 5 Triassic

Acanthotriletes varius Nilsson 1958
Accinctisporites radiatus (Leschik) Schulz 1965
Alisporites circulicorpus Clarke 1965
A. grauvogeli Klaus 1964
Angustisulcites gorpii Visscher 1966
A. grandis (Freudenthal) Visscher 1966
A. klausii Freudenthal 1964
Apiculatasporites plicatus Visscher 1966
Aratrisporites rotundus Mädler 1964
A. saturni (Thiergart) Mädler 1964

Calamospora sp.
Camerosporites secatus Leschik emend. Scheuring 1978
Chasmatosporites magnolioides (Erdtman) Nilsson 1958
Cingulizonates rhaeticus (Reinhardt) Schulz 1967
Classopollis torosus (Reissinger) Balme 1957
Colpectopollis ellipsoideus Visscher 1966
Convolutispora microrugulata Schulz 1967
Cycadopites coxii Visscher 1966
C. subgranulosus (Couper) Clarke 1965
C. trusheimii Visscher 1966
Cyclogranisporites sp.
Cyclotriletes microgranifer Mädler 1964
C. oligogranifer Mädler 1964
Cymatiosphaera polypartita Morbey 1975

Dapcodinium priscum Evitt 1961
Duplicisporites verrucosus Leschik emend. Scheuring 1978

Echinitosporites iliacoides Schulz & Krutzsch 1961
Euestheria minuta (Alberti)

Geopollis zwolinskai (Lund) Brenner 1986
Gliscopollis
G. meyeriana (Klaus) Venkatachala 1966

Illinites chitonoides Klaus 1964
I. kosankei Klaus 1964

Lunatisporites sp.
L. rhaeticus (Schulz) Warrington 1974

Micrhystridium lymense var. *gliscum* Wall 1965

Microreticulatisporites sp.
M. fuscus (Nilsson) Morbey 1975

Ovalipollis pseudoalatus (Thiergart) Schuurman 1976

Perotrilites minor (Mädler) Antonescu & Taugourdeau Lantz 1973
Protocardia rhaetica (Merian)
Protodiploxypinus fastidiosus (Jansonius) Warrington 1974
P. sittleri (Klaus) Scheuring 1970
Punctatisporites triassicus Schulz 1964

Quadraeculina anellaeformis Maljavkina 1949

Retisulcites perforatus (Mädler) Scheuring 1970
Rhaetavicula contorta (Portlock)
Rhaetipollis germanicus Schulz 1967
Rhaetogonyaulax rhaetica (Sarjeant) Loeblich & Loeblich emend. Below 1987
Ricciisporites tuberculatus Lundblad 1954
Rugulatisporites mesozoicus Mädler 1964

Scabratisporites scabratus Visscher 1966
Spinotriletes echinoides Mädler 1964
Stellapollenites thiergartii (Mädler) Clement-Westerhof, van der Eem, van Erve, Klasen, Schuurman & Visscher 1974
Striatoabieites balmei Klaus emend. Scheuring 1978
Sulcatisporites kraeuseli Mädler 1964

Tasmanites sp.
Triadispora crassa Klaus 1964
T. falcata Klaus 1964
T. plicata Klaus emend. Scheuring 1978
Tsugaepollenites oriens Klaus 1964

Vallasporites ignacii Leschik 1955
Verrucosisporites applanatus Mädler 1964
V. jenensis Reinhardt & Schmitz ex. Reinhardt 1964
V. pseudomorulae Visscher 1966
V. thuringiacus Mädler 1964
Veryhachium reductum (Deunff) Jekhowsky 1961
Vesicaspora fuscus (Pautsch) Morbey 1975
Voltziaceaesporites heteromorpha Klaus 1964

146

INDEX

BRITISH GEOLOGICAL SURVEY

Keyworth, Nottingham NG12 5GG
0115 936 3100

Murchison House, West Mains Road, Edinburgh
EH9 3LA   0131-667 1000

London Information Office, Natural History Museum
Earth Galleries, Exhibition Road, London SW7 2DE
0171-589 4090

The full range of Survey publications is available through the
Sales Desks at Keyworth and at Murchison House, Edinburgh,
and in the BGS London Information Office in the Natural
History Museum (Earth Galleries). The adjacent bookshop
stocks the more popular books for sale over the counter. Most
BGS books and reports can be bought from The Stationery
Office and through Stationery Office agents and retailers.
Maps are listed in the BGS Map Catalogue, and can be bought
together with books and reports through BGS-approved
stockists and agents as well as direct from BGS.

*The British Geological Survey carries out the geological survey of Great
Britain and Northern Ireland (the latter as an agency service for the
government of Northern Ireland), and of the surrounding continental
shelf, as well as its basic research projects. It also undertakes
programmes of British technical aid in geology in developing countries
as arranged by the Department for International Development and
other agencies.*

*The British Geological Survey is a component body of the Natural
Environment Research Council.*

Published by The Stationery Office and available from:

The Publications Centre
(mail, telephone and fax orders only)
PO Box 276, London SW8 5DT
General enquiries 0171 873 0011
Telephone orders 0171 873 9090
Fax orders 0171 873 8200

The Stationery Office Bookshops
59–60 Holborn Viaduct, London EC1A 2FD
temporary until mid 1998
(counter service and fax orders only)
Fax 0171 831 1326
68–69 Bull Street, Birmingham B4 6AD
0121 236 9696   Fax 0121 236 9699
33 Wine Street, Bristol BS1 2BQ
0117 9264306 Fax 0117 9294515
9–21 Princess Street, Manchester M60 8AS
0161 834 7201 Fax 0161 833 0634
16 Arthur Street, Belfast BT1 4GD
01232 238451 Fax 01232 235401
The Stationery Office Oriel Bookshop
The Friary, Cardiff CF1 4AA
01222 395548 Fax 01222 384347
71 Lothian Road, Edinburgh EH3 9AZ
(counter service only)

Customers in Scotland may
mail, telephone or fax their orders to:
Scottish Publications Sales
South Gyle Crescent, Edinburgh EH12 9EB
0131 228 4181 Fax 0131 622 7017

The Stationery Office's Accredited Agents
(see Yellow Pages)

and through good booksellers